Quantification and Modelling of Fugitive Greenhouse Gas Emissions from Urban Water Systems

Scientific and Technical Report Series No. 26

Quantification and Modelling of Fugitive Greenhouse Gas Emissions from Urban Water Systems

Edited by
Liu Ye, Jose Porro and Ingmar Nopens

IWA PUBLISHING

Published by **IWA Publishing**
Unit 104–105, Export Building
1 Clove Crescent
London E14 2BA, UK
Telephone: +44 (0)20 7654 5500
Fax: +44 (0)20 7654 5555
Email: publications@iwap.co.uk
Web: www.iwapublishing.com

First published 2022

British Library Cataloguing in Publication Data
A CIP catalogue record for this book is available from the British Library

ISBN: 9781789060454 (Paperback)
ISBN: 9781789060461 (eBook)
ISBN: 9781789060478 (ePUB)

This eBook was made Open Access in April 2022

Contents

Chapter 7

Modelling N_2O production and emissions . . . 167

Mathieu Spérandio, Longqi Lang, Fabrizio Sabba, Robert Nerenberg, Peter Vanrolleghem, Carlos Domingo-Félez, Barth F. Smets, Haoran Duan, Bing-Jie Ni and Zhiguo Yuan

Chapter 8

Modelling of methane production and emissions . . . 197

Keshab Sharma, Oriol Gutierrez, Zhiguo Yuan, Matthijs R. J. Daelman, Mark C. M. van Loosdrecht and Eveline I. P. Volcke

List of Contributors

Ahlem Filali
Université Paris-Saclay, INRAE, PROSE, France

Amanda Lake
Jacobs, Scotland

Antonio Delre
Department of Environmental Engineering, Technical University of Denmark, Denmark

Ariane Coelho Brotto
Jacobs, UK

Barth F. Smets
Department of Environmental Engineering, Technical University of Denmark, Denmark

Bing-Jie Ni
Advanced Water Management Centre, The University of Queensland, Australia

Carlos Domingo-Félez
Department of Environmental Engineering, Technical University of Denmark, Denmark

Chris Sweetapple
Centre for Water Systems, College of Engineering, Mathematics and Physical Sciences, University of Exeter, UK

David Butler
Centre for Water Systems, College of Engineering, Mathematics and Physical Sciences, University of Exeter, UK

Eveline I.P. Volcke
Department of Green Chemistry and Technology, Ghent University, Belgium

Evina Katsou
Department of Civil & Environmental Engineering, Brunel University London, UK

Fabrizio Sabba
Department of Civil and Environmental Engineering and Earth Sciences, University of Notre Dame, USA

Giacomo Bellandi
AM-TEAM, Belgium

Guangtao Fu
Centre for Water Systems, College of Engineering, Mathematics and Physical Sciences, University of Exeter, UK

Haoran Duan
School of Chemical Engineering, The University of Queensland, Australia

Ingmar Nopens
Biomath, Ghent University, Belgium
Captur

Jose Porro
Cobalt Water Global, Inc, USA

Keshab R. Sharma
Advanced Water Management Centre, The University of Queensland, Australia

Krist V. Gernaey
Process and Systems Engineering Centre, Department of Chemical and Biochemical Engineering, Technical University of Denmark, Denmark

Liu Ye
School of Chemical Engineering, The University of Queensland, Australia

Lluís Corominas
Catalan Institute for Water Research, Spain

Longqi Lang
TBI, Université de Toulouse, CNRS, INRAE, INSA, France

Magnus Arnell
Unit of Urban Water Management, RISE Research Institutes of Sweden, Sweden
Division of Industrial Electrical Engineering and Automation, Department of Biomedical Engineering, Lund University, Sweden

Maite Pijuan
Catalan Institute for Water Research, Spain
The University of Girona, Spain

Mark C.M. van Loosdrecht
Department of Biotechnology, Delft University of Technology, The Netherlands

Mathieu Spérandio
TBI, Université de Toulouse, CNRS, INRAE, INSA, France

Matthijs R.J. Daelman
Department of Biotechnology, Delft University of Technology, The Netherlands

Oriol Gutierrez
Catalan Institute for Water Research, Spain
The University of Girona, Spain

Peter A. Vanrolleghem
ModelEAU, Département de génie civil et de génie des eaux, Université Laval, Canada

Robert Nerenberg
Department of Civil and Environmental Engineering and Earth Sciences, University of Notre Dame, USA

Ulf Jeppsson
Division of Industrial Electrical Engineering and Automation, Department of Biomedical Engineering, Lund University, Sw4424

Vanessa Parravicini
TU Wien, Institute for Water Quality, Resource and Waste Management, Austria

Vasileia Vasilaki
Department of Civil & Environmental Engineering, Brunel University London, UK

Xavier Flores-Alsina
Process and Systems Engineering Centre, Department of Chemical and Biochemical Engineering, Technical University of Denmark, Denmark

Yinfen Zhao
School of Chemical Engineering, The University of Queensland, Australia

Ziping Wu
School of Chemical Engineering, The University of Queensland, Australia

Zhiguo YUAN
Advanced Water Management Centre, The University of Queensland, Australia

Foreword

The *IWA Task Group on the use of water quality and process models for minimizing wastewater utility greenhouse gas footprints* and the idea of this book came about because we wanted to fill a gap in knowledge related to fugitive emissions of greenhouse gas from wastewater systems, so that we can then avoid or eliminate them. For instance, in the case of nitrous oxide (N_2O) emissions, we wanted to better understand the N_2O mechanisms and pathways, because the mechanisms were initially thought to be associated with only one of the relevant processes, denitrification, but we later learned that they could include nitrification and multiple pathways within nitrification. Furthermore, we realized that fugitive GHG emissions in wastewater involved other aspects, such as methane (CH_4) generation in sewers, CH_4 emissions related to sludge storage and biosolids from anaerobically digested sludge, and fugitive emissions of both CH_4 and N_2O from water bodies receiving discharges.

The main motivation or rationale behind the Task Group effort was to improve upon the Intergovernmental Panel on Climate Change (IPCC) emission factors or assumptions associated with quantifying GHG emissions from wastewater systems and provide corresponding tools for the industry so that any water reclamation facility in design or being retrofitted or optimized could eliminate these emissions. Specifically, the intent was to develop a more mechanistic approach to understanding the production of greenhouse gases and the use of mechanistic models to mitigate these emissions. However, it was quickly understood that developing a mechanistic model alone may not resolve or address all of the challenges in managing water utility greenhouse gas emissions; therefore, non-mechanistic approaches have also been considered, such as artificial intelligence including knowledge-based approaches and machine learning.

In terms of collaboration, what the Task Group and the overall approach to this book inevitably did was bring people together. It brought together researchers from Europe, Australia, and the United States to build a consensus-based approach, as well as to publish the models that were being developed. When the Task Group began, it was understood that it would take several years to develop consensus-based mechanistic models, as opposed to having one or several models already developed to build consensus around, which is mainly what previous Task Groups have done. The goal for this Task Group was different and tended to be a bit more ambitious by trying to spur model development as much as also providing a consensus approach. Therefore, collaboration to integrate these models, which were mainly based on lab-scale or empirical data, as well as full-scale data-driven techniques was needed and continues today.

Fugitive emissions remain a risk. The water reclamation community is aware of this risk, but has not yet fully taken charge to address it. This is mainly because there has been a lack of tools and guidance on how to address fugitive GHG emissions. Therefore, water technology providers and consultants who implement water reclamation technologies have been making decisions in retrofits without adequate knowledge. This forward-looking book aims to change that. It is an optimistic report hoping to inspire the water community to take the lead in resolving and addressing these emissions themselves. It will provide modelers, process engineers, and water treatment practitioners with the tools and approaches to address the climate risk of GHG emissions. It also provides a science-based framework for regulators wanting to control such emissions. This book could not be more timely with respect to the climate crisis and the urgency for taking climate action. Now it's up to the water industry to put it to use and help make its mark on better protecting the planet.

Sudhir Murthy, Ph.D., P.E.
CEO of NEWhub Corp. and Senior VP of IWA

Acknowledgements

Our sincere thanks to all the reviewers listed below (in alphabetical order) for their contributions to this important work.

- Adrian Oehmen, *The University of Queensland*
- Ben van den Akker, *South Australian Water Corporation*
- David de Haas, *GHD*
- Ignasi Rodriguez-Roda, *University of Girona*
- Imre Takács, *Dynamita*
- Joaquim Comas, *University of Girona and Catalan Institute for Water Research*
- Keshab Sharma, *The University of Queensland*
- Maite Pijuan, *Catalan Institute for Water Research*
- Romain Lemaire, *Veolia*
- Vanessa Parravicini, *TU Wien*
- Yan Zhou, *Nanyang Technological University*

We wish to acknowledge the significant contributions made by all the authors, especially the lead authors in each of the chapters, and their collaborative approach to delivery of this book.

We believe this is a solid step in the journey to the net-zero emission goal that lies ahead of us...

From

Liu Ye (Lead Editor, The University of Queensland)

Jose Porro (Co-Lead Editor, Cobalt Water Global)

Ingmar Nopens (Editor, Ghent University)
March 2022

For additional acknowledgements of Task Group members and supporters, please see *A note from the Task Group GHG* at the end of the book.

doi: 10.2166/9781789060461_001

Chapter 1
Introduction

Liu Ye[1] *and Jose Porro*[2]

[1]*School of Chemical Engineering, The University of Queensland, St Lucia, Brisbane, QLD 4072, Australia. l.ye@uq.edu.au*
[2]*Cobalt Water Global, Inc. 77 Sands Street, Brooklyn, NY 11201, USA, jose.porro@cobaltwater-global.com*

SUMMARY

The *Introduction* chapter explains the motivation for assessing and reducing greenhouse (GHG) emissions from urban water systems and gives an overview of the design for this book. This chapter also provides a summary for the content of each chapter in terms of the state-of-the-art regarding urban water system GHG emissions sources, inventory protocols, quantification methods, and the existing modelling tools. Finally, this chapter explains the scope and objectives of this book, as well as providing a general guide for the use of this book in GHG assessment and reduction efforts.

Keywords: Greenhouse gas, methane, nitrous oxide, and urban water systems

TERMINOLOGY

Term	Definition
Urban water system	An engineered system to decouple the water used by the human community and the natural water environment
N_2O	Nitrous oxide, a potent GHG, with a global warming potential 265-fold stronger than that of carbon dioxide (CO_2)
CH_4	Methane, a potent GHG, with a global warming potential 25-fold stronger than that of carbon dioxide (CO_2)
Biomass	A clump of organic material consisting of living organisms, which live on the substrates in wastewater, or the dead organism debris.
Greenhouse gas	Gas that absorbs and emits radiant energy within the thermal infrared range and contributes to the global warming effect.
Nutrient	Substances such as nitrogenous compounds, phosphate or organic carbon that can be assimilated by microbes to promote the metabolism and growth of microbes in wastewater.
Organic matter	Organic waste of plant or animal origin from homes or industry, or originated from storm water run-offs, and so on., which mainly contains volatile fraction of solids.

Mathematic model	A system of mathematical equations that describes physical and biological processes. It is a simplified representation of the real process.
Wastewater	The used water including solids discharged from communities, businesses, industry or agriculture that flows into a wastewater treatment plant. Storm water, surface water, and groundwater infiltration also may be included.

1.1 CLIMATE CHANGE, SUSTAINABILITY, AND GHG LEGISLATION IN THE WATER SECTOR

The effects of climate change can have a tremendous impact on almost all facets of life on Earth. In many parts of the world, they are already being felt and, to no surprise, the global position on climate change, as evidenced since COP21, is that we must stand and fight to minimize anthropogenic greenhouse gas (GHG) emissions, and move towards net zero emissions to help mitigate climate change impacts. Although the water sector's overall contribution is small compared to the global GHG emissions, urban water management can make up a significant part of a city's GHG emissions inventory. In New York City, for example, urban water management is second only to the buildings sector in GHG emissions and makes up 20% of the city-wide GHG emissions inventory (Bonczak *et al.* 2020). This is largely due to the low emissions per capita resulting from the mass transit system, and the same results are likely for water utilities in other large cities around the world with extensive mass transit systems. However, even if, as a whole, the water sector represents a small fraction of the total GHG emissions, the effects of climate change can be detrimental to the urban water cycle, prompting problems such as water scarcity and stress from droughts, flooding and disruption of service from extreme weather events, combined sewer overflows, and a variety of water quality issues. Since we cannot live without water, and the sustainability and resilience of the urban water cycle is central to human quality of life, there is a physical need and driver for climate change mitigation and minimizing GHG emissions in the water sector. Therefore, the water sector needs to lead the fight against climate change by example.

The physical effects that climate change has on the urban water cycle have also prompted a cultural change among water utilities. Sustainability and resilience are now widely incorporated into the planning and operations of water utilities and are implemented in varying degrees. Therefore, in addition to addressing economic and social factors, minimizing the environmental impact of urban water systems is at the forefront for water utilities with a wish list of activities to be more environmentally friendly, such as resource recovery, minimizing water, chemical, and energy consumption, including minimizing the GHG emissions related to urban water system management and development. There is this idea of the green utility (Welch, 2010). Therefore, there is also a water utility cultural driver for mitigating GHG emissions.

Although there is no argument among water utilities that we should all be striving for environmental stewardship from a sustainability point of view, there are other specific drivers for reducing GHG emissions in the water sector. These include various types of GHG legislation that implicate some water utilities in different parts of the world, such as in the US, UK, Denmark, Australia, and The Netherlands, as well as general voluntary GHG reduction goals that are incorporated into strategies that are implemented at local, regional, and national levels, which also implicate water utilities in various parts of the world. However, there should be no other driver needed than the real and visible climate-related disasters that are happening much more frequently and led to the Climate Crisis being declared in 2019. Hence, clear drivers exist for water utilities to mitigate climate change and action is either already being taken by leaders in the field, or is planned by water utilities in response to these drivers. These drivers are what makes this book relevant, and they provide the main motivation to equip the water sector with knowledge and tools to start taking climate action now by quantifying, modelling, and mitigating urban wastewater system GHG emissions.

Given the urgency in addressing the Climate Crisis, we know that GHG emissions must be reduced in the next decade to minimize the effects of climate change, as the time left for us to take action has been cut by two-thirds. We cannot wait until the middle of the century to achieve the original target (Höhne *et al.*, 2020). This requires efforts from all sectors to reduce the emissions.

The wastewater sector contributes to greenhouse gas emissions not only through its significant energy consumption, but also through direct emissions of fugitive gases such as carbon dioxide (CO_2), methane (CH_4) and nitrous oxide (N_2O). According to the guidelines from the Intergovernmental Panel on Climate Change (IPCC) on national GHG inventories calculation (IPCC, 2006), the GHG emissions are categorized into three scopes:

- Scope 1: direct GHG emissions from the treatment of wastewater;
- Scope 2: indirect emissions from the generation of purchased electricity, heat or steam that is consumed in its owned or controlled equipment or operations;
- Scope 3: indirect GHG emissions from materials and consumables used for the treatment of wastewater – for example chemicals manufacture and transport and the emissions associated with purchased goods and services, including those for capital infrastructure works, waste generated by company operations, as well as employee travel and commuting and so on.

Currently, the wastewater sector is required to report both direct (Scope 1) and indirect (Scope 2) GHG emissions from wastewater systems as part of the waste and energy sectors. The indirect emissions are expected to decrease substantially in the coming decade due to both the increased recovery of energy from wastewaters and the improved energy utilization efficiency for wastewater treatment. Use of onsite wind and solar energy will further reduce Scope 2 emissions. Therefore, the Scope 1 emissions will become the key contributors and are also more difficult to reduce. As most of the CO_2 produced from wastewater treatment processes are biogenic carbon, the majority of the Scope 1 emissions from urban wastewater systems are CH_4 and N_2O. This is because both CH_4 and N_2O are potent GHGs, with global warming potentials 25-fold and 265-fold, respectively, stronger than that of CO_2. N_2O emissions are especially important, contributing up to 80% of the overall carbon footprint of a wastewater treatment plant (WWTP) (Daelman *et al.*, 2013). The pathways and factors leading to biological nitrous oxide and methane formation and emissions from wastewater are also highly complex and site-specific. In 2019, IPCC published a refinement to its 2006 GHG inventory guidelines and a substantial increase of the default emission factor for N_2O was used for WWTPs (IPCC, 2019). It is also noted that the national level GHG methodologies lack the level of detail required to properly quantify N_2O emissions at the asset level because they do not account for the site-specific conditions.

As the water industry has a strong stake in improving environmental performance, sustainability and reducing emissions, the objective of this book is to provide a detailed summary of the current state-of-knowledge for both N_2O and CH_4 generation, quantification and modelling from urban wastewater systems, therefore improving the scientific basis used by the water industry in understanding and mitigating the Scope 1 emissions from wastewater processes.

1.2 OVERVIEW OF GHG EMISSION SOURCES IN URBAN WATER SYSTEMS

The urban water cycle is not only intricate and complex by nature, it is also unique for every city, like a fingerprint. It can include a drinking water stage with supply, treatment, and distribution; a wastewater stage with collection, treatment, and discharge; and everything in between, such as water reuse, rainwater harvesting, and urban drainage. The layout and configuration for each part will have varying impacts on the total urban water cycle carbon footprint depending upon various factors, such as elevations, population, climate, water consumption behaviour, electrical grid energy mix, water resource type (e.g., surface water or groundwater), and how everything is managed and further developed by water utilities. Therefore, climate change mitigation by minimizing GHG emissions from urban water systems requires a multi-faceted and holistic approach to address the different issues related to the different stages/systems of the urban water cycle, and the different sources of GHG emissions in each stage. This holistic approach not only requires looking at planning, design,

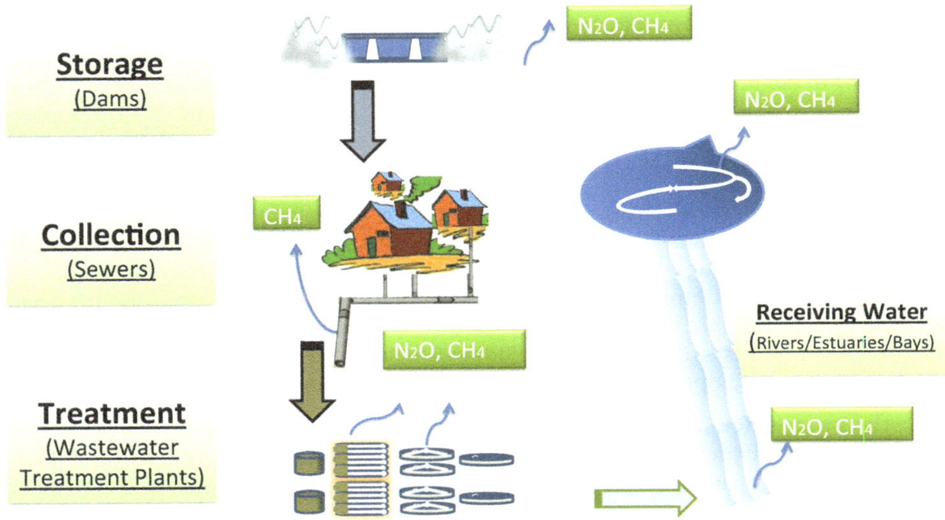

Figure 1.1 Fugitive greenhouse gas (N_2O and CH_4) emissions from urban water systems.

operations, and construction of urban water systems, but also requires consideration of urban water system management and development in an integrated fashion, as the various parts of the urban water cycle are inherently linked and can impact the GHG emissions of each other. However, the focus of this book is on the urban wastewater system.

As a whole, the urban wastewater system is engineered to decouple the water used by the human community and the natural water environment, from water storage to wastewater collection and treatment before the water is safely discharged to receiving water bodies (illustrated in Figure 1.1).

It has been reported that all urban water systems and receiving water bodies (i.e., rivers, estuaries and bays) will produce N_2O and CH_4 (Sturm *et al.*, 2013). This is because N_2O and CH_4 are generated biologically either as a by-product or an obligatory intermediate during the biological nitrogen or carbon cycle (more details can be found in Chapters 2 and 3). Given the amount of nitrogen and carbon inventory in the urban water system, only underground wastewater collection systems (sewers) and the subsequent wastewater treatment plants are within the scope of this book as these are regarded as the key emissions sources of N_2O and CH_4. It should be noted that while the fundamental generation pathways of N_2O and CH_4 are similar in either the natural environment or in urban water systems, the quantification methodologies may vary. This book will focus on the N_2O and CH_4 emission points in sewers and wastewater treatment plants (as shown in Figure 1.2 below) to introduce their associated generation mechanisms and factors (operational or environmental) that may affect their emissions.

Chapter 2 focuses on sources and pathways of N_2O generation. As an environmentally detrimental greenhouse gas, N_2O generated from wastewater treatment systems has attracted a lot of attention due to its important contribution to the overall facility carbon footprint. When aiming for mitigation of N_2O, it is imperative to understand the mechanisms that trigger its formation. Sources and pathways for N_2O production are reviewed and discussed. Despite extensive investigations, the mechanisms for N_2O production continue to undergo extensive academic research because of the complexity of wastewater treatment systems. N_2O generation is influenced by a diversity of interrelated conditions: the different species of microbes with specialized functions, the interactions among these microbes in a mixed system, and the responses of microbes to the different environmental factors and operational

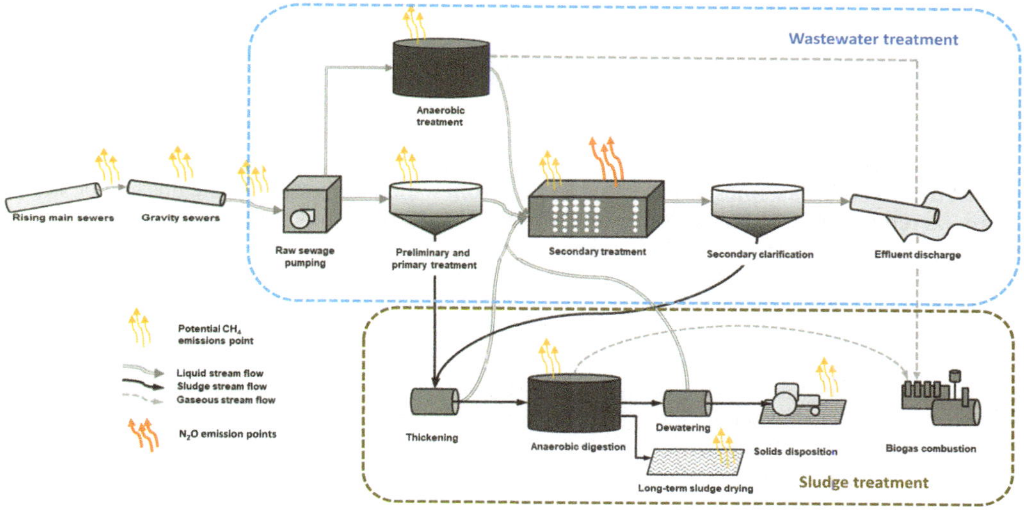

Figure 1.2 Sources of N_2O and CH_4 emission points from wastewater transport and treatment (Adapted from Figure 3.3 in Chapter 3).

conditions. Chapter 2 provides an overview of the N_2O production pathways and mechanisms during the biological nitrogen removal (BNR) process and describes the different factors that have been reported to have an effect on N_2O production.

Chapter 3 provides an overview of the contribution of urban wastewater system methane emissions. CH_4 formation in urban wastewater systems occurs through anaerobic digestion (AD) of organics contained in sewage. AD consists of a sequence of concomitant reactions by which a consortium of microorganisms, in the absence of oxygen, break down biodegradable carbon material producing biogas, a mixture of methane, carbon dioxide and traces of H_2S. In an urban wastewater system, those reactions can occur naturally in sewer systems depleted of oxygen or be artificially promoted in wastewater treatment plants to capture and recover the energy contained in molecules of methane. Specifics of methane generation in sewer biofilms, sewer sediments, anaerobic wastewater treatment and sludge disposal of wastewater treatment plants are presented in this chapter. Identification of CH_4 emission spots in urban wastewater engineered systems represents the initial step for reliable quantification of GHG; this then facilitates development and implementation of effective mitigation strategies.

1.3 GHG EMISSIONS INVENTORY PROTOCOLS

Following a better understanding of N_2O and CH_4 generation for the urban wastewater system, **Chapter 4** provides readers with focused analysis of the accounting methodologies and protocols supporting GHG emissions assessment and reporting of relevance to the urban water system in wastewater treatment of domestic and industrial wastewaters. It summarizes the basis for existing CH_4 and N_2O emission factors, the three-tier approach set out in the internationally accepted IPCC methodology and areas where further work is required. This chapter also summarizes the implications of the 2019 IPCC refinement of top-down emission factors on the magnitude of N_2O emissions from secondary treatment, as well as country-specific emission factors developed through national bottom-up monitoring and reporting guidelines. Finally, this chapter highlights the importance of bottom-up approaches to understand the opportunities to optimize treatment processes and conditions that minimize direct GHG emissions and help move the water industry towards net zero GHG emissions.

1.4 DIRECT MEASUREMENT OF URBAN WATER SYSTEM GHG EMISSIONS

Current reporting guidelines on N_2O and CH_4 emissions from the wastewater sector, whether at international or national level, provide for single emission factors each for N_2O and CH_4 in a top-down approach. Evidence from scientific and industry investigations shows that this is not likely to be accurate as the fixed emissions factor methods are adopted to report all emissions. Therefore, the quantification of direct GHG emissions from sewers and wastewater treatment plants is of great importance to provide science-based emissions baselines and understanding from which mitigation strategies can be developed and emissions reductions can be implemented and sustained in urban water systems.

Chapter 5 provides an overview of the currently available nitrous oxide and methane quantification methods applied at full-scale in sewers and wastewater treatment plants. Since the first measurement campaigns in the early 90 s were based on sporadic grab sampling, quantification methodologies and sampling strategies have evolved significantly, in order to describe the spatial-temporal dynamics of the emissions. The selection of a suitable quantification method is mainly dictated by the objective of the measurement survey and by specific local requirements. Plant-wide quantification methods provide information on the overall emissions of wastewater treatment plants, including unknown sources, which can be used for GHG inventory purposes. To develop on-site mitigation strategies, in-depth analysis of GHG generation pathways and emission patterns is required. In this case, process-unit quantifications can be employed to provide data for developing mechanistic models or to statistically link GHG emissions to operational conditions. With regard to sewers, current available methods are not yet capable of capturing the complexity of these systems, due to their geographical extension and variability of conditions, and only allow monitoring of specific locations where hotspots for GHG formation and emission have been identified.

Chapter 6 reviews and summarizes recent studies from N_2O and CH_4 monitoring campaigns in full-scale WWTPs and sewer networks. The analysis classifies quantified N_2O and CH_4 emissions, triggering operational conditions and formation pathways for different configurations. Control strategies to minimize N_2O emissions are proposed for different process groups. Main reasons for the emission factor (EF) discrepancies in the control strategies (i.e., aeration control), configuration, and operational and environmental conditions that favour the preferred enzymatic pathways are discussed as well.

Compared with N_2O, CH_4 quantification from full-scale WWTPs is less investigated, while it also contributes significantly to the overall plant carbon footprint. The results of full-scale CH_4 quantification studies are summarized in this chapter. Emissions of CH_4 in WWTPs mainly originate from the influent, anaerobic wastewater treatment and anaerobic sludge handling processes. The amount of CH_4 emissions varies greatly with different configurations of WWTPs. For WWTPs without anaerobic sludge handling processes, the CH_4 emissions can mainly be traced back to the CH_4 dissolved in the influent. When anaerobic treatment is applied in WWTPs for wastewater chemical oxygen demand (COD) removal, its CH_4 emissions might substantially increase the overall plant carbon footprint. GHG monitoring campaigns carried out in WWTPs should include the monitoring of fugitive CH_4 emissions. Finally, CH_4 and N_2O emissions reported from sewer networks are also summarized in this chapter.

The last part of the chapter also summarizes some mitigation strategies applied at full-scale to control fugitive CHG emissions from WWTPs and sewers.

1.5 MODELLING TOOLS FOR ASSESSING GHG EMISSIONS FROM URBAN WATER SYSTEMS

Mathematical modelling plays a critical role towards the understanding and the success of mitigation of GHG emissions from urban wastewater systems. Mechanistic models based on biological pathways for both N_2O and CH_4, so called first principle models, have been successfully used to simulate the emissions. Benchmarking has been a useful tool for unbiased comparison of control strategies in wastewater treatment plants in terms of effluent quality, operational cost and risk of suffering microbiology-related total suspended solids (TSS) separation problems. Recently, deep learning has been increasingly applied to urban water management, models based on neural networks and machine learning also develop quickly in this field. In this book, Chapters 7–10 review all the different types of models and their applications in full-scale N_2O and CH_4 simulation and mitigation.

Chapter 7 reviews the current status of the modelling of N_2O emissions from wastewater treatment. The existing mathematical models describing all known microbial pathways for N_2O production are reviewed and discussed. These include N_2O production and consumption by heterotrophic denitrifiers, N_2O production by ammonia-oxidizing bacteria (AOB) through the hydroxylamine oxidation pathway and the AOB denitrification pathway, and the integration of these pathways in single-pathway N_2O models. The two-pathway models are compared to single-pathway models. The calibration and validation of these models using lab-scale and full-scale experimental data is also reviewed. The mathematical modelling of N_2O production, while still being enhanced by new knowledge development, has reached a maturity that facilitates the estimation of site-specific N_2O emissions and the development of mitigation strategies for wastewater treatment plants taking into account the specific design and operational conditions of the plant.

Chapter 8 provides a review of the models available for estimating the production and emission of methane from wastewater collection and treatment systems. The details of a number of mechanistic models as well as the simplified empirical models are summarized. Their limitations are identified and general methods for calibration and validation are presented.

Chapter 9 presents the status of extending the original Benchmark Simulation Model No. 2 (BSM2) towards including greenhouse gas emissions. A mathematical approach based on a set of comprehensive models that estimate all potential on-site and off-site sources of CO_2, CH_4 and N_2O is presented and discussed in detail. Based upon the assumptions built into the model structures, simulation results highlight the potential undesirable effects of increased GHG emissions when carrying out local energy optimization in the activated sludge section and/or energy recovery in the anaerobic digester. Although off-site CO_2 emissions may decrease in such scenarios due to either lower aeration energy requirement or higher heat and electricity production, these effects may be counterbalanced by increased N_2O emissions, especially since N_2O has a 300-fold stronger greenhouse effect than CO_2. The reported results emphasize the importance of using integrated approaches when comparing and evaluating (plant-wide) control strategies in wastewater treatment plants for more informed operational decision-making.

In **Chapter 10**, alternatives to mechanistic modelling, such as knowledge-based and data-driven approaches for assessing and mitigating GHG emissions from urban wastewater systems are detailed. Examples include knowledge-based artificial intelligence (AI), integrating mechanistic modelling and computational fluid dynamics (CFD) with AI, and data-driven and machine learning (ML) methods for assessing and mitigating nitrous oxide emissions from wastewater treatment. Using a knowledge-based AI approach, the expert knowledge of N_2O pathways and influencing factors can be represented and applied to generate a dynamic risk score using the process data and identify mitigation actions. Using data-driven methods like principal component analysis (PCA) and machine learning, specifically support vector machines (SVM), patterns in data can be detected to identify mitigation opportunities and to classify process conditions for guiding monitoring campaigns and minimizing the time needed for monitoring to properly represent the full range of emissions for a specific site. Although the focus is on the use of these approaches for assessing and mitigating N_2O emissions, the same general approach can be applied for assessing and mitigating CH_4.

1.6 MITIGATION OF GHG EMISSIONS FROM URBAN WATER SYSTEMS

The final chapter, **Chapter 11**, summarizes the key knowledge presented in this book. It also discusses the issues, knowledge gaps and perspectives in GHG quantification, reporting guidelines and modelling, and provides the perspectives for future work. This gives us the knowledge and tools for starting to monitor and mitigate emissions and begin to contribute towards net zero plants. Further work is needed and will likely also surface while implementing this knowledge in practice. However, it is of great importance that the already available knowledge is put into practice without delay, in order to initiate efforts to reduce the impact of the water sector on climate change now.

1.7 GENERAL GUIDE FOR USE OF THIS BOOK IN GHG ASSESSMENT AND REDUCTION EFFORTS

As mentioned, the goal of this book is to provide practitioners with the knowledge and tools to start taking climate action now by quantifying, modelling, and mitigating urban wastewater system GHG emissions. Therefore, the following use of this book is suggested:

- Read Chapter 1 for understanding the full context and drivers for GHG emissions quantification, modelling, and mitigation, and an overview of the relevant urban wastewater system GHG emission sources, as well as for understanding the contents of each chapter.
- Depending on the GHG of interest, read either Chapters 2 or 3 to understand mechanisms and pathways.
- Assuming the starting point is GHG accounting and reporting and no measurements or mitigation has begun, read Chapter 4 to understand how emissions can be quantified using best use of the protocols until measurement can be performed.
- Read Chapter 5 for quantification methods through direct measurements.
- Read Chapter 6 to understand the results of previous monitoring campaigns, the emission factors that have been derived from different monitoring strategies, and factors affecting monitoring results.
- Read either Chapters 7 or 8 depending on the GHG of focus to understand how mechanistic models can and should be used.
- Read Chapter 9 to understand results of control benchmarking studies and the types of effects different control strategies can have on WWTP GHG emissions for the specific case of the Benchmark Simulation Model platform.
- Read Chapter 10 to understand what AI and data-driven approaches can be taken.
- Read Chapter 11 for perspectives on where we are and where we are going with quantifying, modelling and mitigating urban wastewater system GHG emissions.

It should be noted however that, depending on the specific objectives, not all chapters may need to be read. For example, if there are mechanistic modelling studies planned, then perhaps it makes sense to start with either Chapters 7 or 8. Similarly, if only quantification is needed at the moment, then the focus can be on Chapters 5 and 6.

ACKNOWLEDGEMENT

Liu Ye acknowledges the funding support through Australian Research Council Discovery Project (180103369), and the University of Queensland Foundation Research Excellence Award.

REFERENCES

Bonczak B., Chua U. and Kontokosta C. (2020). Inventory of New York City Greenhouse Gas Emissions. NYC Mayor's Office of Sustainability. https://nyc-ghg-inventory.cusp.nyu.edu/ (accessed 19 August 2021).

Daelman M. R., van Voorthuizen E. M., van Dongen L., Vockle E. I. and van Loosdrecht M. C. (2013). Methane and nitrous oxide emissions from municipal wastewater treatment – results from a long-term study. *Water Science and Technology*, **67**(10), 2350–2355, https://doi.org/10.2166/wst.2013.109

IPCC. (2006). Guidelines for National Greenhouse Gas Inventories, (Chapter 5, Waste).

IPCC. (2019). Wastewater treatment and discharge. Bartram, D., Short, M. D., Ebie, Y., Farkăs, J., Gueguen, C., Peters, G. M., Zanzottera, N. M., Karthik, M., Masuda, S. In: 2019 Refinement to the 2006 IPCC Guidelines for National Greenhouse Gas Inventories, F. B. Demirok and A. Herold (eds). Vol. **5**, 6.39 (Chapter 6). Available at https://www.ipcc-nggip.iges.or.jp/public/2019rf/vol5.html

Höhne N., den Elzen M., Rogelj J., Metz B., Fransen T., Kuramochi T., Olhoff A., Alcamo J., Winkler H., Fu S., Schaeffer M., Schaeffer R., Peters G. P., Maxwell S. and Dubash N. K. (2020). Emissions: world has four times the work or one-third of the time. *Nature*, **579**, 25–28, https://doi.org/10.1038/d41586-020-00571-x

Sturm K., Yuan Z., Gibbes B. and Grinham A. (2013). Methane and nitrous oxide sources and emissions in a subtropical freshwater reservoir, south east queensland, Australia. *Biogeosciences Discussions*, **10**(12), 19485–19508, https://doi.org/10.5194/bgd-10-19485-2013

Welch C. (2010). The Green Utility: A Practical Guide to Sustainability. American Water Works Association. United States, ISBN: 978-1-61300-000-7

NOMENCLATURE

AD	Anaerobic digestion
AI	Artificial intelligence
AOB	Ammonia oxidizing bacteria
BNR	Biological nutrient removal
BSM	Benchmark simulation model
CFD	Computational fluid dynamics
EF	Emission factor
GHG	Greenhouse gas
GWP	Global warming potential
IPCC	Intergovernmental Panel on Climate Change
ML	Machine learning
PCA	Principal component analysis
SVM	Support vector machines
TSS	Total suspended solids
UK	United Kingdom
USA	United States of America
WWTP	Wastewater treatment plant

doi: 10.2166/9781789060461_011

Chapter 2

Full-scale source, mechanisms and factors affecting nitrous oxide emissions

Maite Pijuan[1,2] and Yingfen Zhao[3]

[1]Catalan Institute for Water Research (ICRA), Emili Grahit 101, 17003 Girona, Spain
[2]The University of Girona, 17003 Girona, Spain. E-mail: mpijuan@icra.cat
[3]School of Chemical Engineering, The University of Queensland, St Lucia, QLD 4072, Australia. E-mail: yingfen.zhao@uq.edu.au

SUMMARY

Biological wastewater treatment is conducted by a wide range of microbes with multiple metabolic pathways that are normally affected by the operational conditions applied. Nitrous oxide (N_2O) can be generated during wastewater treatment either as a by-product or an obligatory intermediate. As an environmentally detrimental greenhouse gas, N_2O generated from wastewater treatment systems has attracted a lot of attention due to its important contribution to the overall carbon footprint. When aiming at its mitigation, it is imperative to understand the mechanisms that trigger its formation. Sources and pathways for N_2O production have been categorized into four types based on previous studies: (i) hydroxylamine (NH_2OH) oxidation during ammonia (NH_3) conversion to nitrite (NO_2^-); (ii) NO_2^- reduction by ammonia oxidizing bacteria (AOB), which is called nitrifier denitrification; (iii) heterotrophic denitrification by denitrifiers; and (iv) hybrid abiotic/biotic N_2O production. Despite extensive investigations, the mechanisms for N_2O production await further clarification because of the complexity of wastewater treatment systems. N_2O generation is influenced by a diversity of interrelated conditions: the different species of microbes with specialized functions, the interactions among these microbes in a mixed system, and the responses of microbes to the different environmental factors and operational conditions. This chapter provides an overview of the N_2O production pathways and mechanisms during the biological nitrogen removal (BNR) process and describes the different factors that have been reported to have an effect on N_2O production.

Keywords: Abiotic/biotic N_2O production; denitrification; emission pathways; nitrification; nitrous oxide

TERMINOLOGY

Term	Definition
Activated sludge	Flocs of sludge particles containing living microbes, mainly bacteria and protozoans, which are formed in the presence of oxygen in aeration tanks.

Activated sludge process	The wastewater treatment process developed around 1912–1914 and applied to deal with sewage and industrial wastewater. It contains three main components: an aeration tank, a settling tank and a return activated sludge line. In the aeration tank, activated sludge is applied to speed up the decomposition of contaminants in wastewater. Oxygen is provided in the aeration tank for the metabolization of activated sludge, to convert contaminants into harmless products. After the aeration tank, the mixed activated sludge goes to a clarifier to separate the sludge and treated water.
Acid	A substance that tends to donate a proton and lower pH, or dissolves in water with the formation of hydrogen ions.
Aeration	The introduction of air into wastewater in order to oxidize organic or nitrogenous compounds by microbes, and also for keeping the activated sludge suspended and well mixed.
Aerobic	Conditions with free oxygen in the wastewater.
Ammonia monooxygenase (AMO)	An enzyme catalyzing NH_4^+ oxidation to NH_2OH.
Anaerobic	Conditions without atmospheric or dissolved molecular oxygen in the wastewater.
Anoxic	Conditions of oxygen deficiency or lacking sufficient oxygen as the electron acceptor in the wastewater. Other electron acceptors such as nitrate and nitrite (NOx) would be used by microbes under these situations.
Biomass	A clump of organic material consisting of living organisms, which live on the substrates in wastewater, or the dead organism debris.
Chemical oxygen demand	An indication of the amount of organic materials in wastewater. It refers to the amount of oxygen equivalent consumed in the chemical oxidation of organic matter by strong oxidants such as potassium dichromate.
Chemical oxygen demand to nitrogen ratio	An index to reflect the carbon source availability during denitrification, which requires organic carbon to provide electrons for the reduction of nitrogenous compounds such as nitrate or NO_2^- in wastewater.
Dissolved oxygen	Molecular oxygen dissolved in wastewater.
Greenhouse gas	Gas that absorbs and emits radiant energy within the thermal infrared range and contributes to the global warming effect.
Heterotrophic denitrification (HDN)	A series of reduction reactions from nitrate to nitrogen gas by heterotrophic denitrifiers under anoxic conditions, with organic carbon as the electron donor for the reactions.
Hybrid abiotic/biotic N_2O production	The reactions for N_2O production with the interactions between microbes and chemical compounds during wastewater treatment process.
Hydrogen sulfide gas	A kind of poisonous gas of no color and with rotten egg odor, produced under anaerobic conditions by sulfide reducing bacteria.
Hydroxylamine oxidation	An intermediate step during ammonium oxidation by aerobic ammonium oxidizing bacteria, which would produce greenhouse gas N_2O.
Hydroxylamine oxidoreductase (HAO)	An enzyme catalyzing NH_2OH oxidation to NO_2^-.
Influent	Untreated or partially treated wastewater, which flows into the treatment system for contaminants removal.
Nitrifier denitrification	Reduction of nitrate or NO_2^- to N_2O by nitrifiers under oxygen limiting conditions, with O_2 or H_2 as the electron donor.
Nitrate reductase (NaR)	An enzyme catalyzing nitrate to NO_2^-.
Nitrite reductase (NiRS or NirK)	An enzyme catalyzing NO_2^- to nitric oxide.

Nitric oxide reductase (NoR)	An enzyme catalyzing nitric oxide to N_2O.
Nitrous oxide reductase (NoS)	An enzyme catalyzing N_2O to nitrogen gas.
Nutrient	Substances such as nitrogenous compounds, phosphate or organic carbon that can be assimilated by microbes to promote the metabolism and growth of microbes in wastewater.
Organic matter	Organic waste of plant or animal origin from homes or industry, or originated from storm water run-offs, and so on., which mainly contains the volatile fraction of solids.
Oxidation	Oxidation is the addition of oxygen, removal of hydrogen, or the removal of electrons from an element or compound. In wastewater treatment, organic matter is oxidized to more stable substances.
pH	An indication of the acidity or alkalinity of solutions.
Reactor	A vessel or tank of different size or design which can hold the mixed microbial sludge to conduct physical, chemical or biological reactions for wastewater treatment processes.
Wastewater	The used water including solids discharged from communities, businesses, industry or agriculture that flow into a wastewater treatment plant. Storm water, surface water, and groundwater infiltration also may be included.

2.1 INTRODUCTION

Nitrous oxide (N_2O) is a potent greenhouse gas with a 310-fold greater potential for global warming effects compared with that of carbon dioxide (CO_2) (Edenhofer *et al.*, 2014). Its atmospheric concentration is at present 21% higher than pre-industrial levels, and since the beginning of the 20th century has been exponentially increasing at a rate of 5% per decade due to the anthropogenic introduction of fixed nitrogen into the environment (European Environmental Agency, https://www.eea.europa.eu/). In addition, N_2O is considered as the most important ozone-depleting substance, with a long lifetime in the atmosphere (116 years), and is expected to remain as the largest contributor to ozone depletion throughout the 21st century if no effective mitigation strategies are implemented (Ravishankara *et al.*, 2009).

The majority of N_2O production originates from microbial mediated nitrification and denitrification processes occurring in both terrestrial and aquatic systems. Nitrification and denitrification are also key activated sludge processes in the biological treatment of wastewater and have been implemented in wastewater treatment plants (WWTPs) around the globe. Nitrification refers to the oxidation of ammonia (NH_3) to nitrate (NO_3^-) in a two-step process: first to nitrite (NO_2^-) and then to NO_3^-. Ammonia oxidizing bacteria (AOB) is the most well-known group of microorganisms capable of conducting the first step in NH_3 oxidation (Figure 2.1). Although N_2O is not part of their key metabolic route, it is known that AOB can produce large quantities of N_2O. Interestingly, their genetic inventory does not possess any homologues of the N_2O reductase genes, suggesting that N_2O is the terminal product of NOx reduction in AOB (Klotz & Stein, 2011).

Also, some archaea (the ammonia oxidizing archaea, AOA) have been shown to oxidize NH_3 to NO_2^-. The extent of their contribution to nitrification in natural and managed ecosystems is still under debate, although a recent study found that their abundance correlated negatively to N_2O emission in four WWTPs (Castellano-Hinojosa *et al.*, 2018). Also, AOA have been confirmed to be unable to conduct nitrifier denitrification, one of the main pathways leading to N_2O production (Kits *et al.*, 2017).

The second step of nitrification is conducted by nitrite oxidizing bacteria (NOB). The activity of NOB is closely linked to that of AOB since NO_2^- is usually very scarce in natural environments and therefore NOB depend on NO_2^- production by AOB. Although there is a report showing that NOB may

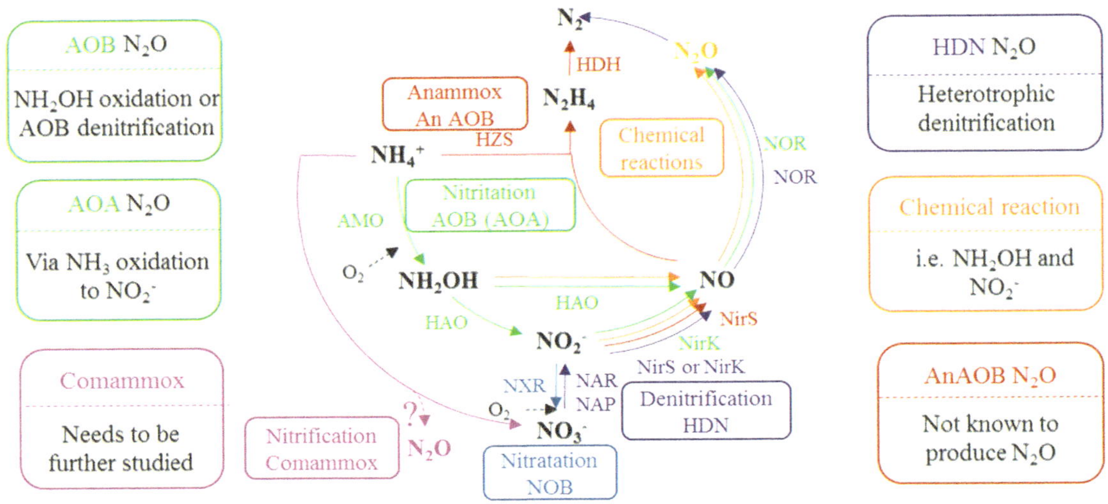

Figure 2.1 Overview of N_2O production and consumption pathways and related microbes during biological nitrogen removal (BNR) (after Desloover *et al.*, 2012). HDH, hydrazine dehydrogenase; HZS, hydrazine synthase; NAP, periplasmic nitrate reductase; NXR, nitrite oxidoreductase.

form N_2O during denitrification of NO_3^- or NO_2^- under anoxic conditions with pyruvate as the electron donor (Freitag *et al.*, 1987), it is widely assumed that their contribution to N_2O emissions is negligible (Law *et al.*, 2012a). In 2015, a complete ammonium oxidizer (comammox) was identified from the genus *Nitrospira*. This bacterium is able to conduct the full oxidation of NH_3 to NO_3^- (Daims *et al.*, 2015; Van Kessel *et al.*, 2015). However, there are very few studies available on the contribution of this process to the overall N_2O emissions. Notably, a pure comammox culture was studied by Kits *et al.* (2019) and they suggested N_2O emitted by *Nitrospira inopinata* was comparable to that from AOA, but much lower than that from AOB. They also demonstrated that this N_2O was originated from abiotic conversion of NH_2OH. This was confirmed by the assumption that the lack of genes for encoding nitrite reductase (NiR) seems to suggest it could only produce N_2O via abiotic pathways (Chen *et al.*, 2020).

On the other hand, denitrification has the potential to produce and consume N_2O. Under anoxic conditions, many different groups of heterotrophic denitrifiers can use NO_3^- or NO_2^- to oxidize a growth substrate. During this process NO_3^- is converted to NO_2^-, which is further reduced to nitric oxide (NO), then to N_2O and finally to nitrogen gas (N_2). The fact that a significant amount of N_2O has been detected in soils carrying out bacterial denitrification suggests that in some cases the last step of denitrification may not be as efficient as the first ones, either because the last step is more sensitive to environmental factors or because the majority of the microbial population does not have the capability to further reduce N_2O to N_2 (Richardson *et al.*, 2009).

Besides the biological reactions for N_2O production, the investigation of the contribution to overall N_2O emissions from abiotic/biotic N-nitrosation production pathway has drawn increasing attention. The abiotic reactions may occur among the different intermediates produced during nitritation (e.g. NH_2OH and NO_2^-) leading to N_2O production, especially in the presence of trace metals (e.g., Fe^{2+}/Fe^{3+}) (Schreiber *et al.*, 2012). However, controversies regarding the overall N_2O contribution caused by this pathway exist: it has been found under some conditions, such as high concentrations of NO_2^- and NH_2OH, the N_2O production from this process can be significant (Soler-Jofra *et al.*, 2016; Terada *et al.*, 2017). But in other studies (Su *et al.*, 2019a), it was found that the abiotic N_2O production pathway contributes little to overall N_2O emissions under the typical pH range in WWTPs.

During the last decade, concerns have increased regarding the direct N_2O emissions produced in WWTPs since they are a nonnegligible source which contributes to approximately 3% of N_2O emissions and represents the sixth largest contributor (Mannina *et al.*, 2018). In order to mitigate N_2O emissions, extensive studies have been conducted in various configurations both at lab-scale and full-scale. The N_2O emission factor reported (amount of N_2O-N emitted relative to the N-load or N-converted) varied from almost negligible emissions to up to 25% (Aboobakar *et al.*, 2013; Ahn *et al.*, 2010; Desloover *et al.*, 2011; Foley *et al.*, 2010; Joss *et al.*, 2009; Vasilaki *et al.*, 2019). In general, higher N_2O emission from lab-scale studies was reported compared with that from full-scale studies. It may be easier for lab-scale studies to focus on a certain parameter affecting N_2O production while controlling other factors, in order to determine the relationship between N_2O production and the factor of interest. In full-scale scenarios, many factors may exert influence simultaneously, so the results from lab-scale studies may not be sufficiently representative of the complexity of full-scale conditions. However, the findings from lab-scale studies could provide some reference and guidance on the impact of different factors for full-scale application. Also, with more and more data collected from lab-scale studies covering all the influencing factors, further insights will be obtained using data-driven technology for the full-scale N_2O emission control applications (Vasilaki *et al.*, 2019).

In the widely applied nitrification-denitrification wastewater treatment process, N_2O can be produced from both anoxic and aerobic zones, but the aerobic zones have been reported to contribute more to N_2O emissions than anoxic zones from BNR reactors. In comparison with conventional nitrification and denitrification process, even more N_2O emissions were found from aerobic zones in novel nitrogen removal processes where partial nitrification (PN) (oxidation of NH_3 to NO_2^-) took place and aeration stripping promoted the N_2O emission (Ahn *et al.*, 2010; Desloover *et al.*, 2012) (Figure 2.2). In the following anammox process, however, anaerobic AOB were not reported to produce N_2O (Desloover *et al.*, 2012). Further, where a partial denitrification (PD) process is proposed

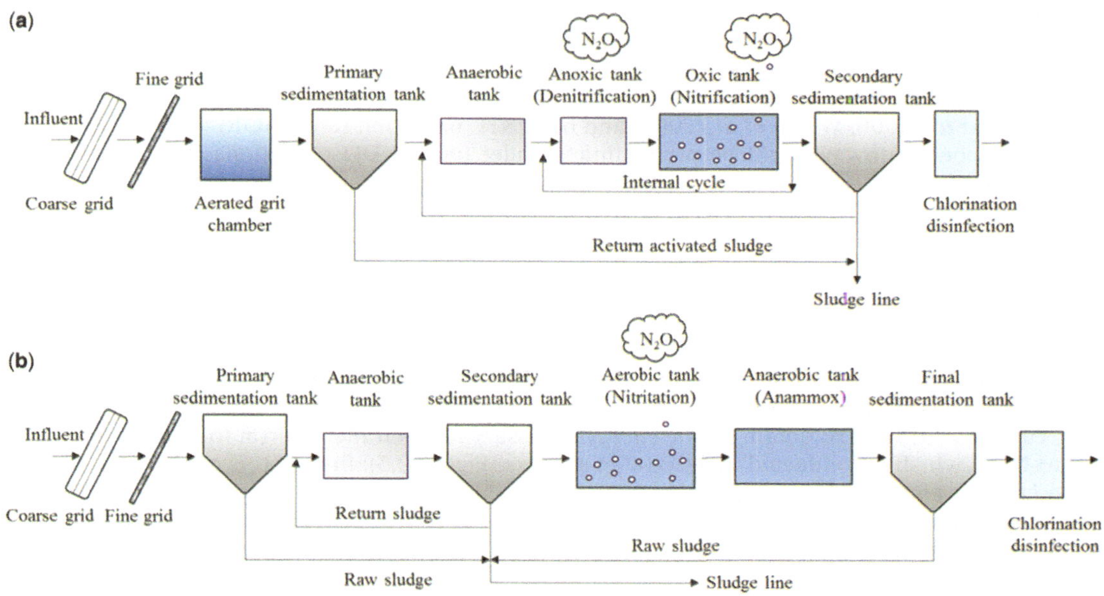

Figure 2.2 Mainstream wastewater treatment scenarios and the units contributing to N_2O emissions: (a) Conventional nitrification-denitrification process; (b) PN-Anammox process.

as an alternative for the PN process to provide NO_2^- for the following anammox, N_2O emissions are believed to be more complicated due to the intensified electron competition among different microbes in these complex systems (Zhou *et al.*, 2020).

Different operational and environmental conditions are applied for nitrification and denitrification processes in WWTPs, including dissolved oxygen (DO), pH, temperature, and so on. These parameters are found to have a close relationship with N_2O emissions (Adouani *et al.*, 2015; Li *et al.*, 2015; Su *et al.*, 2019b; Tumendelger *et al.*, 2014). Also, the substrates or intermediates in the nitrogen removal process are reported to influence N_2O production, such as the nitrogen loading (Frison *et al.*, 2015; Seuntjens *et al.*, 2018), NO_2^-, NH_2OH, NO (Domingo-Félez & Smets, 2019) and organic carbon (Zhu & Chen, 2011). In order to unravel how N_2O generation is influenced by different parameters, a 'black box' approach was first applied with the aim of finding the apparent relationship between N_2O emission and a particular condition (Duan *et al.*, 2017). However, as different conditions are applied simultaneously during the wastewater treatment process, the N_2O production dynamics and how they are influenced under changing conditions deserve further investigation.

This chapter focuses on the pathways leading to N_2O production during nitrification and denitrification and on the main factors regulating its production in wastewater treatment environments. Also, the chemical production of N_2O under wastewater environments is discussed. The effect of several nitrogenous compounds on N_2O production is discussed as well as the effect of easily controllable process parameters such as DO concentration and pH.

2.2 PATHWAYS LEADING TO N_2O PRODUCTION

2.2.1 N_2O production during nitrification

N_2O emissions have been reported from pure culture AOB reactors (Kozlowski *et al.*, 2016a; Liu *et al.*, 2017; Shaw *et al.*, 2005; Yu *et al.*, 2010) and mixed culture ammonium oxidation systems (Kampschreur *et al.*, 2008a; Law *et al.*, 2011; Su *et al.*, 2019a; Terada *et al.*, 2017; Wunderlin *et al.*, 2012). Although there are still some questions to be resolved on the exact mechanisms leading to N_2O formation in AOB, two different metabolic pathways have been proposed as sources of N_2O: (i) the NH_2OH oxidation pathway and (ii) the nitrifier denitrification pathway. The first pathway seems to be favored under high oxygen conditions (Chen *et al.*, 2018; Dundee & Hopkins, 2001; Peng *et al.*, 2014, 2015; Sutka *et al.*, 2006; Wrage *et al.*, 2004) and high NH_3 oxidation rates (AOR) (Law *et al.*, 2012b), while the second pathway might be predominant under limited DO conditions (Peng *et al.*, 2014, 2015; Wrage-Mönnig *et al.*, 2018). However, both pathways seem to occur simultaneously in situ in many cases, and each pathway is regulated differently, even by the same environmental factor, under different operational conditions.

2.2.1.1 NH_2OH oxidation

The production of N_2O via the NH_2OH oxidation pathway is a result of transient NH_2OH accumulation, under conditions where enzyme turnover within the NH_3 oxidation pathway is unbalanced (Cantera & Stein, 2007). The oxidation of accumulated NH_2OH can continue to generate an electron flux that enhances NO_2^- or NO reduction, resulting in increased N_2O production (Domingo-Félez & Smets, 2019; Yu *et al.*, 2018). AOB obtain all the energy necessary for their metabolism from the oxidation of NH_3 to NO_2^- which is conducted in a two-step process (Figure 2.3): first NH_3 is oxidized to NH_2OH by the enzyme ammonia monooxygenase (AMO) and then NH_2OH is further oxidized to NO_2^- by the enzyme NH_2OH oxidoreductase (HAO). N_2O can be produced through biotic or abiotic chemical oxidation. The exact pathway leading to N_2O production from NH_2OH oxidation has been the subject of debate. One accepted model is that HAO oxidizes NH_2OH to NO, which is then reduced to N_2O by NorS, a homologue of nitric oxide reductases (NoR) (Stein *et al.*, 2007; Stein, 2011a) (Figure 2.1). Another possible model is the conversion of NH_2OH by HAO to a nitrosyl radical (NOH), which could then be chemically decomposed to form N_2O (Hynes & Knowles, 1984; Poughon *et al.*, 2000).

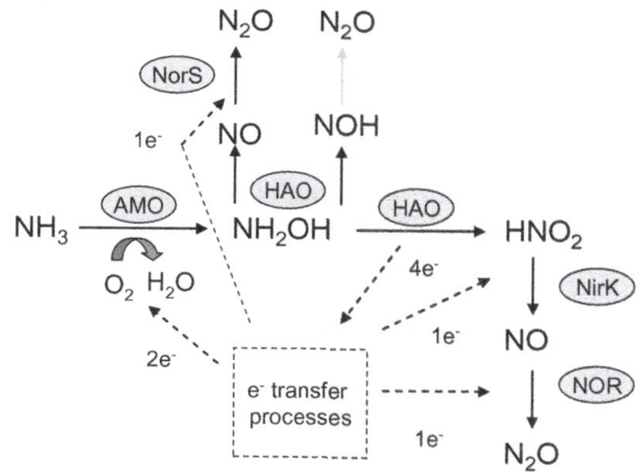

Figure 2.3 Possible nitrogen transformation pathways and enzymes involved in AOB (adapted from Kim *et al.*, 2010). Black arrows represent biological processes; grey arrows represent chemical mediated processes; dashed arrows represent electron fluxes.

However, in contrast to the prevailing view that NH_2OH is the only obligatory intermediate from NH_3 to NO_2^- under the catalysis of AMO and HAO by AOB, a more recent study proposed that NO is an additional obligate intermediate besides NH_2OH in *N. europaea* (Caranto & Lancaster, 2017), predicting participation of a third enzyme in the biological oxidation of NH_3 to NO_2^-, and necessitating more intricate studies of the N_2O production by AOB.

2.2.1.2 Nitrifier denitrification

AOB is also able to reduce NO_2^- to N_2O via NO by NO_2^- and NO reductases (NirK and NoR respectively) without the need for organic carbon. This process is called nitrifier denitrification. Interestingly, homologues to N_2O reductase are absent from the AOB genomes available to date (Klotz & Stein, 2011), indicating the inability of AOB to further reduce the produced N_2O.

Nitrifier denitrification occurs during aerobic conditions together with NH_3 oxidation and it is enhanced under microaerobic conditions (Goreau *et al.*, 1980; Kampschreur *et al.*, 2009; Kozlowski *et al.*, 2016b; Lipschultz *et al.*, 1981; Tallec *et al.*, 2006; Zhu *et al.*, 2013a). The exact function of this pathway in AOB is unclear but several hypotheses have been postulated: (i) energy conservation and production under low DO concentration (Abeliovich & Vonshak, 1992), or electron dissipation when present at high supply rates under high ammonium concentrations (Domingo-Félez & Smets, 2019; Hink *et al.*, 2017), (ii) a decrease in competition for oxygen by removing the substrate for NOB (Poth & Focht, 1985), (iii) a detoxification mechanism to remove the excess NO_2^- (Beaumont *et al.*, 2002; Stein & Arp, 1998; Wrage-Mönnig *et al.*, 2018), and (iv) an electron sink to speed up the oxidation of NH_2OH during aerobic metabolism (Cantera & Stein, 2007; Domingo-Félez & Smets, 2019; Yu *et al.*, 2018). It still remains unknown if only one or a combination of these hypotheses are controlling the activation of this pathway.

The predominance of each of these pathways during nitrification in wastewater treatment systems seems to be influenced by the concentration of the different nitrogen species present in the mixed liquor. Wunderlin *et al.* traced the N_2O sources in their experiments conducted with nitrifying sludge (Wunderlin *et al.*, 2013). They analyzed the nitrogen isotope fractionation of N_2O, which was compared to the isotopic signatures of published pure-culture investigations where the active pathways producing N_2O are known. They found that nitrifier denitrification was the dominant pathway for N_2O production by AOB in their pilot plant treating domestic wastewater. However, during periods of high NH_3 and low NO_2^- concentration, the NH_2OH oxidation pathway became increasingly relevant.

They also conducted some experiments where NH_2OH was added and no NH_3 was present. In that case, almost all the N_2O production originated from the NH_2OH oxidation pathway. However, the exact mechanisms of N_2O formation from this pathway could not be determined. It was suggested that higher ammonium concentrations promote N_2O production due to enhanced ammonium oxidation rates, which will further increase the ammonium turnover into NH_2OH. NH_2OH concentrations in AOB pure cultures or engineered systems are typically low, that is <0.2 mg N/L (Liu *et al.*, 2017), suggesting its continuous oxidation during this process. This would result in an increased flux of available electrons, which would be further dissipated to the anabolic pathways and used for the reduction of NO_2^- or NO, explaining the increasing N_2O production (Domingo-Félez & Smets, 2019). In another study, Law *et al.* attributed the majority of the N_2O detected in their system to the chemical breakdown of the nitrosyl radical formed during NH_2OH oxidation to NO_2^- (Law *et al.*, 2012a). They studied an enriched AOB culture adapted to high levels of NH_4^+ and NO_2^- (~500 mg N/L) and low DO concentrations (0.5–0.8 mg O_2/L) in a lab-scale sequencing batch reactor (SBR) performing partial nitritation of synthetic reject wastewater (1 g N-NH_4^+/L). These conditions might have triggered the predominance of this particular N_2O production pathway. Also, it is possible that different operational conditions favor the development of different AOB strains, which, although being closely related, might possess different N_2O production pathways (Stein, 2011b). More research on the biochemistry and microbial ecology of nitrifying systems is still needed to verify if N_2O production is mainly driven by environmental factors or if it is also related to the predominant AOB strains.

2.2.2 N_2O production during denitrification

Denitrification is performed by a very diverse group of microorganisms which couple oxidation of organic or inorganic substrates to reduction of NO_3^-, NO_2^-, NO, N_2O and then to N_2 under anoxic conditions. Four different enzymes are involved in the process: nitrate reductase (NaR), NiR, NoR and N_2O reductase (NoS) (Figure 2.4) (Zumft, 1997). Each enzyme uses a redox active metal cofactor, such as molybdenum for NO_3^- reduction, iron or copper for NO_2^- reduction, iron for NO reduction, and copper for N_2O reduction (Richardson *et al.*, 2009).

There are several scenarios that can lead to an incomplete reduction of N_2O to N_2, resulting in N_2O accumulation including: (i) the prevalence of microbes harboring incomplete (truncated) denitrification pathways, notably 'N_2O producers' whose denitrifier genomes lack the genes for NoS (Hallin *et al.*, 2018). Such denitrifying communities only conduct the first steps of denitrification, having as an end product N_2O instead of N_2 (Gao *et al.*, 2019). This is not a common problem in mixed microbial environments such as WWTPs, where a high degree of diversity exists, and other N_2O reducing communities would also be present in tandem with N_2O producing communities. (ii) environmental factors that either impose higher inhibition effects on N_2O reductase than other upstream nitrogen reductases, or make N_2O reductase less competitive for substrates (Pan *et al.*, 2012, 2013a; Wang *et al.*, 2014). This leads to an imbalanced rate between N_2O production and reduction during denitrification, which normally further results in N_2O accumulation under these conditions.

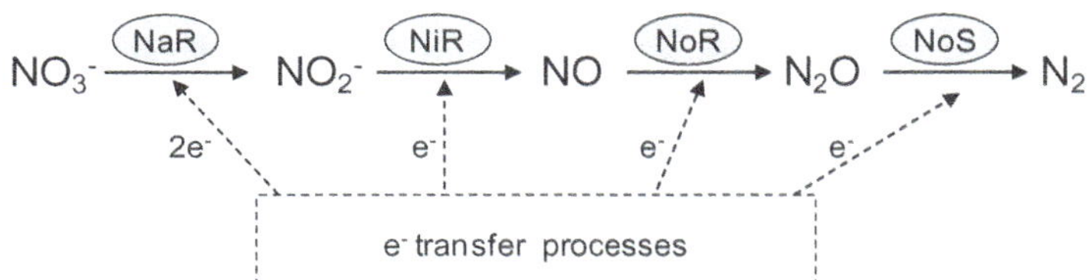

Figure 2.4 Nitrogen reduction steps and enzymes associated with denitrification.

2.2.3 N_2O production through abiotic pathways

In BNR systems, abiotic reactions for N_2O production were reported to occur among reactive nitrogen intermediates such as NH_2OH and free nitrous acid (HNO_2) (Terada *et al.*, 2017). These reactions include: the oxidation of NH_2OH by HNO_2, O_2 and Fe^{3+}, respectively; the reduction of HNO_2 by Fe^{2+}, and the disproportionation of NH_2OH, and so on. (Equations (2).(1) to (2).(5)) (Su *et al.*, 2019b). Chemical reactions among redox active metals, such as iron and manganese, and organics such as humic and fulvic acids can also lead to N_2O production (Zhu-Barker *et al.*, 2015). In soils it has been reported that Fe (III) and Mn (IV) are able to oxidize NH_2OH, producing N_2O. Also, chemical denitrification of NO_2^- coupled with Fe (II) oxidation can result in N_2O formation as shown by Kampschreur *et al.* (2011). Although in previous studies chemical reactions leading to N_2O production were ignored or deemed unimportant in wastewater treatment systems (Law *et al.*, 2011; Zhu-Barker *et al.*, 2015), their contribution could increase when dealing with wastewater with high levels of metals and when NO_2^- accumulates (Harper *et al.*, 2015; Kampschreur *et al.*, 2011; Zhu-Barker *et al.*, 2015). For example, some recent findings reported that hybrid N_2O production was a dominant pathway in a PN SBR, accounting for about 51% of the total N_2O production. Also, Su *et al.* (2019a) suggested that abiotic reaction contributions would become important at acidic pH (≤5). Further investigation of the contribution to overall N_2O emissions from the abiotic/biotic N-nitrosation production pathway is warranted, especially when more novel nitrogen removal technologies are developed for wastewater treatment.

(1) The oxidation of NH_2OH by HNO_2:

$$NH_2OH + HNO_2 \rightarrow N_2O + 2H_2O \tag{2.1}$$

(2) The oxidation of NH_2OH by O_2:

$$2NH_2OH + O_2 \rightarrow N_2O + 3H_2O \tag{2.2}$$

(3) The disproportionation of NH_2OH:

$$4NH_2OH \rightarrow 2NH_3N_2O + 3H_2O \tag{2.3}$$

(4) The reduction of HNO_2 by Fe^{2+}:

$$2HNO_2 + 4Fe^{2+} + 4H^+ \rightarrow 4Fe^{3+} + N_2O + 3H_2O \tag{2.4}$$

(5) The oxidation of NH_2OH by Fe^{3+}:

$$4Fe^{3+} + 2NH_2OH \rightarrow 4Fe^{2+} + N_2O + H_2O + 4H^+ \tag{2.5}$$

2.3 FACTORS AFFECTING N_2O PRODUCTION

2.3.1 Factors influencing N_2O production during nitrification

2.3.1.1 Role of the N compounds on N_2O production by AOB

In nitrification, nitrogenous compounds such as substrates (NH_3) and intermediates/products (NH_2OH, NO_2^- and NO) could all affect N_2O production amounts and pathways by AOB. However, their exact effect on N_2O production and the optimal levels for N_2O minimization in nitrifying systems are not fully understood. The following section summarizes some of the studies that have inferred the correlation between N_2O emissions and these N compounds.

2.3.1.1.1 NH_3

NH_3/ammonium (NH_4^+) has been reported as an important factor affecting N_2O and NO emissions in AOB under aerobic and anaerobic conditions (Kampschreur *et al.*, 2008b; Wunderlin *et al.*, 2012; Yu *et al.*, 2010). The effect of pulse NH_4^+ additions on N_2O production under aerobic conditions was first reported by Kampschreur *et al.* (2008a). An increase of N_2O emissions was found each time that NH_4^+ was added. NO was also emitted but only when NH_4^+ was present, and was not affected by the concentration of NH_4^+.

Wunderlin *et al.* observed N_2O production as soon as NH_4^+ was added in a batch test conducted with nitrifying sludge (Wunderlin *et al.*, 2012), reaching its maximum when NH_4^+ was still high (~20 mg N/L) and NO_2^- was low (~0.5 mg N/L). They attributed this N_2O production to a shift in the AOB metabolism from a low specific activity (periods without NH_4^+) towards the maximum specific activity (after a pulse of NH_4^+). This was previously suggested by Yu *et al.* who studied the effect of transient conditions in a pure culture of *Nitrosomonas europaea* (Yu *et al.*, 2010). They detected a peak of N_2O during the transition from anoxia to aerobic conditions, when NH_4^+ was present. Based on the experimental data they concluded that the N_2O peak detected was due to a shift in the AOB metabolism from periods of low metabolic rates (during anoxia) to periods of high nitrogen flux through the catabolic pathways (during aerobic conditions).

An exponential correlation between AOR and N_2O emission rate was reported in a study conducted with an enriched AOB population treating synthetic reject wastewater (Law *et al.*, 2012b). These authors also indicated that under controlled DO and pH conditions, both AOR and N_2O emission rates were constant despite NH_4^+ and NO_2^- concentrations varying from 250–550 and 450–750 mg N/L, respectively. Higher NH_3 concentration promotes higher AOR, potentially leading to higher *amo* gene expression. This might result in NH_2OH accumulation which, as explained in the following section, leads to N_2O formation through the NH_2OH oxidation pathway (Chandran *et al.*, 2011).

In addition to the impact of NH_4^+ feeding patterns and concentrations, NH_4^+ loading variation was also reported to affect N_2O emissions. A positive linear relationship was found between the NH_4^+ loading rate (ranging from 746 to 2988 mg/L/d) and the N_2O emissions in an expanded granular sludge bed reactor, with the N_2O conversion rates increasing from 5.5% to 8.5% (Fang *et al.*, 2020). Such promoted N_2O emissions may be explained by the increased NH_3 turnover to NH_2OH under high NH_4^+ loading, which would cause NH_2OH accumulation, with the oxidation of NH_2OH providing more electrons for NO_2^- or NO (Domingo-Félez & Smets, 2019). Therefore, decreasing the NH_4^+ shock loads has been proposed as one of the strategies for N_2O mitigation (Peng *et al.*, 2017). However, these studies were based on temporary effects of increased NH_4^+ loading on N_2O emissions. Contrasting results were found with an *N. europaea* chemostat enrichment under repeated exposure to NH_3 pulse loadings on an hourly basis every day. The authors observed the maximum N_2O emission at 160 ppm on day 1 and it decreased to around 60 ppm on day 18 (Chandran *et al.*, 2011). It was suggested that further studies be conducted on the possible adaptive response in N_2O emissions to long-term NH_3-N loading shock.

2.3.1.1.2 NH_2OH

NH_2OH is one of the key intermediates in the catabolism of AOB. NH_2OH oxidation is an energy generating step responsible for the proton gradient formation in AOB (Arp & Stein 2003; Casciotti *et al.*, 2003; Domingo-Félez & Smets, 2019). Also, NH_2OH is highly toxic for many bacteria and although AOB seem to be more tolerant to this compound than other microorganisms, its accumulation can cause a decrease in their NH_3 oxidation rate (Böttcher & Koops, 1994; Xu *et al.*, 2012). These characteristics make its build-up unfavorable for AOB. It has been postulated that NH_2OH only accumulates under transient conditions where enzyme turnover within the AOB pathway is unbalanced (Cantera & Stein, 2007; Yu *et al.*, 2010) or in those systems with high conversion of NH_3, such as high-strength wastewater systems (Schreiber *et al.*, 2012), particularly when NO_2^- accumulates, since NH_2OH oxidation can be inhibited by NO_2^- (Yu & Chandran, 2010).

As discussed previously, NH_2OH accumulation enhances N_2O production via the NH_2OH oxidation pathway. So, when NH_2OH is externally added in an AOB culture, N_2O production will be enhanced (Stein, 2011a). However, despite this clear link between NH_2OH and N_2O production, there are only a few studies reporting on the effect of NH_2OH on N_2O emission from wastewater treatment systems. Wunderlin *et al.* explored the effect of NH_2OH addition in a nitrifying culture (Wunderlin *et al.*,

2012). Two batch tests were conducted at different DO concentrations (1.1 and 2.2 mg O_2/L) where NH_2OH was added as a pulse, resulting in a concentration of 10 mg N-NH_2OH/L. In these tests, it was observed that 6.9–8.5% of the oxidized NH_2OH was converted to N_2O, which was much higher than the N_2O emitted in those experiments where NH_3 instead of NH_2OH was added (1.3–3.8%). In another study, Rodriguez-Caballero and Pijuan explored the N_2O emission dynamics of a nitritation SBR treating synthetic reject wastewater (Rodriguez-Caballero & Pijuan, 2013). They observed that the presence of only NH_2OH at the beginning of the settling phase, when the DO concentration was zero, triggered production of N_2O, which was emitted at the beginning of the subsequent cycle.

Terada *et al.* applied $^{15}NH_2OH$ to an enriched AOB culture from a PN SBR to study the NH_2OH interactions with NO_2^- for N_2O production (Terada *et al.*, 2017). They found that under NO_2^- concentration of 400 mg N/L for each batch test, the N_2O production rates positively correlated with the initial $^{15}NH_2OH$ concentrations (1, 5, 10 and 20 mg N/L). It is however important to consider that the addition of NH_2OH, even in small quantities, creates conditions which might differ from those occurring in real systems and may artificially enhance the production of N_2O in AOB cells which are forced to remove the electrons generated from the oxidation of this external supply of NH_2OH.

2.3.1.1.3 NO_2^-

NO_2^- is the toxic product of aerobic NH_3 oxidation in AOB and it is considered as one of the key parameters affecting N_2O emissions in these bacteria, by increasing their nitrifier denitrification activity. Higher N_2O generation has been associated with higher NO_2^- concentrations in wastewater treatment systems (Foley *et al.*, 2010). In general, N_2O emissions are reported to be higher in those systems performing PN (where NH_3 is mainly oxidized to NO_2^-) compared to full nitrification or PN-anammox systems where generally NO_2^- does not accumulate (Ahn *et al.*, 2010; Desloover *et al.*, 2011; Kampschreur *et al.*, 2008b; Okabe *et al.*, 2011).

Rodriguez-Caballero *et al.* measured the N_2O emissions from a PN lab-scale reactor which was converted to a full nitrification system by adding enriched NOB biomass (Rodriguez-Caballero *et al.*, 2013). They observed a reduction of more than 50% of the N_2O emissions when operating the reactor under full nitrification conditions, which was attributed to avoiding having NO_2^- accumulation. Similar results had been previously reported by Ahn *et al.* (2011), detecting an increase in N_2O emissions when transforming a full nitrification system to a PN reactor. Interestingly, in their case N_2O emissions during the PN period were reduced after 75 days of operation at that particular mode and they suggested that transition periods rather than the modes themselves could be responsible for the increase in N_2O production observed.

However, contrasting results were reported by Law and co-workers (Law *et al.*, 2013). They determined that the N_2O production rate was the highest at NO_2^- concentrations below 50 mg N/L using an enriched AOB biomass from a partial nitritation reactor treating synthetic reject wastewater. When NO_2^- was increased, the N_2O production rate gradually decreased. In their study, higher NO_2^- concentrations resulted in lower N_2O emissions, suggesting that exceedingly high NO_2^- concentrations in nitritation systems is not necessarily related to an increase in N_2O production.

This is also somehow contradictory to the results reported by Kampschreur *et al.*, who found higher N_2O production when adding NO_2^- in a step wise mode (NO_2^- pulses of 5 and 15 mg N/L) during aerobic NH_4^+ oxidation in a full nitrification system (Kampschreur *et al.*, 2008a). Interestingly, they also reported a proportional relationship between the NO_2^- concentration and the NO concentration measured in the concentration range from 2.5 to 25 mg N-NO_2^-/L. More recently, Castro-Barros *et al.* reported that NO_2^- pulses resulted in an increase in N_2O and NO emissions in a nitrifying lab-scale reactor fed with low strength ammonium wastewater. These emissions decreased to the original levels when NO_2^- was completely oxidized to NO_3^- (Castro-Barros *et al.*, 2016).

These differences could be related to the fact that different AOB strains possess different adaptation strategies to high NO_2^- environments. This was suggested by Cua and Stein (2011) who tested the response of 3 different AOB strains to NO_2^-. Their results indicated that each strain evolves its own set of genetic and physiological adaptations to high NO_2^- environments and showed that NO_2^- have physiological and genetic effects that vary among different strains. Therefore, it is possible that the same NO_2^- concentration triggers different N_2O production dynamics depending on the type of AOB. Another explanation could refer to the adaptation of AOB to different environments with different NO_2^- concentrations. In the study by Law *et al.*, AOB were adapted to 500 mg N-NO_2^-/L (Law *et al.*, 2013) whereas the AOB from Kampschreur *et al.* (2008a) were developed in a full nitrification reactor, where NO_2^- accumulated up to 15 mg N/L but was quickly oxidized to NO_3^- afterwards (Kampschreur *et al.*, 2008b). More research is needed to determine whether the same AOB strains can respond differently to certain environmental factors, producing more or less N_2O, depending on their adaptation capabilities.

2.3.1.1.4 NO

Production of NO during NH_3 oxidation by AOB has been reported in pure culture studies (Kester *et al.*, 1997; Yu *et al.*, 2010), lab-scale systems (Kampschreur *et al.*, 2008a; Ribera-Guardia & Pijuan, 2017; Rodriguez-Caballero *et al.*, 2013) and full-scale nitrifying reactors (Gustavsson & la Cour Jansen, 2011; Kampschreur *et al.*, 2008b). Under oxic conditions small quantities of NO are produced during the oxidation of NH_2OH to NO_2^- by HAO (Domingo-Félez & Smets, 2019; Hooper & Terry, 1979) or via the reduction of NO_2^- through the nitrifying denitrification pathway (Goreau *et al.*, 1980; Starkenburg *et al.*, 2006; Wrage-Mönnig *et al.*, 2018). This NO has been suggested to trigger N_2O production as a way for AOB to avoid nitrosative stress (Klotz & Stein, 2011). However, some studies suggest that NO is an essential intermediate in NH_3 oxidation (Caranto & Lancaster, 2017), enhancing NH_3 oxidation when present and suppressing AOB growth when it is stripped from the media (Zart *et al.*, 2000).

Few studies have reported the combined emission of NO and N_2O from mixed nitrifying systems at lab and full-scale, which makes it difficult to establish a clear relationship between the two compounds. NO emissions are usually one order of magnitude lower than N_2O emissions from the same system: 0.03% of the N-converted emitted as NO and 2.8% as N_2O in a nitrifying lab-scale reactor (Kampschreur *et al.*, 2008a); 0.05% NO and 0.83% N_2O in a nitritation lab-scale reactor (Rodriguez-Caballero & Pijuan, 2013); and 0.4% NO and 3.4% N_2O in a full-scale nitritation reactor (Kampschreur *et al.*, 2008b). When NH_4^+ is added as a pulse, NO slightly peaks and then decreases to a lower but constant concentration that finally drops when NH_4^+ is consumed (Kampschreur *et al.*, 2008a; Rodriguez-Caballero *et al.*, 2013). These emissions, however, increase by one order of magnitude when the nitrifying reactor is subjected to anoxic conditions. Kampschreur *et al.* (2008a) observed an immediate NO peak when air was switched to N_2 in a mixed nitrifying reactor (Kampschreur *et al.*, 2008a). Interestingly, this NO increase was followed by an increase in N_2O. A similar increase in N_2O and NO was also observed by Rodriguez-Caballero and Pijuan during the transition from aerobic to anoxic conditions in an enriched AOB reactor (Rodriguez-Caballero & Pijuan, 2013). In their case an NO peak at the start of anoxia was observed, which slowly decreased afterwards. To explain this decrease, they suggested that a compound (maybe nitrogen tetroxide, N_2O_4) necessary for NO production in anoxia, was being consumed. The N_2O production in their system, however, remained constant during the 20 minutes of imposed anoxic conditions. This was also observed by Ribera-Guardia *et al.*, when subjecting an enriched AOB culture to anaerobic conditions for more than an hour (Ribera-Guardia & Pijuan, 2017). On the contrary, Yu *et al.* only reported N_2O generation during the transition from anoxic to aerobic conditions in a pure culture of *Nitrosomonas europaea* (Yu *et al.*, 2010). In their experiments, the switch from aerobic to anoxic conditions only triggered the production of NO.

It seems clear that NO is not only a simple intermediate product in N_2O formation by AOB. What is still unknown is if its production enhances the production of N_2O in a direct or indirect way.

2.3.1.2 Role of controllable process parameters during nitrification: DO and pH

DO and pH have been recognized as factors affecting N_2O emissions. Both parameters can be modified to a certain extent and controlled to the levels desired in wastewater treatment systems, with consideration given to achieving balance between desirable nitrogen removal performance and N_2O minimization. However, the optimum operational threshold for these parameters is still unknown. A summary of the studies that have reported on the effect of these two parameters on N_2O emissions from nitrifying cultures is presented below.

2.3.1.2.1 DO

DO is considered an important parameter affecting N_2O emissions (Kampschreur *et al.*, 2009), with lower DO concentrations generally increasing N_2O emissions. Under oxygen limiting conditions, AOB would use NO_2^- instead of oxygen as the terminal electron acceptor, producing N_2O. However, it is still unclear if an optimum DO concentration threshold to minimize N_2O emissions can be established for nitrifying systems. This is most likely because other compounds (i.e., N compounds discussed above) are also having a simultaneous effect on N_2O.

It is however possible to find a correlation between N_2O emissions and DO concentration in experiments conducted in the same reactor. Zheng *et al.* investigated the effect of having different DO concentrations in a mixed nitrifying reactor (Zheng *et al.*, 1994). They observed a maximum N_2O production of 7% (N_2O/N-converted) at 0.2 mg O_2/L which decreased at concentrations lower and higher than this DO level. Similar results were also found by Tallec *et al.* (2006). They found that the N_2O emission factor in their experiments reached a maximum of 0.4% at DO concentrations around 1 mg O_2/L. At DO concentrations higher than 2 mg O_2/L and lower than 1 mg O_2/L the N_2O decreased. They repeated the same experiments adding NO_2^- and observed that N_2O emissions significantly increased but displayed the same trend.

In another study, Law *et al.* studied the dependency of AOR and N_2O production of an enriched AOB culture on varying DO concentration (Law *et al.*, 2012a). They found that N_2O production was minimal at low DO concentrations (0.05–0.2 mg O_2/L) and increased when DO was increased (up to 2.4 mg O_2/L). However, in this study, the increase of the DO caused an increase in the AOR, which was shown by the same authors to be exponentially correlated with the N_2O emission rate. Therefore, in this case, the increase in N_2O emissions detected at higher DO could not be solely attributed to this parameter.

A study on a full-scale nitrifying activated sludge treatment plant reported higher N_2O emissions at DO concentrations lower than 1 mg O_2/L (Aboobakar *et al.*, 2013). These emissions decreased with increasing DO concentrations. A similar profile was found in a granular airlift pilot reactor (150 L) conducting full nitritation from reject wastewater (Pijuan *et al.*, 2014). This reactor operated continuously having similar concentrations of NO_2^- and NH_4^+ throughout the experimental period. A clear dependency of N_2O emissions on DO concentration was found in the range of 1 to 4.5 mg O_2/L, increasing within this range as DO values were lowered. At higher DO concentrations, N_2O emissions remained constant. The strong dependency of N_2O emissions at relatively high DO concentrations found in this pilot plant might be due to the fact that nitrifying granular sludge was predominant in this reactor. Granular biomass is expected to display larger oxygen gradients, implying that some of the nitrifying biomass experiences low or even zero oxygen levels, triggering more N_2O emissions (Kampschreur *et al.*, 2008b).

It is possible that the optimal DO level for minimal N_2O emissions will have to be established for each system, considering also the concentration of other compounds that are affecting these emissions. In this sense, Peng *et al.* reported the combined effect of DO and NO_2^- concentrations on the N_2O production of a nitrifying culture (Peng *et al.*, 2015). Results showed that at each DO level, as NO_2^- concentration increased so did the N_2O production rate and emission factor (Figure 2.5). On the other hand, at each NO_2^- level, N_2O production rate and emission factor decreased as DO concentrations increased. Further studies are needed to assess the combined effects of several parameters on N_2O emissions to provide more reference for the full-scale N_2O mitigation practice.

2.3.1.2.2 pH

Changes in pH have also been reported to influence N_2O production in AOB. An early study by Hynes and Knowles reported that the rates of production of N_2O were changing when changing pH within

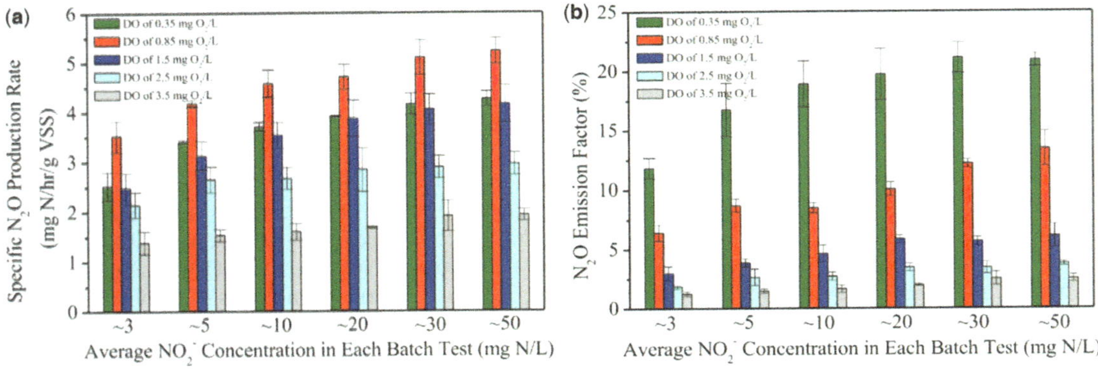

Figure 2.5 The combined effect of NO_2^- and DO concentrations on N_2O production rate and emission factor (from Peng *et al.*, 2015). (VSS refers to volatile suspended solids).

the range of 5.4 to 9.5, having a maximum N_2O production at pH 8.5 in a pure culture of *Nitrosomonas europaea* (Hynes & Knowles, 1984). However, when changing pH, other parameters such as free ammonia (FA) concentration and free nitrous acid (FNA) concentration were also changing. Shiskowski and Mavinic suggested that FNA rather than NO_2^- was the actual electron acceptor for the nitrifier denitrification pathway in AOB (Shiskowski & Mavinic, 2006). They observed a reduction in N_2O production rate when pH was increased, contradicting the study of Hynes and Knowles, which they attributed to the lower availability of FNA to AOB cells.

In another study, Law *et al.* conducted several batch tests with an enriched AOB culture where the pH effect on N_2O production was assessed within the range of 6.0 to 8.5 (Law *et al.*, 2011). The N_2O production rate increased when pH increased from 7 to 8 but started to decrease at pH 8.5. Changes of pH at levels below 7 did not affect the N_2O production rate. A similar profile was obtained for the AOR in their culture at different pH set points. They suggested that the pH effect on N_2O production could be indirect, increasing the AOR, which has been reported to be exponentially correlated with N_2O production as discussed previously. In their study, they also exposed the AOB culture to a range of FA and FNA concentrations under pH-controlled conditions and showed that changes in FA and FNA did not result in significant changes in N_2O production. Similar results of the pH effect on N_2O production were also found in a more recent study (Su *et al.*, 2019b). Two intermittently-fed SBRs were operated in parallel to achieve nitritation. Different pH set points (from 6.5 to 8.5, with an increment of 0.5) were applied to test the pH effect on nitrogenous compounds conversion and N_2O production. For all the pH levels studied, the NH_4^+ removal rate remained nearly constant, and no significant changes in NO_2^- accumulation rate were measured. The net N_2O production rate showed a clear relationship with the pH variations, which increased with the pH from 6.5 to 8, then decreased slightly when pH increased to 8.5. The same trend was found for the net N_2O yield of NH_4^+ removed (Figure 2.6). A best-fit model was applied in this study to simulate the nitrogen conversion and N_2O production pathways under these pH levels, and the nitrifier denitrification pathway was dominant for the N_2O production at all pH levels. A reduction of up to seven-fold N_2O production was suggested by the authors if the nitritation reactors were operated at slightly acidic or neutral pH levels. Such pH control, however, should be further evaluated in terms of economic and environmental applicability before being implemented in full-scale WWTPs. In order to unravel the mechanism of the pH effect on N_2O production, more microbial information should also be investigated at a deeper level. The effect of pH on N_2O production from the perspective of functional enzymatic processes, pathways and microbial activities was reviewed (Blum *et al.*, 2018). In this paper they suggested the pH optima

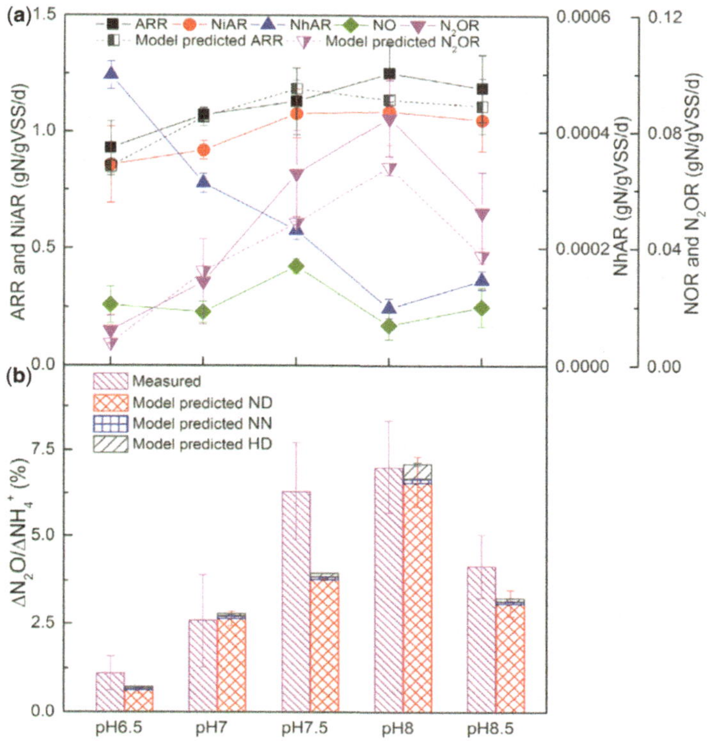

Figure 2.6 Specific nitrogenous compounds conversion rates (a); the net N_2O yield of NH_4^+ removed (b) at different pH set-points (experimental and simulation results) (from Su *et al.*, 2019b). Note: abbreviations can be found in the nomenclature at the end of this chapter.

of different N-converting enzymes based on previous literature. Since the pH optimum varies among different functional enzymes in nitrification, the imbalance of enzymatic activities under a certain pH point could promote N_2O accumulation.

While the studies on the effect of pH on N_2O production have thus far only focused on high-strength wastewater, studies on how N_2O production is affected by AOB in mainstream conditions are still lacking. In addition, long-term experiments would help to clarify if AOB can adapt to the new conditions and minimize their N_2O production, especially in those scenarios where pH changes are linked to an increase in the AOR.

2.3.2 Factors influencing N_2O production during denitrification

N_2O is an intermediate compound in the denitrification process and its accumulation is strictly linked to the activity of the NoS enzyme. N_2O can accumulate due to two main reasons: (i) when the majority of the denitrifying community does not possess the gene encoding for NoS, therefore having N_2O as the end product of denitrification; or (ii) when NoR is affected by a certain environmental or operational factor, becoming lower than the NaR or NiR. Several factors which have been reported to lead to N_2O accumulation during denitrification to date, will be discussed in this section.

2.3.2.1 Role of electron donors and acceptors under denitrification

2.3.2.1.1 External carbon source

In the denitrification process, organic carbon sources are required to provide electron donors for heterotrophic denitrifying bacteria. Methanol, ethanol, acetate, sludge fermentates, and real wastewater have been used as carbon sources for denitrification (Law *et al.*, 2012b; Weissbach *et al.*, 2018). Microbes have different metabolic pathways for different carbon sources. The metabolism and utilization of carbon sources with different biodegradabilities are different during denitrification. Therefore, different organic carbon sources have different electron donor capabilities, which would affect denitrification performance and N_2O production. Generally, the denitrification rate with easily biodegradable small organic matter as a carbon source is higher than that of large organic matter. For example, Lu and Chandran investigated the emissions of N_2O in two different denitrification reactors using easily biodegradable organic carbon sources: methanol and ethanol, respectively (Lu & Chandran, 2010). Better denitrification efficiency and lower N_2O release were found (less than 0.2% of total nitrogen in water) in the denitrification process (Lu & Chandran, 2010), though the authors suggested that emissions were different depending on the carbon source used, and they concluded that N_2O emissions could not be generalized for all carbon sources. This is because the microbial structure in different systems varied as a result of the carbon sources applied. For example, some microbes may not have genes encoding NoS, or the expression of NoS of some microbes varies under different carbon source conditions, resulting in different denitrification performance and N_2O production (Law *et al.*, 2012a). Another study performed by Belmonte *et al.* explored the N_2O emissions using acetate and swine wastewater as carbon sources during the denitrification process and the results showed different N_2O production depending on the carbon source, with more emissions observed for the swine wastewater (Belmonte *et al.*, 2012). However, even though these correlations were found between N_2O emissions and different carbon types, it is still unclear if the type of carbon source can have an effect on the N_2O reduction rate and what the mechanism behind it is.

2.3.2.1.2 Internal carbon source

Under the conditions of insufficient external organic carbon, denitrifiers will store internal carbon sources such as polyhydroxyalkanoates (PHA) to use for endogenous respiration in the biological nitrogen and phosphorus removal process. Previous studies have reported the accumulation of N_2O in those systems where denitrification was conducted using PHA, such as in biological reactors containing denitrifying phosphorus accumulating organisms (dPAOs) or denitrifying glycogen accumulating organisms (dGAOs) (Wang *et al.*, 2011; Zeng *et al.*, 2003). For example, Schalk *et al.* observed that when external COD was limited and PHA served as the growth substrate, N_2O started to accumulate (Schalk-Otte *et al.*, 2000). PHA consumption is a rate-limiting step (Beun *et al.*, 2002; Murnleitner *et al.*, 1997), which may trigger competition for electrons between the denitrifying enzymes, which is a possible mechanism to explain N_2O emissions by microorganisms growing on storage compounds. Using a mathematical model calibrated and validated by experimental denitrifying systems, Liu *et al.* demonstrated the linear relationship between N_2O accumulation and polyhydroxybutyrate (PHB; the main PHA fraction) production during denitrification (Liu *et al.*, 2015). And N_2O started to accumulate when PHB was consumed as the sole electron donors. They suggested that when PHB was used as a carbon source, N_2O accumulation would increase as a result of the relatively low N_2O reduction rate under increased PHB consumption.

2.3.2.1.3 Electron competition

As organic carbon is required to provide the electron donor for the N reductases in denitrification, their capability to compete for electrons is important to reduce nitrogenous compounds. This is also related to the availability of organic carbon and the abundance of electrons provided. The negative effect of the simultaneous presence of different nitrogen oxides (NO_3^-, NO_2^- and N_2O) on their reduction rates during denitrification was first reported under low chemical oxygen demand per nitrogen (COD/N) ratios for ordinary heterotrophic denitrifiers that metabolized externally available carbon sources as the electron donor (VonSchulthess & Gujer, 1996). This concept, known as electron competition, was comprehensively evaluated in a denitrifying culture using methanol as the sole carbon source (Pan *et al.*, 2013b). In order to demonstrate the electron competition among the four nitrogenous reductases as well as the effect of the availability of electrons provided by organic carbon on electron competition, two series of batch tests were conducted, namely, under non-carbon-limiting conditions and carbon-limiting conditions, respectively. During each series, a single electron acceptor (NO_3^-, NO_2^-, N_2O) and a combination of their mixtures were applied for denitrification to measure the reduction rates and calculate the electron consumption rates of each nitrogen reductase (NaR, NiR, NoR, NoS), with the electron consumption rate representing the electron competition capability. They found that under non-carbon limiting conditions, the highest electron consumption rates of NaR, NiR and NoS were achieved when NO_3^-, NO_2^- and N_2O were the single electron acceptors, respectively. The electron consumption rates of all the denitrification reductases decreased when a mixture of electron acceptors was applied in comparison to when single nitrogen oxides were added (Figure 2.7). The results for the scenario under carbon-limiting conditions are in agreement with those from non-carbon-limiting conditions, where the electron consumption rates of each nitrogen reductase decreased when mixed nitrogen oxides were added as electron acceptors as compared with when single electron acceptors were applied, suggesting that electron competition also existed in non-carbon-limiting conditions. This further led the authors to conclude that the key for N_2O accumulation was electron competition among nitrogen reductases rather than simply the COD/N ratio.

Similar results were found in a denitrifying culture with glucose as the external carbon source. Under the COD/N ratio of 2, 4 and 8, respectively, the authors found the electron competition intensified among NaR, NiR, NoR and NoS when the COD/N ratio increased, by calculating the electron distribution ratio and competition rates among the four reductases (Zhao *et al.*, 2018). This competition for the electrons in the reduction steps of denitrification was also observed in other populations adapted to other substrates which can have higher denitrification rates than methanol, such as acetate or ethanol (Ribera-Guardia *et al.*, 2014).

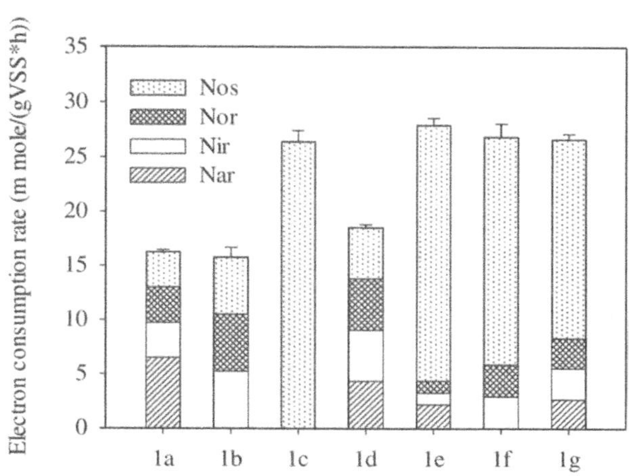

Figure 2.7 Electron consumption rate of NaR, NiR, NoR and NoS during denitrification. Note: 1 refers to batch series under the non-carbon-limiting condition; a–g represents different scenarios of electron acceptors applied: a. NO_3^-; b. NO_2^-; c. N_2O; d. NO_3^- + NO_2^-; e. NO_3^- + N_2O; f. NO_2^- + N_2O; g: NO_3^- + NO_2^- + N_2O (from Pan *et al.*, 2013a).

2.3.2.1.4 NO_2^-/FNA

FNA is an important factor affecting microbial activity and metabolic characteristics and therefore also N_2O production. It is the protonated product of NO_2^- under acidic conditions, with its concentration determined by temperature, pH and NO_2-N concentration (Anthonisen *et al.*, 1976). FNA is a cytotoxin that can lead to the formation of active nitrogen oxides in the cytoplasm which will result in toxicity to microbial cells, inhibiting the growth and productivity of a variety of microorganisms including denitrifying bacteria and denitrifying phosphorus accumulating organisms (dPAOs) (Vadivelu *et al.*, 2006; Zhou *et al.*, 2007). On the other hand, FNA also directly inhibits enzyme activity in microbial cells. Zhou *et al.* (2007) studied the effect of FNA on N_2O reduction by dPAOs, and the study showed that the inhibition degree of N_2O reduction was related to the concentration of FNA. When the concentration of FNA was 0.004 mg N/L, the reduction of N_2O was completely inhibited. The following possible mechanism of FNA inhibiting N_2O reduction is suggested: NoS contains two metal centers, one of which is CuA, a binuclear copper center. It can accept electrons from water-soluble electron donors. The other metal center is a quad-nucleated copper-sulfur center located at the active site (Rasmussen *et al.*, 2005). FNA can bind to the active site containing copper NoS, resulting in competitive inhibition of N_2O reduction and an increase in N_2O accumulation during the reaction (Zhou *et al.*, 2007). Gao *et al.* studied the inhibition mechanism of FNA on *Pseudomonas Aeruginosa*, a denitrifying strain, and pointed out that when the concentration of FNA was 0.1 mg/L, the transcription degree of genes encoding NoS decreased, providing a reasonable explanation for the accumulation of N_2O (Gao *et al.*, 2016). However, several studies have suggested that the presence of NO_2^- in the anoxic period could lead to N_2O accumulation. For example, Pijuan and Yuan showed a higher accumulation of N_2O when NO_2^- rather that NO_3^- was present in the anoxic phase of an SBR reactor treating nutrient-rich abattoir wastewater (Pijuan & Yuan, 2010). While Zhou *et al.* demonstrated that FNA rather than NO_2^- was the compound responsible for the inhibition detected in the N_2O reduction of an enriched dPAO culture (Zhou *et al.*, 2008). Further research on the influencing mechanism of FNA/NO_2^- on microbial activity and N_2O generation in biological denitrification processes is still needed in order to obtain appropriate strategies for N_2O mitigation from denitrification.

2.3.2.2 Role of controllable process parameters under denitrification: DO and pH

2.3.2.2.1 DO

DO is known to inhibit both the synthesis and activity of denitrification enzymes (Lu & Chandran, 2010; VonSchulthess *et al.*, 1994). Also, it is known that NoS is more sensitive to oxygen than the other reductases. Therefore, NoS is more susceptible to inactivation by DO than other upstream enzymes during denitrification, leading to the slower N_2O reduction rate than NO_3^- and NO_2^-, and thus an accumulation of N_2O (Wunderlin *et al.*, 2012). Although oxygen is not expected to be present in the anoxic parts of a WWTP, an over aeration in the aerobic tanks linked with a high internal recirculation might lead to the detection of certain concentrations of oxygen in the anoxic reactor, causing inhibition to the reduction of N_2O.

2.3.2.2.2 pH

pH is known to have an effect on N_2O emissions. Hanaki *et al.* determined that N_2O accumulated at low pH in a lab-scale denitrifying culture using acetate and yeast extract as electron donors and NO_3^- as the final electron acceptor (Hanaki *et al.*, 1992). N_2O production at pH of 6.5 was significantly higher than that at pH of 7.5, although pH of 7.5 and 8.5 showed less difference. Later Thörn and Sörensson (1996) determined a N_2O maximum production when the pH was between 5 and 6 in a pilot plant, which was run as a nitrogen removal system with pre-denitrification in an anoxic basin followed by sedimentation. Similarly, Pan *et al.* determined that substantial N_2O accumulation was observed at low pH levels (6.0–6.5) during denitrification, likely due to electron competition among the four denitrification steps when the electron supply from carbon oxidation was limited (Pan *et al.*, 2012).

The pH effect on N_2O production from denitrification is interpreted from the perspective of the enzyme structure (Blum *et al.*, 2018). As NoS contains two Cu-centers (CuA and CuZ), the N_2O reduction capability is influenced by the electron transfer between the two sites. In the pH range of 4–8, the electron transfer between CuA and CuZ appears to be rate limiting for N_2O reduction (Gorelsky *et al.*, 2006), and it was suggested that pH 7–8 is the optimum range to minimize N_2O production from denitrification (Blum *et al.*, 2018; Fujita & Dooley, 2007).

2.3.2.3 Role of other typical compounds under denitrification

2.3.2.3.1 Hydrogen sulfide or sulfide (H_2S)

H_2S is produced biologically in sewer pipes and could be introduced to the denitrification tank via the influent wastewater. H_2S is known to affect microbial activity in general since it is usually toxic to bacteria. Schönharting *et al.* suggested that H_2S in sewage could alter the activity of heterotrophic denitrification and lead to N_2O accumulation during biological wastewater treatment (Schönharting *et al.*, 1998). In 2013, Pan *et al.* studied the potential inhibitory effects of H_2S on NO_3^-, NO_2^-, and N_2O reduction with a methanol-utilizing denitrifying culture (Pan *et al.*, 2013a). H_2S was found to be strongly inhibitory to N_2O reduction, with 50% inhibition (Figure 2.8). They also observed N_2O accumulation during NO_3^- and NO_2^- reduction when concentrations were above 0.5 and 0.2 mg H_2S-S/L, respectively. Finally, they revealed that the protonated form of H_2S was likely the true inhibitor of N_2O reduction, and the inhibitory effect was reversible. The effect of H_2S on N_2O accumulation during denitrification was also revealed by a mathematical model, with the consideration of electron competition capability of different nitrogen reductases under various H_2S levels (Pan *et al.*, 2019). It was suggested that the N_2O accumulation was due to the reduced electron competition ability of NoS, compared to that of NaR and NiR.

2.3.2.3.2 Copper

A deficiency of copper can lead to N_2O accumulation, since copper is necessary for the production of the enzyme NoS (Richardson *et al.*, 2009). This has also been reported for soils. For example, copper oxide (CuO) nanoparticles were used to study the performance and N_2O production from soil denitrification (Zhao *et al.*, 2020). Copper ions were suggested to be the dominant toxic composition presenting as CuO, which has a severe inhibition effect on denitrification, and N_2O emission rates

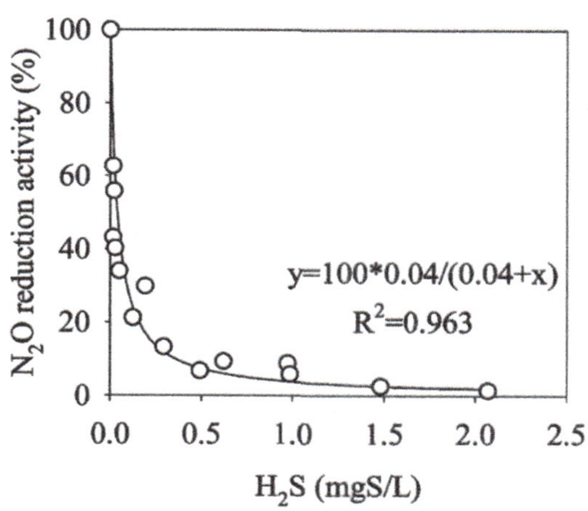

Figure 2.8 The effect of H_2S concentrations on N_2O reduction activity in a methanol-utilizing denitrifying system (from Pan *et al.*, 2013b).

decreased by 10–24% under CuO concentrations of 500 mg/kg. In wastewater treatment systems, the copper concentration is expected to be much lower. Chen *et al.* studied the long-term effect of copper nanoparticles on nitrogen removal and N_2O generation from activated sludge systems, with copper concentrations varying from 0.1–10 mg/L. They found enhanced total nitrogen removal with decreased N_2O production at all concentrations of copper applied (Chen *et al.*, 2012). Similarly, Zhu *et al.* proposed that the N_2O emissions from the denitrification process could be minimized by controlling copper ion concentrations (Zhu *et al.*, 2013b). A reduction of 55–73% for N_2O production was found with the addition of copper at 10–100 μg/L. They also demonstrated that the N_2O reducing denitrifiers decreased after the addition of copper ions by polymerase chain reaction (PCR) assays. However, N_2O production will be stimulated if the copper concentrations are too high to cause the inhibition of NoS (Zhan *et al.*, 2018). Besides the potential effect on N_2O emissions, the effect of copper introduced to wastewater treatment systems on the subsequent sludge handling process should also be considered.

2.3.3 Effect of environmental conditions on N_2O production during nitrification and denitrification

The effect of environmental conditions on BNR processes is unavoidable. This is because the growth of the functional microbes in nitrogen removal benefits from optimal conditions, while the environmental conditions vary with place and time, so it is hard to maintain them at optimum conditions for microbes. Temperature plays one of the most important roles among different environmental conditions. Since temperature control is not feasible in WWTPs, the growth and metabolism of microbes will inevitably undergo diurnal and seasonal temperature variations and thus have different performance levels and N_2O emissions. Besides temperature, salinity is another environmental condition causing disturbance to wastewater treatment processes. Salinity (mostly NaCl) contained in municipal wastewater mainly comes from the intrusion of seawater or saline groundwater in the sewer system, industrial activity and toilet flushing with seawater (De Graaff *et al.*, 2020). This will affect the microbial growth and performance at different levels, and therefore N_2O production. The following sections describe the effect of these factors on microbial growth and N_2O production from nitrification and denitrification processes in detail.

2.3.3.1 Temperature

Temperature could be one of the key factors contributing to N_2O emissions during nitrification and denitrification processes, especially in areas with distinct seasonal temperature variations (Gruber *et al.*, 2020). Temperature could affect both the physicochemical process and biological interactions in wastewater treatment, therefore affecting N_2O production and emission directly or indirectly. The physicochemical processes that are affected include N_2O solubility, mass transfer and chemical equilibrium, and so on., while biological processes mainly include microbial species, growth rate and enzyme activity under different temperatures (Van Hulle *et al.*, 2010).

The apparent relationship between N_2O emissions and temperature has been investigated by some studies. The N_2O solubility decreases as a result of increased temperature (being about two times lower at 25°C than at 5°C), therefore the liquid phase N_2O is more easily stripped into the gas phase, leading to the enhancement of N_2O emissions (Van Hulle *et al.*, 2010). A granular sludge airlift reactor conducting PN at mainstream conditions was operated to demonstrate the relationship between N_2O emissions and temperature. Three different temperature set points (10, 15 and 20°C) were applied during different stages. The authors found that the N_2O gas emissions relative to the oxidized ammonium at 20°C was 2.5 times higher than that at 10°C (emission factor $3.7 \pm 0.5\%$ vs $1.5 \pm 0.3\%$). It was proposed in this study that temperatures higher than 15°C increased N_2O emissions in the PN process. As suggested by Chen *et al.* (2020), temperature disturbances will cause an imbalance between the oxidation reaction of NH_4^+ and NO_2^-. It is possible that a higher temperature may facilitate the NH_4^+ oxidation, which would produce more NH_2OH and thus more N_2O via the NH_2OH oxidation pathway. However, contradictory results were found in a full-scale study of WWTPs, where higher

N_2O emissions were reported at lower temperatures (10–15°C), while at 15–20°C, N_2O emissions were much lower (Gruber *et al.*, 2020). The mechanisms for such N_2O emission dynamics in response to temperature variations were not clear in this study, and the authors postulated that either the accumulation of NO_2^- induced by failure of nitrification, or the reduced denitrification capacity at lower temperatures, may cause the variation of N_2O emission, which awaits further validation.

Studies were also carried out focusing on the effect of temperature on microbial growth and activity during nitrification and denitrification. Zhang *et al.* found an increased AOB activity, which was 148.3% higher at elevated temperature (20°C) than it was at 15°C, in an acid paddy soil. By qPCR (polymerase chain reaction) analysis, the microbial community was transformed from AOA to AOB with elevated temperatures, but the dominant AOB species was not altered during this process (Zhang *et al.*, 2019). Another study of the temperature effect on AOB activity was conducted in a PN SBR treating high strength landfill leachate (Gabarró *et al.*, 2012). In this case, the equilibrium of free ammonium (FA) and free nitrous acid (FNA) concentrations were also considered, where the temperature variations would lead to obvious changes in FA and FNA concentrations under extremely high NH_4^+ concentrations. The authors suggested that the AOB activity was inhibited by both FA and FNA, as a result of the different FA and FNA concentrations induced by temperature variations. From the gene and enzyme level, the relationship between N_2O accumulation ratio and functional gene abundance during denitrification was reported (Zhang *et al.*, 2019). The lowest (NirK+NirS)/NoSZ value occurred at a temperature of 25°C, which suggested that the NoS expression was stimulated and the activity was higher than that of NiR, leading to decreased N_2O accumulation. However, the mechanism of temperature effect on N_2O emissions awaits further investigation to distinguish it from other interrelated factors. Especially, it may be worth investigating if the N_2O emission variations at different temperatures are caused by changes to the biological processes or the physicochemical processes, since both are simultaneously affected by temperature variations.

2.3.3.2 Salinity

Salinity is another factor contributing to N_2O emissions from BNR. High salinity would stimulate N_2O production both directly and indirectly. The direct effect mainly includes the changes of N_2O production pathways under different salinity conditions, while the indirect effect derives from the different inhibitory effects of salinity on microbial properties.

In a study concerning N_2O emissions from a single stage PN/A reactor treating ammonia-rich saline wastewater (Yan *et al.*, 2016), where increasing salinity was applied from 0 to 20 g NaCl/L (in increments of 5 g/L), the authors found that the highest N_2O emission occurred at relatively low NaCl concentrations (5 g/L, with an emission factor of 0.75% N_2O-N/TN influent), while the lowest emission (0.16%) was found at the highest NaCl concentration (20 g/L). The authors suggested that at all salinities studied, the NH_2OH oxidation pathway was the dominant pathway for N_2O production. This was deduced from the exponential relationship between the ammonium oxidation rate (AOR) and the N_2O production rate at all salinities studied. NaCl was also proposed as a selecting factor which can help achieve partial-denitrification (NO_3^- to NO_2^-) in order to provide NO_2^- for the subsequent anammox process (Li *et al.*, 2018). Progressive increases of the salinity (0–5%) in the influent were applied to a denitrifying up flow sludge bed reactor. An average, an NO_2^- accumulation ratio (NO_2^- converted from NO_3^-) of 75% was achieved under the salinity concentration of 5%. Despite the fact that the N_2O emission was not assessed in this study, it is highly possible that the N_2O emission would be promoted in this system, not only by the inhibitory effect of salinity on denitrification reductases (especially NoS), but also by the accumulated NO_2^-. The effect of NaCl on N_2O emissions was shown in mainstream BNR systems as well (Vieira *et al.*, 2019). N_2O emissions were quantified in correlation with process conditions and the periods of infiltration of seawater in a full-scale biological aerated filter. The authors suggested that increased seawater infiltration at high tide led to the augmentation of the daily N_2O production and emission to 13.78 g N_2O-N/kg of NH_4-N removed, compared with the average daily N_2O emissions (6.16 g N_2O-N/kg of NH_4-N removed) monitored. Therefore, the authors proposed the high influent conductivity (salinity) as the indicator of possible increased N_2O emissions for the wastewater treatment plant. However, studies regarding the mechanisms of the effect of salinity on the regulation of the N_2O production pathway are very rare. To improve understanding of these mechanisms, further studies need to be conducted, and the combination of molecular methods and isotope-based technology are recommended.

Besides the direct effect on the N_2O related production pathway, salinity could also affect microbial properties, such as gene expression, enzyme activity, growth rate and microbial activity, and so on. High salinity has been proven to cause significant differences in gene expression of freshwater microorganisms (Völker *et al.*, 1994). Some genes were activated under high salinity conditions, suggesting their possible roles in balancing the osmotic stress (Marin *et al.*, 2004). But studies are still lacking with regard to the nitrogen conversion genes, especially the NoS encoding gene expression under various salinities. Despite this, extensive studies have been conducted on the effect of salinity on microbial activity. For example, it was reported that salinity had a stronger inhibition effect on NOB activity than on AOB activity (Mosquera-Corral *et al.*, 2005; Ye *et al.*, 2009), and such difference was adopted as a strategy to inhibit NOB and facilitate NO_2^- accumulation for PN. For example, it was found that AOB activity was inhibited by 32% at 20 g/L of NaCl, while NOB activity was inhibited by 100% during the steady-state operation of an aerobic granular sludge SBR (Pronk *et al.*, 2014). The decreased NOB activity would facilitate NO_2^- accumulation, which could further promote N_2O production.

2.4 CONCLUDING REMARKS

The research conducted till now has enabled identification of the key metabolic pathways leading to N_2O production in wastewater treatment systems, as well as the key conditions and environmental factors triggering the activation of these pathways. In some cases, quantitative relationships between some of these parameters and the N_2O emissions have been found. Although this is an important step towards the development of new models which can predict N_2O emissions from WWTPs under different scenarios, there are still many unknowns regarding the combined effect of several factors. Also, some literature seems to suggest that microbial communities can adapt to certain environmental conditions, thus reducing their N_2O production.

This review has mainly focused on N_2O emission pathways and reported how several factors affect the nitrifiers and denitrifiers that are involved in N_2O production in wastewater treatment systems. It is important to remark that the complexity increases in real systems where nitrifiers and denitrifiers coexist, making it more difficult to interpret how one group can directly or indirectly affect the N_2O production potential.

ACKNOWLEDGEMENTS

Maite Pijuan acknowledges the support from the Economy and Knowledge Department of the Catalan Government through a Consolidated Research Group (ICRA-TECH – 2017 SGR 1318) – Catalan Institute for Water Research and the Spanish Government through the Salvador de Madariaga mobility program (PRX19/00051). Yingfen Zhao acknowledges the scholarship support from the China Scholarship Council (CSC), the ARC Discovery Project (180103369), and the University of Queensland.

REFERENCES

Abeliovich A. and Vonshak A. (1992). Anaerobic metabolism of Nitrosomonas europaea. *Archives of Microbiology*, **158**(4), 267–270, https://doi.org/10.1007/BF00245243

Aboobakar A., Cartmell E., Stephenson T., Jones M., Vale P. and Dotro G. (2013). Nitrous oxide emissions and dissolved oxygen profiling in a full-scale nitrifying activated sludge treatment sludge treatment plant. *Water Research*, **47**(2), 524–534, https://doi.org/10.1016/j.watres.2012.10.004

Adouani N., Limousy L., Lendormi T. and Sire O. (2015). N_2O and NO emissions during wastewater denitrification step: influence of temperature on the biological process. *Comptes Rendus Chimie*, **18**(1), 15–22, https://doi.org/10.1016/j.crci.2014.11.005

Ahn J. H., Kim S. P., Park H. K., Rahm B., Pagilla K. and Chandran K. (2010). N_2O emissions from activated sludge processes, 2008–2009: results of a national monitoring survey in the United States. *Environmental Science & Technology*, **44**(12), 4505–4511, https://doi.org/10.1021/es903845y

Ahn J. H., Kwan T. and Chandran K. (2011). Comparison of partial and full nitrification processes applied for treating high-strength nitrogen wastewaters: microbial ecology through nitrous oxide production. *Environmental Science & Technology*, **45**(7), 2734–2740, https://doi.org/10.1021/es103534g

Anthonisen A. C., Loehr R. C., Prakasam T. B. S. and Srinath E. G. (1976). Inhibition of nitrification by ammonia and nitrous acid. Journal of the Water Pollution Control Federation, **48**(5), 835–852.

Arp J. and Stein L. (2003). Metabolism of inorganic N compounds by ammonia-oxidizing bacteria. *Critical Reviews in Biochemistry and Molecular Biology*, **38**(6), 471–495, https://doi.org/10.1080/10409230390267446

Beaumont H., Hommes N., Sayavedra-Soto L., Arp D .J., Arciero D. M., Hooper A. B., Westerhoff H. V. and van Spanning R. J. M. (2002). Nitrite reductase of Nitrosomonas europaea is not essential for production of gaseous nitrogen oxides and confers tolerance to nitrite. *Journal of Bacteriology*, **184**(9), 2557–2560, https://doi.org/10.1128/JB.184.9.2557-2560.2002

Belmonte M., Vázquez-Padín J., Figueroa M., Campos J. L., Méndez R., Vidal G. and Mosquera-Corral A. (2012). Denitrifying activity via nitrite and N_2O production using acetate and swine wastewater. *Process Biochemistry*, **47**(7), 1202–1206, https://doi.org/10.1016/j.procbio.2012.04.012

Beun J. J., Dircks K., van Loosdrecht M. C. M. and Heijnen J. J. (2002). Poly-beta-hydroxybutyrate metabolism in dynamically fed mixed microbial cultures. *Water Research*, **36**(5), 1167–1180, https://doi.org/10.1016/S0043-1354(01)00317-7

Blum J. M., Su Q., Ma Y., Valverde-Pérez B., Domingo-Félez C., Jensen M. M. and Smets B. F. (2018). The pH dependency of N-converting enzymatic processes, pathways and microbes: effect on net N_2O production. *Environmental Microbiology*, **20**(5), 1623–1640, https://doi.org/10.1111/1462-2920.14063

Böttcher B. and Koops H. (1994). Growth of lithotrophic ammonia-oxidizing bacteria on hydroxylamine. *FEMS Microbiology Letters*, **122**(3), 263–266, https://doi.org/10.1111/j.1574-6968.1994.tb07178.x

Cantera J. and Stein L. (2007). Role of nitrite reductase in the ammonia-oxidizing pathway of Nitrosomonas europaea. *Archives of Microbiology*, **188**(4), 349–354, https://doi.org/10.1007/s00203-007-0255-4

Caranto J. D. and Lancaster K. M. (2017). Nitric oxide is an obligate bacterial nitrification intermediate produced by hydroxylamine oxidoreductase. *Proceedings of the National Academy of Sciences of the United States of America*, **114**(31), 8217–8222, https://doi.org/10.1073/pnas.1704504114

Casciotti K., Sigman D. and Ward B. (2003). Linking diversity and stable isotope fractionation in ammonia-oxidizing bacteria. *Geomicrobiology Journal*, **20**(4), 335–353, https://doi.org/10.1080/01490450303895

Castellano-Hinojosa A., Maza-Marquez P., Melero-Rubio Y., Gonzalez-Lopez J. and Rodelas B. (2018). Linking N_2O emissions to population dynamics of nitrifying and denitrifying prokaryotes in 4 full-scale WWTP. *Chemosphere*, **200**, 57–66, https://doi.org/10.1016/j.chemosphere.2018.02.102

Castro-Barros C., Rodriguez-Caballero A., Volcke E. and Pijuan M. (2016). Effect of nitrite on the N_2O and NO production on the nitrification of low-strength ammonium wastewater. *Chemical Engineering Journal*, **287**, 269–276, https://doi.org/10.1016/j.cej.2015.10.121

Chandran K., Stein L., Klotz M. and van Loosdrecht M. (2011). Nitrous oxide production by lithotrophic ammonia-oxidizing bacteria and implications for engineered nitrogen-removal systems. *Biochemical Society Transactions*, **39**(part 6), 1832–1837, https://doi.org/10.1042/BST20110717

Chen Y., Wang D., Zhu X., Zheng X. and Feng L. (2012). Long-term effects of copper nanoparticles on wastewater biological nutrient removal and N_2O generation in the activated sludge process. *Environmental Science & Technology*, **46**(22), 12452–12458, https://doi.org/10.1021/es302646q

Chen X., Yuan Z. and Ni B. J. (2018). Nitrite accumulation inside sludge flocs significantly influencing nitrous oxide production by ammonium-oxidizing bacteria. *Water Research*, **143**, 99–108, https://doi.org/10.1016/j.watres.2018.06.025

Chen H., Zeng L., Wang D., Zhou Y. and Yang X. (2020). Recent advances in nitrous oxide production and mitigation in wastewater treatment. *Water Research*, **184**, 116168, https://doi.org/10.1016/j.watres.2020.116168

Cua L. and Stein L. (2011). Effects of nitrite on ammonia-oxidizing activity and gene regulation in three ammonia-oxidizing bacteria. *FEMS Microbiology Letters*, **319**(2), 169–175, https://doi.org/10.1111/j.1574-6968.2011.02277.x

Daims H., Lebedeva E., Pjevac P., Han P., Herbold C., Albertsen M., Jehmlich N., Palatinszky M., Vierheilig J., Bulaev A., Kirkegaard R., von Bergen M., Rattei T., Bendinger B., Nielsen P. and Wagner M. (2015). Complete nitrification by Nitrospira bacteria. *Nature*, **528**(7583), 504–509, https://doi.org/10.1038/nature16461

De Graaff D. R., van Loosdrecht M. C. and Pronk M. (2020). Biological phosphorus removal in seawater-adapted aerobic granular sludge. *Water Research*, **172**, 115531, https://doi.org/10.1016/j.watres.2020.115531

Desloover J., De Clippeleir H., Boeckx P., Du Laing G., Colsen J., Verstraete W. and Vlaeminck S. E. (2011). Floc-based sequential partial nitritation and anammox at full scale with contrasting N_2O emissions. *Water Research*, **45**(9), 2811–2821, https://doi.org/10.1016/j.watres.2011.02.028

Desloover J., Vlaeminck S., Clauwaert P., Verstraete W. and Boon N. (2012). Strategies to mitigate N_2O emissions from biological nitrogen removal systems. *Current Opinion in Biotechnology*, **23**(3), 474–482, https://doi.org/10.1016/j.copbio.2011.12.030

Domingo-Félez C. and Smets B. F. (2019). Regulation of key N_2O production mechanisms during biological water treatment. *Current Opinion in Biotechnology*, **57**, 119–126, https://doi.org/10.1016/j.copbio.2019.03.006

Duan H., Ye L., Erler D., Ni B. J. and Yuan Z. (2017). Quantifying nitrous oxide production pathways in wastewater treatment systems using isotope technology–a critical review. *Water Research*, **122**, 96–113, https://doi.org/10.1016/j.watres.2017.05.054

Dundee L. and Hopkins W. (2001). Different sensitivities to oxygen of nitrous oxide production by Nitrosomonas europaea and Nitrosolobus multiformis. *Soil Biology and Biochemistry*, **33**(11), 1563–1565, https://doi.org/10.1016/S0038-0717(01)00059-1

Edenhofer O., Pichs-Madruga R., Sokona Y., Farahani E., Kadner S., Seyboth K., Adler A., Baum I., Brunner S. and Eickemeier P. J. C. (2014). IPCC, 2014: Summary for Policymakers. In: Climate Change 2014, Mitigation of Climate Change.

Fang F., Li H., Jiang X., Deng X., Yan P., Guo J., Chen Y. and Yang J. J. J. (2020). Significant N2O emission from a high rate granular reactor for completely autotrophic nitrogen removal over nitrite. *Journal of Environmental Management*, **266**, 110586, https://doi.org/10.1016/j.jenvman.2020.110586

Foley J., de Haas D., Yuan Z. and Lant P. (2010). Nitrous oxide generation in full scale BNR wastewater treatment plants. *Water Research*, **44**(3), 831–844, https://doi.org/10.1016/j.watres.2009.10.033

Freitag A., Rudert M. and Bock E. (1987). Growth of Nitrobacter by dissimilatoric nitrate reduction. *FEMS Microbiology Letters*, **48**(1–2), 105–109, https://doi.org/10.1111/j.1574-6968.1987.tb02524.x

Frison N., Chiumenti A., Katsou E., Malamis S., Bolzonella D. and Fatone F. (2015). Mitigating off-gas emissions in the biological nitrogen removal via nitrite process treating anaerobic effluents. *Journal of Cleaner Production*, **93**, 126–133, https://doi.org/10.1016/j.jclepro.2015.01.017

Fujita K. and Dooley D. M. (2007). Insights into the mechanism of N_2O reduction by reductively activated N_2O reductase from kinetics and spectroscopic studies of pH effects. *Inorganic Chemistry*, **46**(3), 613–615, https://doi.org/10.1021/ic061843f

Gabarró J., Ganigué R., Gich F., Ruscalleda M., Balaguer M. D. and Colprim J. (2012). Effect of temperature on AOB activity of a partial nitritation SBR treating landfill leachate with extremely high nitrogen concentration. *Bioresource Technology*, **126**, 283–289, https://doi.org/10.1016/j.biortech.2012.09.011

Gao S. H., Fan L., Peng L., Guo J., Agulló-Barceló M., Yuan Z. and Bond P. L. (2016). Determining multiple responses of Pseudomonas aeruginosa PAO1 to an antimicrobial agent, free nitrous acid. *Environmental Science & Technology*, **50**(10), 5305–5312, https://doi.org/10.1021/acs.est.6b00288

Gao H., Mao Y., Zhao X., Liu W. T., Zhang T. and Wells G. (2019). Genome-centric metagenomics resolves microbial diversity and prevalent truncated denitrification pathways in a denitrifying PAO-enriched bioprocess. *Water Research*, **155**, 275–287, https://doi.org/10.1016/j.watres.2019.02.020

Goreau T., Kaplan W., Wofsy S., McElroy M., Valois F. and Watson S. (1980). Production of NO_2^- and N_2O by nitrifying bacteria at reduced concentrations of oxygen. *Applied and Environmental Microbiology*, **40**(3), 526–532, https://doi.org/10.1128/aem.40.3.526-532.1980

Gorelsky S. I., Ghosh S. and Solomon E. I. (2006). Mechanism of N_2O reduction by the μ4-S tetranuclear CuZ cluster of nitrous oxide reductase. *Journal of the American Chemical Society*, **128**(1), 278–290, https://doi.org/10.1021/ja055856o

Gruber W., Villez K., Kipf M., Wunderlin P., Siegrist H., Vogt L. and Joss A. (2020). N_2O emission in full-scale wastewater treatment: proposing a refined monitoring strategy. *Science of the Total Environment*, **699**, 134157, https://doi.org/10.1016/j.scitotenv.2019.134157

Gustavsson D. and la Cour Jansen J. (2011). Dynamics of nitrogen oxides emission from a full-scale sludge liquor treatment plant with nitritation. *Water Science and Technology*, **63**(12), 2828–2845, https://doi.org/10.2166/wst.2011.487

Hallin S., Philippot L., Löffler F. E., Sanford R. A. and Jones C. M. (2018). Genomics and ecology of novel N_2O-reducing microorganisms. *Trends in Microbiology*, **26**(1), 43–55, https://doi.org/10.1016/j.tim.2017.07.003

Hanaki K., Hong Z. and Matsuo T. (1992). Production of nitrous oxide gas during denitrification of wastewater. *Water Science and Technology*, **26**(5–6), 1027–1036, https://doi.org/10.2166/wst.1992.0544

Harper W. F., Takeuchi Y., Riya S., Hosomi M. and Terada A. (2015). Novel abiotic reactions increase nitrous oxide production during partial nitrification: modeling and experiments. *Chemical Engineering Journal*, **281**, 1017–1023, https://doi.org/10.1016/j.cej.2015.06.109

Hink L., Lycus P., Gubry-Rangin C., Frostegård Å., Nicol G. W., Prosser J. I. and Bakken L. R. (2017). Kinetics of NH_3-oxidation, NO-turnover, N_2O-production and electron flow during oxygen depletion in model bacterial and archaeal ammonia oxidisers. *Environmental Microbiology*, **19**(12), 4882–4896, https://doi.org/10.1111/1462-2920.13914

Hooper A. and Terry K. (1979). Hydroxylamine oxidoredcutase of Nitrosomonas: production of nitric-oxide from hydroxylamine. *Biochimica et Biophysica Acta*, **571**(1), 12–20, https://doi.org/10.1016/0005-2744(79)90220-1

Hynes R. and Knowles R. (1984). Production of nitrous oxide by nitrosomonas europaea: effects of acetylene, pH and oxygen. *Canadian Journal of Microbiology*, **30**(11), 1397–1404, https://doi.org/10.1139/m84-222

Joss A., Salzgeber D., Eugster J., König R., Rottermann K., Burger S., Fabijan P., Leumann S., Mohn J. and Siegrist H. (2009). Full-scale nitrogen removal from digester liquid with partial nitritation and anammox in one SBR. *Environmental Science & Technology*, **43**(14), 5301–5306, https://doi.org/10.1021/es900107w

Kampschreur M., Tan N., Kleerebezem R., Picioreanu C., Jetten M. and van Loosdrescht M. (2008a). Effect of dynamic process conditions on nitrogen oxides emission from a nitrifying culture. *Environmental Science & Technology*, **42**(2), 429–435, https://doi.org/10.1021/es071667p

Kampschreur M., van der Star W., Wielders H., Mulder J., Jetten M. and van Loosdrecht M. (2008b). Dynamics of nitric oxide and nitrous oxide emission during full-scale reject water treatment. *Water Research*, **42**(3), 812–826, https://doi.org/10.1016/j.watres.2007.08.022

Kampschreur M., Temmink H., Kleerebezem R., Jetten M. and van Loosdrecht M. (2009). Nitrous oxide emission during wastewater treatment. *Water Research*, **43**(17), 4093–4103, https://doi.org/10.1016/j.watres.2009.03.001

Kampschreur M., Kleerebezem R., de Vet W. and van Loosdrecht M. (2011). Reduced iron induced nitric oxide and nitrous oxide emission. *Water Research*, **45**(18), 5945–5952, https://doi.org/10.1016/j.watres.2011.08.056

Kester R., de Boer W. and Laanbroek H. (1997). Production of NO and N_2O by pure cultures of nitrifying and denitrifying bacteria during changes in aeration. *Applied and Environmental Microbiology*, **63**, 3872–3877, https://doi.org/10.1128/aem.63.10.3872-3877.1997

Kim S., Miyahara M., Fushinobu S., Wakagi T. and Shoun H. (2010). Nitrous oxide emission from nitrifying activated sludge dependent on denitrification by ammonia-oxidizing bacteria. *Bioresource Technology*, **101**(11), 3958–3963, https://doi.org/10.1016/j.biortech.2010.01.030

Kits K. D., Sedlacek C. J., Lebedeva E. V., Han P., Bulaev A., Pjevac P., Daebeler A., Roman S., Albertsen M., Stein L. Y., Daims H. and Wagner M. (2017). Kinetic analysis of a complete nitrifier reveals an oligotrophic lifestyle. *Nature*, **549**(7671), 269–272, https://doi.org/10.1038/nature23679

Kits K. D., Jung M. Y., Vierheilig J., Pjevac P., Sedlacek C. J., Liu S., Herbold C., Stein L. Y., Richter A., Wissel H., Bruggemann N., Wagner M. and Daims H. (2019). Low yield and abiotic origin of N_2O formed by the complete nitrifier Nitrospira inopinata. *Nature Communications*, **10**(1), 1–12, https://doi.org/10.1038/s41467-018-07882-8

Klotz M. and Stein L. (2011). Chapter 4: genomics of ammonia-oxidizing bacteria and insights into their evolution. In: B. Ward, D. Arp and M. Klotz (eds), Nitrification. American Society of Microbiology, Washington DC, pp. 57–94.

Kozlowski J. A., Kits K. D. and Stein L. Y. (2016a). Comparison of nitrogen oxide metabolism among diverse ammonia-oxidizing bacteria. *Frontiers in Microbiology*, **7**, 1090, https://doi.org/10.3389/fmicb.2016.01090

Kozlowski J. A., Stieglmeier M., Schleper C., Klotz M. G. and Stein L. Y. (2016b). Pathways and key intermediates required for obligate aerobic ammonia-dependent chemolithotrophy in bacteria and Thaumarchaeota. *The ISME Journal*, **10**(8), 1836–1845, https://doi.org/10.1038/ismej.2016.2

Law Y., Lant P. and Yuan Z. (2011). The effect of pH on N_2O production under aerobic conditions in a partial nitritation system. *Water Research*, **45**(18), 5934–5944, https://doi.org/10.1016/j.watres.2011.08.055

Law Y., Ni B. J., Lant P. and Yuan Z. (2012a). N_2O production rate of an enriched ammonia-oxidizing bacteria culture exponentially correlates to its ammonia oxidation rate. *Water Research*, **46**(10), 3409–3419, https://doi.org/10.1016/j.watres.2012.03.043

Law Y., Ye L., Pan Y. and Yuan Z. (2012b). Nitrous oxide emissions from wastewater treatment processes. *Philosophical Transactions of the Royal Society*, **367**(1593), 1265–1277, https://doi.org/10.1098/rstb.2011.0317

Law Y., Lant P. and Yuan Z. (2013). The confounding effect of nitrite on N_2O production by an enriched ammonia-oxidizing culture. *Environmental Science & Technology*, **47**(13), 7186–7194, https://doi.org/10.1021/es4009689

Li P., Wang S., Peng Y., Liu Y. and He J. (2015). The synergistic effects of dissolved oxygen and pH on N_2O production in biological domestic wastewater treatment under nitrifying conditions. *Environmental Technology*, **36**(13), 1623–1631, https://doi.org/10.1080/09593330.2014.1002862

Li W., Li H., Liu Y. D., Zheng P. and Shapleigh J. P. (2018). Salinity-aided selection of progressive onset denitrifiers as a means of providing nitrite for anammox. *Environmental Science & Technology*, **52**(18), 10665–10672, https://doi.org/10.1021/acs.est.8b02314

Lipschultz F., Zafiriou O. C., Wofsy S., McElroy M. B., Valois F. W. and Watson S. W. (1981). Production of NO and N_2O by soil nitrifying bacteria. *Nature*, **294**(5842), 641–643, https://doi.org/10.1038/294641a0

Liu Y., Peng L., Guo J., Chen X., Yuan Z. and Ni B. J. (2015). Evaluating the role of microbial internal storage turnover on nitrous oxide accumulation during denitrification. *Scientific Reports*, **5**, 15138, https://doi.org/10.1038/srep15138

Liu S., Han P., Hink L., Prosser J. I., Wagner M. and Bruggemann N. (2017). Abiotic conversion of extracellular NH_2OH contributes to N_2O emission during ammonia oxidation. *Environmental Science & Technology*, **51**(22), 13122–13132, https://doi.org/10.1021/acs.est.7b02360

Lu H. and Chandran K. (2010). Factors promoting emissions of nitrous oxide and nitric oxide from denitrifying sequencing batch reactors operated with methanol and ethanol as electron donors. *Biotechnology and Bioengineering*, **106**(3), 390–398.

Mannina G., Butler D., Benedetti L., Deletic A., Fowdar H., Fu G., Kleidorfer M., McCarthy D., Steen Mikkelsen P., Rauch W., Sweetapple C., Vezzaro L., Yuan Z. and Willems P. (2018). Greenhouse gas emissions from integrated urban drainage systems: where do we stand? *Journal of Hydrology*, **559**, 307–314, https://doi.org/10.1016/j.jhydrol.2018.02.058

Marin K., Kanesaki Y., Los D. A., Murata N., Suzuki I. and Hagemann M. (2004). Gene expression profiling reflects physiological processes in salt acclimation of synechocystis sp. Strain PCC 6803. *Plant Physiology*, **136**(2), 3290–3300, https://doi.org/10.1104/pp.104.045047

Mosquera-Corral A., González F., Campos J. L. and Méndez R. (2005). Partial nitrification in a SHARON reactor in the presence of salts and organic carbon compounds. *Process Biochemistry*, **40**(9), 3109–3118, https://doi.org/10.1016/j.procbio.2005.03.042

Murnleitner E., Kuba T., van Loosdrecht M. C. M. and Heijnen J. J. (1997). An integrated metabolic model for the aerobic and denitrifying biological phosphorus removal. *Biotechnology and Bioengineering*, **54**(5), 434–450, https://doi.org/10.1002/(SICI)1097-0290(19970605)54:5<434::AID-BIT4>3.0.CO;2-F

Okabe S., Oshiki M., Takahashi Y. and Satoh H. (2011). N_2O emission from a partial nitrification–anammox process and identification of a key biological process of N_2O emission from anammox granules. *Water Research*, **45**(19), 6461–6470, https://doi.org/10.1016/j.watres.2011.09.040

Pan Y., Ye L., Ni B.-J. and Yuan Z. (2012). Effect of pH on N_2O reduction and accumulation during denitrification by methanol utilizing denitrifiers. *Water Research*, **46**(15), 4832–4840, https://doi.org/10.1016/j.watres.2012.06.003

Pan Y., Ni B. J., Bond P. L., Ye L. and Yuan Z. (2013a). Electron competition among nitrogen oxides reduction during methanol-utilizing denitrification in wastewater treatment. *Water Research*, **47**(10), 3273–32781, https://doi.org/10.1016/j.watres.2013.02.054

Pan Y., Ye L. and Yuan Z. (2013b). Effect of H_2S on N_2O reduction and accumulation during denitrification by methanol utilizing denitrifiers. *Environmental Science & Technology*, **47**(15), 8408–8415.

Pan Y., Liu Y., Wang D. and Ni B. J. (2019). Modeling effects of H_2S on electron competition among nitrogen oxide reduction and N_2O accumulation during denitrification. *Environmental Science: Water Research & Technology*, **5**(3), 533–542, https://doi.org/10.1039/C8EW00873F

Peng L., Ni B. J., Erler D., Ye L. and Yuan Z. (2014). The effect of dissolved oxygen on N_2O production by ammonia-oxidizing bacteria in an enriched nitrifying sludge. *Water Research*, **66**, 12–21, https://doi.org/10.1016/j.watres.2014.08.009

Peng L., Ni B. and Ye L. (2015). The combined effect of dissolved oxygen and nitrite on N_2O production by ammonia oxidizing bacteria in an enriched nitrifying sludge. *Water Research*, **73**, 29–36, https://doi.org/10.1016/j.watres.2015.01.021

Peng L., Carvajal-Arroyo J. M., Seuntjens D., Prat D., Colica G., Pintucci C. and Vlaeminck S. E. J. (2017). Smart operation of nitritation/denitritation virtually abolishes nitrous oxide emission during treatment of co-digested pig slurry centrate. *Water Research*, **127**, 1–10, https://doi.org/10.1016/j.watres.2017.09.049

Pijuan M., Tora J., Rodriguez-Caballero A., Cesar E., Carrera J. and Perez J. (2014). Effect of process parameters and operational mode on nitrous oxide emissions from a nitritation reactor treating reject wastewater. *Water Research*, **49**, 23–33, https://doi.org/10.1016/j.watres.2013.11.009

Pijuan M. and Yuan Z. (2010). Development and optimization of a sequencing batch reactor for nitrogen and phosphorus removal from abattoir wastewater to meet irrigation standards. *Water Science and Technology*, **61**(8), 2105–2112, https://doi.org/10.2166/wst.2010.973

Poth M. and Focht D. (1985). 15N Kinetic analysis of N_2O production by Nitrosomonas europaea: an examination of nitrifier denitrification. *Applied and Environmental Microbiology*, **49**(5), 1134–1141, https://doi.org/10.1128/aem.49.5.1134-1141.1985

Poughon L., Dussap C. and Gros J. (2000). Energy model and metabolic flux analysis fro autotrophic nitrifiers. *Biotechnology and Bioengineering*, **72**(4), 416–433, https://doi.org/10.1002/1097-0290(20000220)72:4<416::AID-BIT1004>3.0.CO;2-D

Pronk M., Bassin J. P., De Kreuk M. K., Kleerebezem R. and Van Loosdrecht M. C. M. (2014). Evaluating the main and side effects of high salinity on aerobic granular sludge. *Applied Microbiology and Biotechnology*, **98**(3), 1339–1348, https://doi.org/10.1007/s00253-013-4912-z

Rasmussen T., Brittain T., Berks B. C., Watmough N. J. and Thomson A. J. (2005). Formation of a cytochrome c-nitrous oxide reductase complex is obligatory for N_2O reduction by Paracoccus pantotrophus. Dalton Transactions, 3501–3506.

Ravishankara A., Daniel J. and Portmann R. (2009). Nitrous oxide (N_2O): the dominant ozone-depleting substance emitted in the 21st century. *Science*, **326**(5949), 123–125, https://doi.org/10.1126/science.1176985

Ribera-Guardia A. and Pijuan M. (2017). Distinctive NO and N_2O emisión patterns in ammonia oxidizing bacteria: effect of ammonia oxidation rate, DO and pH. *Chemical Engineering Journal*, **321**, 358–365, https://doi.org/10.1016/j.cej.2017.03.122

Ribera-Guardia A., Kassotaki E., Gutierrez O. and Pijuan M. (2014). Effect of carbon source and competition for electrons on nitrous oxide reduction in a mixed denitrifying microbial community. *Process Biochemistry*, **49**(12), 2228–2234, https://doi.org/10.1016/j.procbio.2014.09.020

Richardson D., Felgate H., Watmough N., Thomson A. and Baggs E. (2009). Mitigating release of the potent greenhouse gas N_2O from the nitrogen cycle- could enzymatic regulation hold the key? *Trends in Biotechnology*, **27**(7), 388–397, https://doi.org/10.1016/j.tibtech.2009.03.009

Rodriguez-Caballero A. and Pijuan M. (2013). N_2O and NO emissions from a partial nitrification sequencing batch reactor: exploring dynamics, sources and minimization mechanisms. *Water Research*, **47**(9), 3131–3140, https://doi.org/10.1016/j.watres.2013.03.019

Rodriguez-Caballero A., Ribera A., Balcázar J. L. and Pijuan M. (2013). Nitritation versus full nitrification of ammonium-rich wastewater: comparison in terms of nitrous and nitric oxide emissions. *Bioresource Technology*, **139**, 195–202, https://doi.org/10.1016/j.biortech.2013.04.021

Schalk-Otte S., Seviour R. J., Kuenen J. G. and Jetten M. (2000). Nitrous oxide (N_2O) production by Alcaligenes Faecalis during feast and famine regimes. *Water Research*, **34**(7), 2080–2088, https://doi.org/10.1016/S0043-1354(99)00374-7

Schönharting B., Rehner R., Metzger J., Krauth K. and Rizzi M. (1998). Release of nitrous oxide (N_2O) from denitrifying activated sludge caused by H_2S-containing wastewater: quantification and application of a new mathematical model. *Water Science and Technology*, **38**(1), 237–426, https://doi.org/10.2166/wst.1998.0057

Schreiber F., Wunderlin P., Udert K. and Wells G. (2012). Nitric oxide and nitrous oxide turnover in natural and engineered microbial communities: biological pathways, chemical reactions and novel technologies. *Frontiers in Microbiology*, **3**, 372, https://doi.org/10.3389/fmicb.2012.00372

Seuntjens D., Han M., Kerckhof F. M., Boon N., Al-Omari A., Takacs I., … and Murthy S. (2018). Pinpointing wastewater and process parameters controlling the AOB to NOB activity ratio in sewage treatment plants. *Water Research*, **138**, 37–46, https://doi.org/10.1016/j.watres.2017.11.044

Shaw L., Nicol G., Smith Z., Fear J., Prosser J. and Baggs E. (2005). Nitrosospira spp. Can produce nitrous oxide via a nitrifier denitrificaion pathway. *Environmental Microbiology*, **8**(2), 214–222, https://doi.org/10.1111/j.1462-2920.2005.00882.x

Shiskowski D. and Mavinic D. (2006). The influence of nitrite and pH (nitrous oxide) on aerobic-phase, autotrophic N_2O generation in a wastewater treatment bioreactor. *Journal of Environmental Sciences*, **5**(4), 273–283.

Soler-Jofra A., Stevens B., Hoekstra M., Picioreanu C., Sorokin D., van Loosdrecht M. C. and Pérez J. (2016). Importance of abiotic hydroxylamine conversion on nitrous oxide emissions during nitritation of reject water. *Chemical Engineering Journal*, **287**, 720–726, https://doi.org/10.1016/j.cej.2015.11.073

Starkenburg S. R., Chain P. S., Sayavedra-Soto L. A., Hauser L., Land M. L., Larimer F. W., ... and Hickey W. J. (2006). Genome sequence of the chemolithoautotrophic nitrite-oxidizing bacterium Nitrobacter winogradskyi Nb-255. *Applied and Environmental Microbiology*, **72**(3), 2050–2063, https://doi.org/10.1128/AEM.72.3.2050-2063.2006

Stein L. Y. (2011a). Chapter 5: heterotrophic nitrification and nitrifier denitrification. In: Nitrification, B. Ward, D. Arp and M. Klotz (eds), American Society of Microbiology, Washington DC, pp. 95–114.

Stein L. (2011b). Surveying N_2O producing pathways in bacteria. *Methods Enzymol*, **486**, 131–152, https://doi.org/10.1016/B978-0-12-381294-0.00006-7

Stein L. and Arp D. (1998). Loss of ammonia monooxygenase activity in Nitrosomonas europaea upon exposure to nitrite. *Applied and Environmental Microbiology*, **64**(10), 4098–4102, https://doi.org/10.1128/AEM.64.10.4098-4102.1998

Stein L., Arp D., Berube P. M., Chain P. S. G., Hauser L., Jetten M., Klotz M., Larimer F. W., Norton J. M., Op den Camp H. J. M., Shin M. and Wei X. (2007). Whole genome analysis of the ammonia oxidizing bacterium, Nitrosomonas eutropha C91: implications for niche adaptation. *Environmental Microbiology*, **9**(12), 2993–3007, https://doi.org/10.1111/j.1462-2920.2007.01409.x

Su Q., Domingo-Félez C., Jensen M. M. and Smets B. F. (2019a). Abiotic nitrous oxide (N_2O) production is strongly pH dependent, but contributes little to overall N_2O emissions in biological nitrogen removal systems. *Environmental Science & Technology*, **53**(7), 3508–3516, https://doi.org/10.1021/acs.est.8b06193

Su Q., Domingo-Felez C., Zhang Z., Blum J. M., Jensen M. M. and Smets B. F. (2019b). The effect of pH on N_2O production in intermittently-fed nitritation reactors. *Water Research*, **156**, 223–231, https://doi.org/10.1016/j.watres.2019.03.015

Sutka R. L., Ostrom N. E., Ostrom P. H., Breznak J. A., Gandhi H., Pitt A. J. and Li F. (2006). Distinguishing nitrous oxide production from nitrification and denitrification on the basis of isotopomer abundances. *Applied and Environmental Microbiology*, **72**(1), 638–644, https://doi.org/10.1128/AEM.72.1.638-644.2006

Tallec G., Garnier J., Billen G. and Gousailles M. (2006). Nitrous oxide emissions from secondary activated sludge in nitrifying conditions of urban wastewater treatment plants: effect of oxygenation level. *Water Research*, **40**(15), 2972–2980, https://doi.org/10.1016/j.watres.2006.05.037

Terada A., Sugawara S., Hojo K., Takeuchi Y., Riya S., Harper Jr, W. F., ... and Suwa Y. (2017). Hybrid nitrous oxide production from a partial nitrifying bioreactor: hydroxylamine interactions with nitrite. *Environmental Science & Technology*, **51**(5), 2748–2756, https://doi.org/10.1021/acs.est.6b05521

Thörn M. and Sörensson F. (1996). Variation of nitrous oxide formation in the denitrification basin in a wastewater treatment plant with nitrogen removal. *Water Research*, **30**(6), 1543–1547, https://doi.org/10.1016/0043-1354(95)00327-4

Tumendelger A., Toyoda S. and Yoshida N. (2014). Isotopic analysis of N_2O produced in a conventional wastewater treatment system operated under different aeration conditions. *Rapid Communications*, **28**(17), 1883–1892, https://doi.org/10.1002/rcm.6973

Vadivelu V. M., Yuan Z., Fux C. and Keller J. (2006). The inhibitory effects of free nitrous acid on the energy generation and growth processes of an enriched Nitrobacter culture. *Environmental Science & Technology*, **40**(14), 4442–4448, https://doi.org/10.1021/es051694k

Van Hulle S. W. H., Vandeweyer H. J. P., Meesschaert B. D., Vanrolleghem P. A. and Dejans A. (2010). Dumoulin engineering aspects and practical application of autotrophic nitrogen removal from nitrogen rich streams. *Chemical Engineering Journal*, **162**(1), 1–20, https://doi.org/10.1016/j.cej.2010.05.037

Van Kessel M. A. H. J., Speth D. R., Albertsen M., Nielsen P. H., Op den Camp H. J. M., Kartal B., Jetten M. and Lucker S. (2015). Complete nitrification by a asingle microorganism. *Nature*, **528**(7583), 555–559, https://doi.org/10.1038/nature16459

Vasilaki V., Massara T. M., Stanchev P., Fatone F. and Katsou E. (2019). A decade of nitrous oxide (N_2O) monitoring in full-scale wastewater treatment processes: a critical review. *Water Research*, **161**, 392–412, https://doi.org/10.1016/j.watres.2019.04.022

Vieira A., Marques R., Galinha C., Povoa P., Carvalho G., Oehmen A. J. E. S. and Research P. (2019). Nitrous oxide emissions from a full-scale biological aerated filter (BAF) subject to seawater infiltration. *Environmental Science and Pollution Research International*, **26**(20), 20939–20948, https://doi.org/10.1007/s11356-019-05470-x

VonSchulthess R. and Gujer W. (1996). Release of nitrous oxide (N_2O) from denitrifying activated sludge: verification and application of a mathematical model. *Water Research*, **30**(3), 521–530, https://doi.org/10.1016/0043-1354(95)00204-9

VonSchulthess R., Wild D. and Gujer W. (1994). Nitric and nitrous oxide from denitrifying activated sludge at low oxygen concentration. *Water Science and Technology*, **30**(6), 123–132, https://doi.org/10.2166/wst.1994.0259

Völker U., Engelmann S., Maul B., Riethdorf S., Völker A., Schmid R., … and Hecker M. (1994). Analysis of the induction of general stress proteins of Bacillus subtilis. *Microbiology*, **140**(4), 741–752, https://doi.org/10.1099/00221287-140-4-741

Wang Y., Geng J., Guo G., Wang C. and Liu S. (2011). N_2O production in anaerobic/anoxic denitrifying phosphorus removal process: the effects of carbon sources shock. *Chemical Engineering Journal*, **172**(2–3), 999–1007, https://doi.org/10.1016/j.cej.2011.07.014

Wang Y., Zhou S., Ye L., Wang H., Stephenson T. and Jiang X. (2014). Nitrite survival and nitrous oxide production of denitrifying phosphorus removal sludges in long-term nitrite/nitrate-fed sequencing batch reactors. *Water Research*, **67**, 33–45, https://doi.org/10.1016/j.watres.2014.08.052

Wrage N., Velthof G., Oenema O. and Laanbroek H. (2004). Acetylene and oxygen as inhibitors of nitrous oxide production in Nitrosomonas europaea and Nitrosospira briensis: a cautionary tale. *FEMS Microbiology Ecology*, **47**(1), 13–18, https://doi.org/10.1016/S0168-6496(03)00220-4

Wrage-Mönnig N., Horn M. A., Well R., Müller C., Velthof G. and Oenema O. (2018). The role of nitrifier denitrification in the production of nitrous oxide revisited. *Soil Biology and Biochemistry*, **123**, A3–A16, https://doi.org/10.1016/j.soilbio.2018.03.020

Wunderlin P., Mohn J., Joss A., Emmenegger L. and Siegrist H. (2012). Mechanims of N_2O production in biological wastewater treatment under nitrifying and denitrifying conditions. *Water Research*, **46**(4), 1027–1037, https://doi.org/10.1016/j.watres.2011.11.080

Wunderlin P., Lehmann M., Siegrist H., Tuzson B., Joss A., Emmenegger L. and Mohn J. (2013). Isotope signatures of N_2O in a mixed microbial population system: constraints on N_2O producing pathways in wastewater treatment. *Environmental Science & Technology*, **47**(3), 1339–1348.

Weissbach M., Thiel P., Drewes J. E. and Koch K. (2018). Nitrogen removal and intentional nitrous oxide production from reject water in a coupled nitritation/nitrous denitritation system under real feed-stream conditions. *Bioresource Technology*, **255**, 58–66, https://doi.org/10.1016/j.biortech.2018.01.080

Xu G., Xu X., Yang F., Liu S. and Gao Y. (2012). Partial nitrification adjusted by hydroxylamine in aerobic granules under high DO and ambient temperature and subsequent anammox for low C/N wastewater treatment. *Chemical Engineering Journal*, **213**, 338–345, https://doi.org/10.1016/j.cej.2012.10.014

Yan Y., Wang Y., Chen Y., Lin X., Wu M. and Chen J. (2016). Single-stage PN/A technology treating saline ammonia-rich wastewater: finding the balance between efficient performance and less N_2O and NO emissions. *RSC Advances* **6**(114), 113152–113162, https://doi.org/10.1039/C6RA24109C

Ye L., Peng C.-Y., Tang B., Wang S.-Y., Zhao K.-F. and Peng Y.-Z. (2009). Determination effect of influent salinity and inhibition time on partial nitrification in a sequencing batch reactor treating saline sewage. *Desalination*, **246**(1–3), 556–566, https://doi.org/10.1016/j.desal.2009.01.005

Yu R. and Chandran K. (2010) Strategies of Nitrosomonas europaea 19718 to counter low dissolved oxygen and high nitrite concentrations. *BMC Microbiol.* **10** (1), 70, https://doi.org/10.1186/1471-2180-10-70

Yu R., Kampschreur M., van Loosdrecht M. and Chandran K. (2010). Mechanisms and specific directionality of autotrophic nitrous oxide and nitric oxide generation during transient anoxia. *Environmental Science & Technology*, **44**(4), 1313–1319, https://doi.org/10.1021/es902794a

Yu R., Perez-Garcia O., Lu H. and Chandran K. (2018). Nitrosomonas europaea adaptation to anoxic-oxic cycling: insights from transcription analysis, proteomics and metabolic network modeling. *Science of the Total Environment*, **615**, 1566–1573, https://doi.org/10.1016/j.scitotenv.2017.09.142

Zart D., Schmidt I. and Bock E. (2000). Significance of gaseous NO for ammonia oxidation by Nitrosomonas eutropha. *Antonie van Leeuwenhoek*, **77**(1), 49–55, https://doi.org/10.1023/A:1002077726100

Zeng R. J., Lemaire R., Yuan Z. and Keller J. (2003). Simultaneous nitrification, denitrification, and phosphorus removal in a lab-scale sequencing batch reactor. *Biotechnology and Bioengineering*, **84**(2), 170–178, https://doi.org/10.1002/bit.10744

Zhan X., Hu Z. and Wu G. (2018). Greenhouse Gas Emission and Mitigation in Municipal Wastewater Treatment Plants. IWA Publishing, London.

Zhang Q., Li Y., He Y., Brookes P. C. and Xu J. (2019). Elevated temperature increased nitrification activity by stimulating AOB growth and activity in an acidic paddy soil. *Plant and Soil*, **445**(1–2), 71–83, https://doi.org/10.1007/s11104-019-04052-7

Zhao Y., Miao J., Ren X. and Wu G. (2018). Effect of organic carbon on the production of biofuel nitrous oxide during the denitrification process. *International Journal of Environmental Science and Technology*, **15**(2), 461–470, https://doi.org/10.1007/s13762-017-1397-9

Zhao S., Su X., Wang Y., Yang X., Bi M., He Q. and Chen Y. (2020). Copper oxide nanoparticles inhibited denitrifying enzymes and electron transport system activities to influence soil denitrification and N_2O emission. *Chemosphere*, **245**, 125394, https://doi.org/10.1016/j.chemosphere.2019.125394

Zheng H., Hanaki K. and Matsuo T. (1994). Production of nitrous oxide gas during nitrification of wastewater. *Water Science and Technology*, **30**(6), 133–141, https://doi.org/10.2166/wst.1994.0260

Zhou Y., Pijuan M. and Yuan Z. (2007). Free nitrous acid inhibition on anoxic phosphorus uptake and denitrification by poly-phosphate accumulating organisms. *Biotechnology and Bioengineering*, **98**(4), 903–912, https://doi.org/10.1002/bit.21458

Zhou Y., Pijuan M., Zeng R. J. and Yuan Z. (2008). Free nitrous acid inhibition on nitrous oxide reduction by a denitrifying-enhanced biological phosphorus removal sludge. *Environmental Science & Technology*, **42**(22), 8260–8265, https://doi.org/10.1021/es800650j

Zhou X., Song J., Wang G., Yin Z., Cao X. and Gao J. (2020). Unravelling nitrogen removal and nitrous oxide emission from mainstream integrated nitrification-partial denitrification-anammox for low carbon/nitrogen domestic wastewater. *Journal of Environmental Management*, **270**, 110872, https://doi.org/10.1016/j.jenvman.2020.110872

Zhu X. and Chen Y. (2011). Reduction of N_2O and NO generation in anaerobic – aerobic (low dissolved oxygen) biological wastewater treatment process by using sludge alkaline fermentation liquid. *Environmental Science & Technology*, **45**(6), 2137–2143, https://doi.org/10.1021/es102900h

Zhu X., Burger M., Doane T. A. and Horwath W. R. (2013a). Ammonia oxidation pathways and nitrifier denitrification are significant sources of N2O and NO under low oxygen availability. *Proceedings of the National Academy of Sciences of the United States of America*, **110**(16), 6328–6333, https://doi.org/10.1073/pnas.1219993110

Zhu X., Chen Y., Chen H., Li X., Peng Y. and Wang S. (2013b). Minimizing nitrous oxide in biological nutrient removal from municipal wastewater by controlling copper ion concentrations. *Applied Microbiology and Biotechnology*, **97**(3), 1325–1334, https://doi.org/10.1007/s00253-012-3988-1

Zhu-Barker X., Cavazos A. R., Ostrom N. E., Horwarth W. R. and Glass J. B. (2015). The importance of abiotic reactions for nitrous oxide production. *Biogeochemistry*, **126**(3), 251–267, https://doi.org/10.1007/s10533-015-0166-4

Zumft W. G. (1997). Cell biology and molecular basis of denitrification. *Microbiology and Molecular Biology Reviews*, **61**(4), 533–616.

NOMENCLATURE

AMO	Ammonia monooxygenase
AOA	Ammonia oxidizing archaea
AOB	Ammonia oxidizing bacteria
AOR	Ammonia oxidation rate
ARR	Ammonium removal rate
BNR	Biological nitrogen removal
CO_2	Carbon dioxide
Comammox	Complete ammonium oxidizer
CuO	Copper oxide
dGAO	Denitrifying glycogen accumulating organisms
DO	Dissolved oxygen
dPAO	Denitrifying polyphosphate accumulating organism
FA	Free ammonia
FNA (HNO_2)	Free nitrous acid

H_2S	Hydrogen sulfide
HAO	Hydroxylamine oxidoreductase
HD	Heterotrophic denitrification
HDH	Hydrazine dehydrogenase
HZS	Hydrazine synthase
N_2	Nitrogen gas
NAP	Periplasmic nitric reductase
NaR	Nitrate reductase
ND	Nitrifier denitrification
NhAR	Hydroxylamine accumulation rate
NH_2OH	Hydroxylamine
NH_3	Ammonia
NiAR	Nitrite accumulation rate
NiR	Nitrite reductase
NN	Nitrifier nitrification
NO	Nitric oxide
NO_2^-	Nitrite
NO_3^-	Nitrate
NOB	Nitrite oxidizing bacteria
NOH	Nitrosyl radical
NoR	Nitric oxide reductase
NOR	Nitric oxide reductase
NOR	Net NO production rate
NoS	Nitrous oxide reductase
N_2O	Nitrous oxide
N_2O_4	Nitrogen tetroxide
N_2OR	Net N_2O production rate
NXR	Nitrite oxidoreductase
PCR	Polymerase chain reaction
PHA	Polyhydroxyalkanoate
PHB	Polyhydroxybutyrate
PN	Partial nitrification
SBR	Sequencing batch reactor
WWTPs	Wastewater treatment plants

doi: 10.2166/9781789060461_043

Chapter 3

Mechanisms, source, and factors that affect methane emissions

Oriol Gutierrez[1,2], Haoran Duan[3,4], Ziping Wu[3] and Keshab R. Sharma[4]

[1]ICRA, Catalan Institute for Water Research, H2O Building, Scientific and technological park of the UdG, Emili Grahit 101, Girona 17003, Spain. E-mail: ogutierrez@icra.cat

[2]University of Girona, UdG, Girona, 17003, Spain

[3]School of Chemical Engineering, The University of Queensland, St Lucia, QLD 4072, Australia. E-mail: h.duan@uq.edu.au; ziping.wu@uqconnect.edu.au

[4]Advanced Water Management Centre, The University of Queensland, St Lucia, QLD 4072, Australia. E-mail: k.sharma@awmc.uq.edu.au

SUMMARY

Urban wastewater systems (UWSs) are contributing to global greenhouse gas (GHG) increasing concentrations with direct emissions of methane (CH_4). Despite its lower contribution when compared to other sources of anthropogenic methane such as extraction-delivery of fossil fuels and animal-agricultural practices, methane emitted from UWSs can be up to 5% of the total global CH_4 emissions. This chapter provides an overview of the contribution of UWSs to anthropogenic methane emissions.

CH_4 formation in UWSs occurs through anaerobic digestion (AD) of organics contained in sewage. AD consists of a sequence of concomitant reactions by which a consortium of microorganisms, in the absence of oxygen, break down biodegradable carbon material producing biogas, a mixture of methane, carbon dioxide and traces of H_2S. In UWSs, those reactions can occur naturally in sewer systems depleted of oxygen or be artificially promoted in wastewater treatment plants to capture and recover the energy contained in molecules of methane.

Specifics of methane generation in sewer biofilms, sewer sediments, anaerobic wastewater treatment and sludge disposal of wastewater treatment plants are presented in this chapter. Identification of CH_4 emission spots in urban wastewater engineered systems represents the initial step for reliable quantification of GHG, to later develop and establish effective mitigation strategies.

Keywords: Anaerobic processes, methane, sewers, urban wastewater systems, wastewater treatment plants

TERMINOLOGY

Term	Definition
Anaerobic digestion	A sequence of biological processes by which microorganisms break down biodegradable organics in the absence of oxygen.
Anaerobic condition	The anaerobic condition is when the treatment system operates under the absence of the oxygen.
Carbon footprint	The total greenhouse gas emissions, expressed as carbon dioxide equivalent.
Greenhouse gas	Gas that absorbs and emits radiant energy within the thermal infrared range.
Hydraulic retention time (HRT)	A measure of the average length of time that a volume of wastewater remains in a given sewer section or a process unit.
Long-term sludge drying	Drying for a period exceeding 1 month.
Mass transfer	In this chapter, mass transfer refers to the liquid-to-gas transport process of methane. The rate of mass transfer is proportional to the difference between the equilibrium concentration and the concentration of concern. The rate of transfer reduces to zero when equilibrium is reached.
Solids retention time (SRT)	The average length of time that the sludge solids remain in the treatment system.
Sewer	A network of artificial underground conduits that convey and transport wastewater and/or stormwater from its origin to its treatment point.
Sewer rising main pipes	A rising main is a type of drain or sewer through which sewage and/or surface water runoff is pumped from a pumping station to an elevated point. Rising main pipes are fully pressurized, and anaerobic conditions prevail on those sections of sewers.
Sewer gravity pipes	Opposite to rising main pipes, gravity sewer pipes are conduits that use a difference in elevation points, from high to low, and gravity to transport wastewater. Gravity pipes have a liquid and gas phase which implies a certain reaeration of wastewater.
Wastewater	The used water and solids from a community that flows into a treatment plant. Stormwater, surface water, and groundwater infiltration also may be included.

3.1 INTRODUCTION AND CONTEXT

Methane (CH_4) is a chemically and radiatively active gas that is produced from a wide variety of anaerobic (oxygen deficient) processes. Methane is an important greenhouse gas, able to absorb and emit radiant energy within the thermal infrared range. Together with carbon dioxide, nitrous oxide and ozone, it accounts for almost one-tenth of 1% of the earth's atmosphere and has an appreciable greenhouse effect (Foley *et al.*, 2011a, b; IPCC, 2019). Over 100 years CH_4 has a global warming potential of 34 compared to CO_2 (potential of 1). Currently, CH_4 accounts for 20% of the total radiative forcing from all the long-lived and globally mixed greenhouse gases (IPCC, 2019).

Earth's natural greenhouse effect is critical to supporting life by keeping the planetary temperatures at liveable ranges. However, human activities, mainly the burning of fossil fuels and clearcutting of forests, have accelerated the greenhouse effect and are causing rising global warming.

Increasing methane emissions are a major contributor to the rising concentration of greenhouse gases in the earth's atmosphere and are responsible for up to one-third of near-term climate warming (IPCC, 2006). From 2015 to 2019 sharp rises in levels of atmospheric methane have been recorded. During the year 2019, about 360 million tons of methane (60%) were released globally through human activities, while natural sources contributed about 230 million tons (40%). In addition, in February 2020, it was reported that methane emissions from the fossil fuel industry may still have been significantly underestimated (IPCC, 2019).

According to the 2019 report refinement of the Intergovernmental Panel on Climate Change (IPCC) (2019), about 33% of anthropogenic methane emissions are related to the extraction and delivery of fossil fuels. Animal agricultural practices are the second largest source (30%) and the flow related to human-consumer waste (including landfill and wastewater treatment) has grown to become the third major source category (18%). Within this last category, the global CH_4 emissions from municipal and industrial wastewater management have been estimated to contribute to about 5% of the total global CH_4 emissions.

Wastewater utilities are aware of their contribution to global warming through their methane fugitive emissions. In fact, this is one of the main drivers to shift their model from just wastewater treatment plants to water resource recovery facilities (WRRFs) (Kehrein *et al.*, 2020; Puyol *et al.*, 2017). WRRFs aim to improve the efficiency of wastewater treatment installations to extract value from previously unexploited streams using biochemical, physical, physicochemical, and biological conversion processes. CH_4 is a combustible gas (heat capacity of 35.69 J/K·mol) that, when properly harvested, can be used as fuel and subsequently to produce electricity-energy. Within urban wastewater systems (UWSs), streams with high organic content can be fermented in heated biodigesters to produce biogas, which contains about 70% methane, being one of the most robust and valuable resource streams to recover (Metcalf & Eddy, 2003). However, there are sections of the UWS where CH_4 generation and direct emission to the atmosphere still occur.

The present chapter gives a general overview of methane sources from urban wastewater systems, including sewers and wastewater treatment plants. Starting from the fundamentals of the biological processes that lead to the formation of CH_4, it then presents the different units of the UWS where it can be produced and released.

3.2 BIOLOGICAL PROCESSES INVOLVED IN METHANE GENERATION

Formation of CH_4 occurs through the process called anaerobic digestion, a sequence of concomitant reactions by which a consortium of microorganisms, in the absence of oxygen, break down biodegradable carbon material to obtain biogas, a mixture of methane, carbon dioxide and traces of H_2S. CH_4 is the main component of biogas accounting for about 60–70% of its composition in volume/volume.

Biogenic formation of methane is a form of anaerobic respiration in which the terminal electron acceptor is not oxygen but carbon compounds of low molecular weight. In contrast to sulfate reducers, methanogens use a limited number of substrates for growth and energy production. Quantitatively, hydrogen, carbon dioxide and acetate are the most important and best-known substrates for methanogens (Muyzer & Stams, 2008). The equations below represent the process of methane production.

Methanogenic reactions	ΔG0¢ (kJ/reaction)	
$4H_2 + HCO_3^- + H^+ \rightarrow CH_4 + 3H_2O$	−135.6	(3.1)
$CO_2 + 4H_2 \rightarrow CH_4 + 2H_2O$	−130.7	(3.2)
$CH_3COO^- + H_2O \rightarrow CH_4 + HCO_3^-$	−31.0	(3.3)

Wastewater streams can contain high concentrations of organic carbon, up to 200–500 mg chemical oxygen demand (COD)/L in domestic sewage and much higher in industrial streams. Also, oxygen-depleted sections of the UWS are common due to the configuration of wastewater transport and treatment. The occurrence of those two conditions leads to the decay of organic matter and subsequent formation of CH_4 in several locations in the UWS.

In anaerobic respiration, a consortium of microorganisms utilizes organic material that is transformed into intermediate end products such as primarily alcohols, aldehydes, and organic acids, plus carbon dioxide. In the presence of specialized methanogens, the intermediates are converted to the 'final' end products of methane, carbon dioxide, and trace levels of hydrogen sulfide. The overall process of anaerobic digestion can be separated into four key stages which involve hydrolysis, acidogenesis, acetogenesis and methanogenesis (Figure 3.1).

Hydrolysis: the first necessary step of anaerobic digestion where complex organic molecules such as carbohydrates, proteins, and lipids, are broken down into simple sugars, amino acids, and volatile fatty acids. Wastewater contains a wide range of high molecular weight organic polymers that bacteria break down into smaller constituent parts that are readily available to other bacteria. Hydrolysis is carried out mainly by fermentative bacteria and bacteria from the group of relative anaerobes of genera like Streptococcus and Enterobacterium (Metcalf & Eddy, 2003). Acetate and hydrogen produced in this first stage can be used directly by methanogens. But other molecules such as volatile fatty acids, with a chain length greater than that of acetate, must first be catabolized into simpler compounds before they can be directly used by methanogens.

Acidogenesis: the second step in anaerobic digestion where simple monomers are converted into volatile fatty acids (VFAs). Acidogenesis results in further breakdown of the remaining organic components and is carried out by acidogenic-fermentative bacteria. VFAs are created, along with ammonia, carbon dioxide, and hydrogen sulfide, as well as other by-products.

Acetogenesis: the third stage of anaerobic digestion where simple molecules created through the acidogenesis phase are further digested by acetogens to produce largely acetic acid, as well as carbon dioxide and hydrogen.

Methanogenesis: the final step in the decay of organic matter. In this stage, methanogens use the intermediate products of the preceding stages and convert them into methane and carbon dioxide. Organisms capable of producing methane have been identified only from the domain Archaea, a group phylogenetically distinct from both eukaryotes and bacteria.

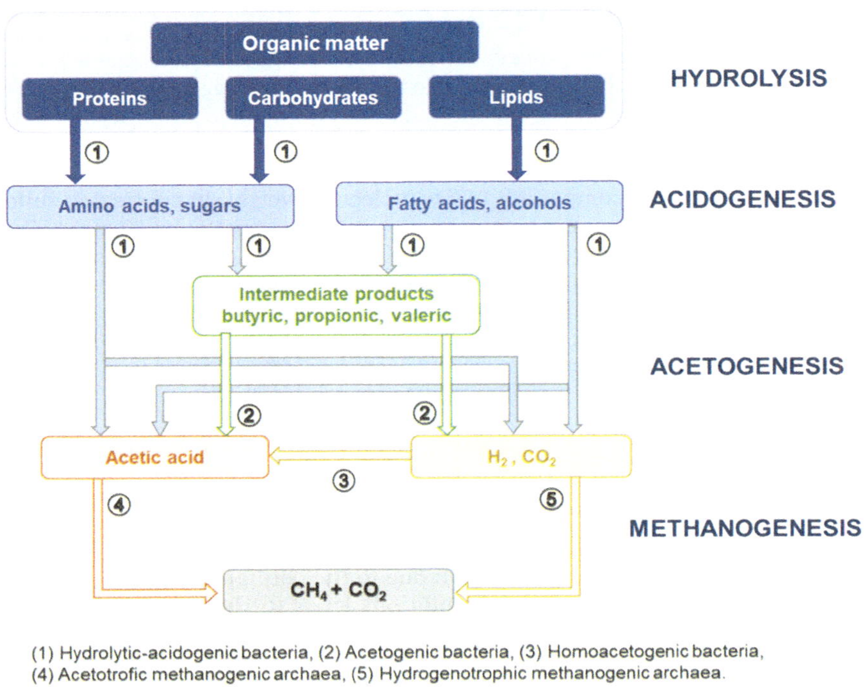

Figure 3.1 Schematic diagram of methane generation pathways.

3.3 METHANE EMISSIONS IN URBAN WASTEWATER SYSTEMS

Urban wastewater systems are the infrastructure that deals with polluted waters from domestic and industrial origins. Their main objectives are to collect and transport wastewater from its generation points to the wastewater treatment plants where it is properly treated before being returned to the environment with acceptable sanitary conditions (Metcalf & Eddy, 2003).

UWSs are composed of two different connected parts: sewer systems and wastewater treatment plants (WWTPs). Sewer systems, also simply known as sewers, are the complex underground infrastructure that conveys wastewater or surface runoff (stormwater, rainwater, meltwater) using an extensive network of pipes, drains, manholes, pumping stations, storm overflows, and screening chambers. Sewer systems are crucial in protecting public health as these prevent the spread of diseases by avoiding population exposure to contaminated wastewater. Sewer systems end at the entry to wastewater treatment plants. WWTPs are centralized facilities in which a combination of various processes (e.g., physical, chemical, and biological) are used to treat wastewater and remove pollutants (Figure 3.2).

Within UWSs, it is quite common to have sections or hotspots where the conditions for CH_4 formation occur. Figure 3.3 presents a UWS and points out locations of potential CH_4 emissions. We can distinguish two different categories of CH_4 production points in UWS:

Natural CH_4-occurrence points: These consist of sections of the UWS where anaerobic conditions naturally prevail, and the microorganisms presented in Section 3.2 can thrive and carry out their metabolism. Amongst these, we can find rising main sewers, some sections of gravity sewers and inlets of primary treatments in the WWTPs. In natural occurrence points, methane production tends to be equal to methane emissions. All methane produced in sewer sections is

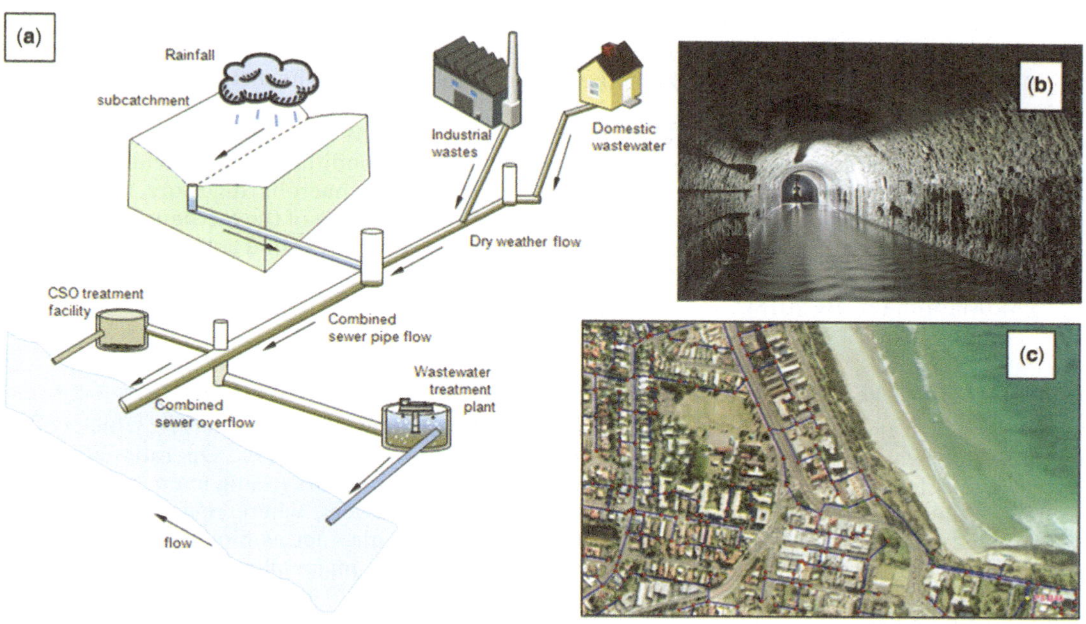

Figure 3.2 (a) Scheme of the urban sewer system. (b) Picture of sewer trunk main in Paris (France), courtesy of sub-urban.com; and (c) Distribution of a sewer network in the domestic suburb of Burleigh Heads, courtesy of Gold Coast city council, (Australia).

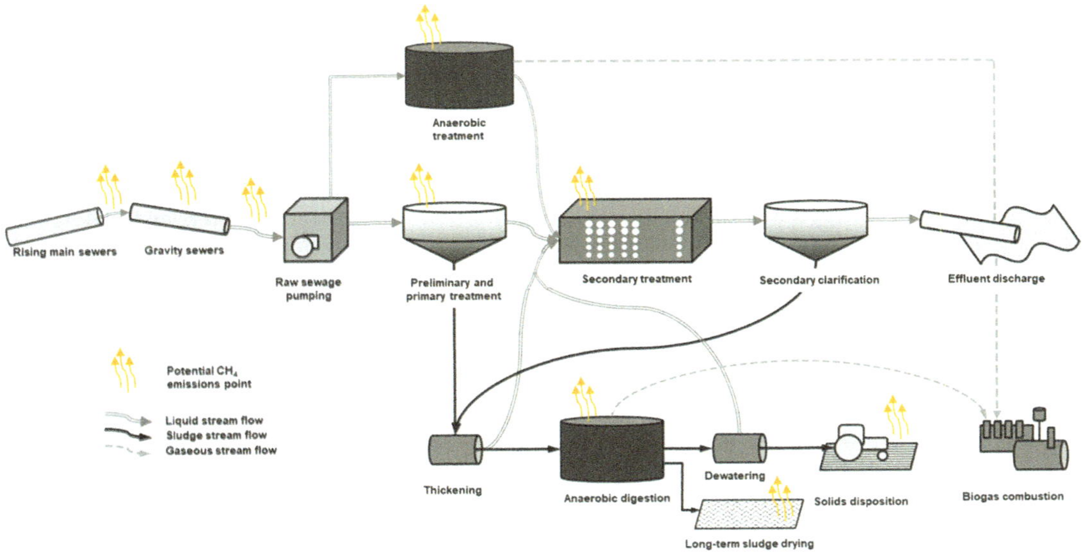

Figure 3.3 Sources of CH_4 emissions from wastewater transport and treatment. Adapted from Willis *et al.* (2016).

usually emitted directly to the atmosphere due to flow turbulence that produces CH4 stripping, widespread distribution of the network, and the difficulty to simultaneously control and collect gas in all emission points (see Section 3.4). Methane also enters the plant from outside via the influent that contains methane that has been formed in the sewer. The methane load is estimated as 1% of the influent COD load.

Engineered CH_4-occurrence points: These consist of units of the WWTP to treat sewage and sludge while intentionally generating methane under controlled conditions and benefiting from the energy contained in CH_4. These are anaerobic treatments and anaerobic digestors. In contrast to natural occurrence points, in the engineered sections the majority of the methane produced can be harvested and thus direct emissions are minimal. However, leaks and collateral streams with high dissolved methane can produce partial emissions of CH_4 to the atmosphere (see Section 3.5) (Mannina *et al.*, 2018).

3.4 METHANE EMISSIONS FROM SEWER SYSTEMS: FACTORS AND SOURCES

Sewer systems are an important and integral component of urban water infrastructure, which collects and transports wastewater from residential houses or industry to WWTPs. Operationally, sewer systems can be divided into two categories: (i) fully filled pressure sewers (rising main sewers), which are predominantly anaerobic, and (ii) partially filled gravity sewers, where re-aeration processes can take place. In addition to transporting wastewater, sewers also act as biological reactors with various microbial processes occurring. Commonly, there are five major phases in a sewer pipe: (i) the suspended wastewater phase, (ii) the wetted sewer biofilms, (iii) the sediments, (iv) the sewer air phase, and (v) the biofilm on pipe surface exposed to sewer air, with the latter two being present in gravity sewers only (Figure 3.4). In-sewer microbial processes mainly take place in biofilms and sediments, with little contribution from the suspended biomass in the water phase or the gas phase (Mohanakrishnan *et al.*, 2009). Wetted anaerobic biofilms are mainly present in rising main sewers,

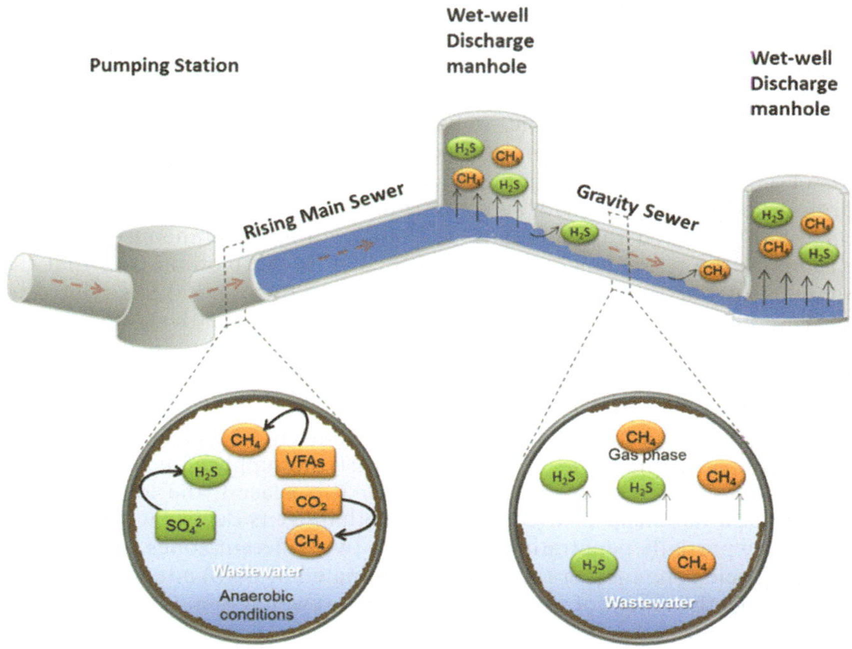

Figure 3.4 Configuration of sewer systems depending on their functioning.

usually with a thickness of a few hundred micrometres. In gravity sewers, both biofilms and sediments below the water surface are in partially anaerobic or fully anaerobic conditions even when oxygen is present in the bulk wastewater, due to limited penetration of the oxygen (Gutierrez *et al.*, 2008). Therefore, processes like methane generation, anaerobic fermentation and sulfate reduction using organic matter or sulfate as electron acceptors occur in deeper layers of sewer biofilms and sediments (Hvitved-Jacobsen, 2002).

Anaerobic conditions and processes in rising main sewers, gravity sewers, and pumping stations have the potential to convert chemical oxygen demand (COD) to CH_4 which would be released to the atmosphere from manholes, and pumping station wet wells. In rising main sewers, methane can be produced and accumulated even beyond saturation concentrations because in-sewer pressure is in excess of atmospheric pressure (Guisasola *et al.*, 2008). When sewage flows from an enclosed anaerobic sewer pipe and is discharged into a ventilated space a large proportion of dissolved methane is stripped off to the atmosphere under turbulence, resulting in significant emissions. This occurs at structures exposed to the atmosphere such as pumping stations, wet wells, and influent works of WWTPs (see Figure 3.2). Some studies observed that while methane emission occurred in gravity sewers following sewage discharge from an upstream rising main, a significant proportion of methane still remained in the liquid phase, and eventually was emitted at the downstream gravity section or inlet headworks of the downstream WWTP.

Methane production from sewers plays an important role in contributing to overall methane emissions over the entire wastewater systems. For instance, Foley *et al.* (2011a) showed that methane production in three rising main pipes (UC09, CO16 and C27) contributed around 18% of the total greenhouse gas (GHG) emission during wastewater handling and treatment of the Pimpama system on the Gold Coast, Australia. Those particular sewers were only a small part of a much larger network, hence more methane production was expected when the wastewater was transported through the remaining parts of the network before reaching the WWTP (Pikaar *et al.*, 2014).

3.4.1 Methane production in anaerobic sewer biofilms

In sewer anaerobic biofilms, methanogenic archaea (MA) compete with other anaerobes, such as sulfate reducing bacteria (SRB) and fermentative bacteria, for the available common substrates. In the presence of sulfate in excess, sulfate reducers compete with methanogens for the common substrates hydrogen and acetate and with syntrophic methanogenic communities (Dar *et al.*, 2008). Guisasola *et al.* (2009) showed that methane and sulfide are simultaneously produced in sewer systems, which implies the coexistence of MA and SRB in sewer biofilms and that these bacteria function simultaneously. The simultaneous functioning of the SRB and MA is related to the spatial arrangement of these bacteria in sewer biofilms. Sewer biofilms are relatively thick (several hundred micrometres; Mohanakrishnan *et al.*, 2009) and the sulfate/organic matter ratio (S/COD) shows spatial variation inside the biofilm, being relatively high near the surface in contact with the bulk liquid and close to zero in the inner zone adjacent to the pipe surface (Sun *et al.*, 2014). Figure 3.5 depicts a schematic view of this hypothesis, which is supported by the sulfide profile measured in sewer biofilms by Mohanakrishnan *et al.* (2009). In contrast, the supply of methanogenesis precursors (VFA) is unlikely to be limiting within the biofilm. For this reason, the lower affinity of MA for these precursors is not a handicap to the growth of methanogens deeper within the biofilm. With sulfate most likely only partially penetrating the biofilm, two different zones appear in the biofilm: a sulfate-reducing anaerobic zone (nearer the surface, dominated by SRB) and a deeper anaerobic zone dominated by MA. Thus, the extent of methanogenesis in a sewer system is inversely proportional to the sulfate penetration length into the biofilm.

3.4.2 Methane production in sewer sediments

Although some studies have suggested that activities which take place in the biofilm walls of sewers systems are the most important microbial transformations (Hvitved-Jacobsen, 2002) other studies have revealed significant methane production also occurs from biologically active gravity sewer sediments (Liu *et al.*, 2015a).

Sewer sediments are deposits found in the invert of combined sewers and predominantly consist of granular mineral particles. The accumulation of these deposits has been observed in discontinuities of sewers, for example, depressions and obstacles in pipe inverts, branches, connections, and poor pipe joints. Sewer sediments normally contain a certain amount of organic material (typically 1% to >20%), which may lead to an organic binding, creating a cohesive-like sediment bed. Sediment deposition rates and characteristics are highly variable both temporally and spatially (Ashley *et al.*, 2005; Ashley and Verbanck, 1996), which is expected to result in heterogeneity of methane and sulfide production from different sewer sediments under varied operational conditions (flow or static).

Biofilms and sediments in gravity sewers typically have a shallow aerobic zone at the surface followed by an anaerobic zone deep in the profile (Gutierrez *et al.*, 2008). The average methane production rate found in sediments 1.56 ± 0.14 g $CH_4/m^2 \cdot d$ was comparable to the areal rate of 1.26 $gCH_4/m^2 \cdot d$ from biofilms in a rising main sewer pipe (Foley *et al.*, 2009). Further studies showed a large variability of methane production in sewer sediments, that is, varying from 0.13 to 2.09 $gCH_4/m^2 \cdot d$. Sampling results from Liu *et al.* (2016) indicated that the main methane production zone was located near the sediment surface (0–2 cm), but extended deeper than the sulfide production zone (0–0.5 cm) with minimal methane production activity in the deeper layer of the sediment (2–3.5 cm) due to limited penetration of fermentable COD (Figure 3.6). As shown in Section 3.4.1, again a clear stratification of the microbial community occurred with sediment depth. SRB dominated at ca. 0–0.5 cm and coexisted with MA at ca. 0.5–1.0 cm. Below this depth, MA dominated the microbial populations (Liu *et al.*, 2016).

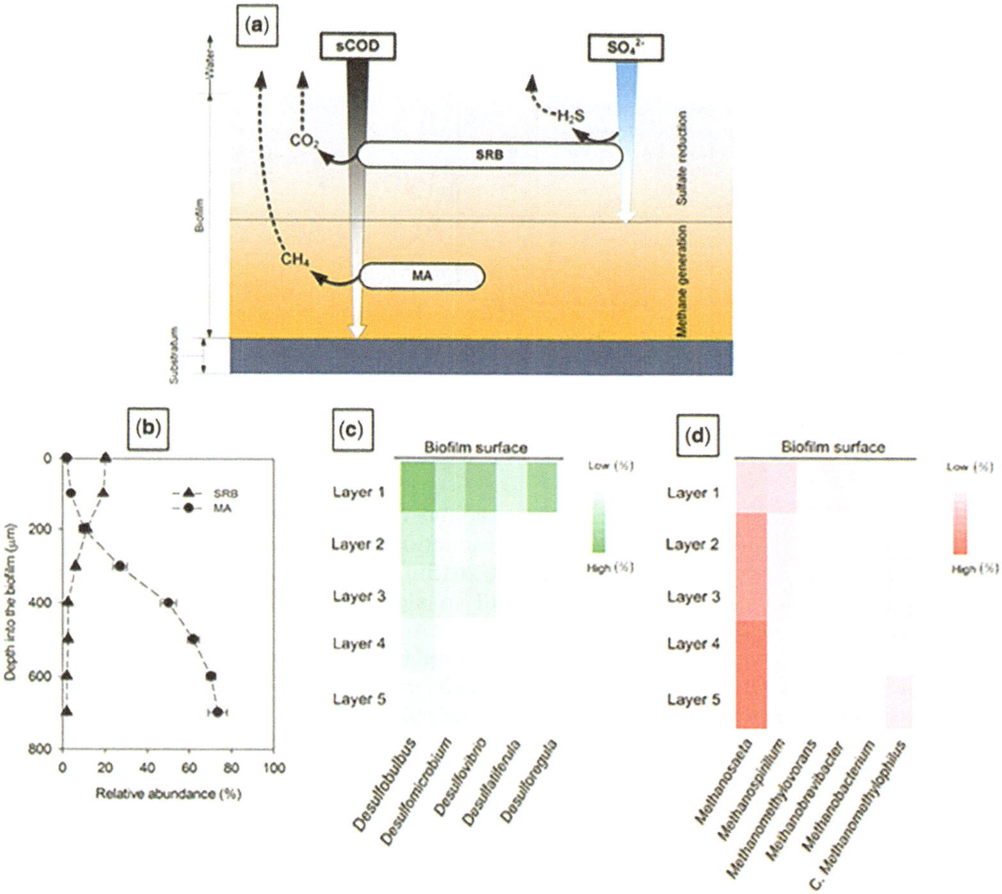

Figure 3.5 (a) Conceptual stratified biofilm model under anaerobic conditions including SRB (sulfate-reducing bacteria) and MA (methanogenic archaea). (b) The SRB and MA proportions of total microorganisms (bacteria and archaea) detected by fluorescence in situ hybridization (FISH) within the sewer biofilms. (c) Heatmap displaying the abundance and distribution of the predominant SRB genera in different sewer biofilm layers from the biofilm surface to the bottom (layer 1 to layer 5). (d) Heatmap displaying the abundance and distribution of the predominant MA genera in different sewer biofilm layers from the biofilm surface to the bottom (layer 1 to layer 5) (Sun *et al.*, 2014).

3.4.3 Factors affecting methane production and emission in sewers

Some key factors regulating methane production and emission in sewers have been identified and listed below. Although the list refers particularly to sewers, they are common in other sections of the UWS and thus can be useful to identify potential CH_4 formation hotspots in the overall UWS.

Dissolved Oxygen (DO): First and foremost, methane production can only take place under anaerobic conditions. Those are predominant in rising main sewers and gravity sewers, even when oxygen is present in the bulk wastewater, biofilms and sediments below the water surface are in partially anaerobic or fully anaerobic conditions due to limited penetration of the oxygen (Gutierrez *et al.*, 2008) and can still generate methane.

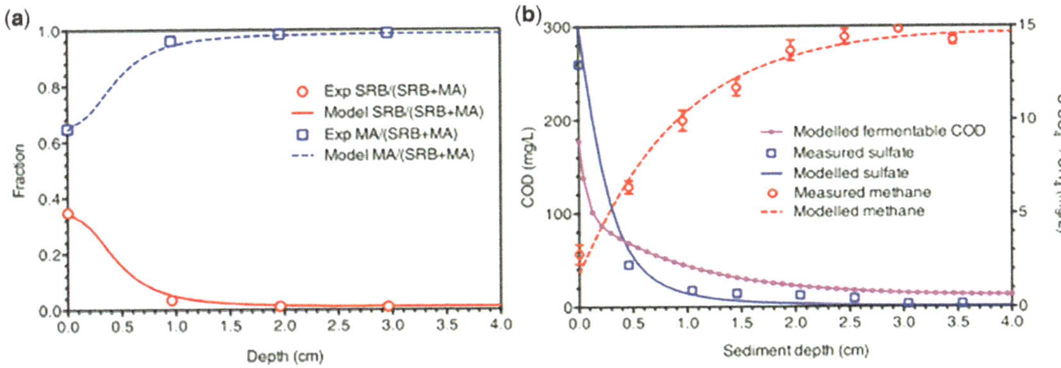

Figure 3.6 (a) Fraction of SRB and MA. (b) Depth profiles of sulfate, methane and fermentable COD in the sewer sediment, adapted from Liu *et al.* (2016). The surface of the sediment was defined as depth 0 cm.

COD content in wastewater: Presence of degradable organics is necessary for the methanogenic metabolisms. The dependency of methane production in sewer sediments is mainly related to fermentable COD concentrations (Liu *et al.*, 2015b). Also, trade waste containing high COD discharged into domestic sewers was found to significantly increase methane production (Sudarjanto & Yuan, 2011).

Hydraulic Retention Time (HRT): HRT is the average length of time that wastewater remains in a pipe or treatment unit. Dissolved methane concentration is positively correlated with HRT in sewers and increases along the length of sewer monitored in field studies (Foley *et al.*, 2009; Guisasola *et al.*, 2008). Liu *et al.* (2015a, 2015b) observed a clear diurnal pattern, with higher dissolved CH_4 concentrations overnight and lower concentrations during the day, likely caused by the diurnal fluctuation in HRT in the network (Sharma *et al.*, 2013). Also, Chaosakul *et al.* (2014) detected a higher methane concentration in both the liquid and gas phase in a gravity sewer during periods of lengthy HRT. The equivalent in a WWTP is contact time.

Area to Volume (A/V) ratio: A/V is the amount of surface area per unit volume in a sewer pipe. In sewers, the surface A/V ratio is an important factor for reactivity, the rate at which the biological reaction will occur, as biofilms grow attached to the surface of sewer inner walls. A higher A/V ratio enables more biofilm growth per unit volume of the wastewater and thus gives a higher methane production rate. Both Guisasola *et al.* (2009) and Foley *et al.* (2009) revealed that higher A/V ratio resulted in higher methane production.

Temperature: Liu *et al.* (2015a, 2015b) showed that temperature also plays an important role in methane production in sewers. A higher methane production rate was observed in warm temperatures corresponding to summer periods (15–25°C) as compared with winter (5–15°C). Results from pumping stations in the USA showed that the concentration of CH_4 in the gas phase was, in 80% of cases, higher in summer than in winter.

High methane concentrations are also expected at sulfide 'hot spots' with severe odour and/or corrosion. Since dissolved sulfide concentration has a positive correlation with HRT, A/V ratio, COD, and temperature, it is also most likely correlated with methane concentration (Sharma *et al.*, 2008). Liu *et al.* (2015a) reported that the dissolved CH_4 and sulfide profiles in a rising main had a strong positive correlation and Guisasola *et al.* (2008) reported similar trends with methane and sulfide profiles, both displaying positive correlation with HRT. This could provide a convenient way of locating likely areas with high methane levels in a sewer network.

3.5 METHANE EMISSIONS FROM WWTPs INCLUDING ANAEROBIC PROCESSES FOR WASTEWATER AND SLUDGE TREATMENT

Anaerobic processes have been applied to treat wastewater and biosolids for more than a century. Organic pollutants in wastewater or biosolids are removed in anaerobic processes via degradation into methane-rich biogas. Unlike in sewer systems, methane generation is intentionally encouraged in wastewater treatment through the organic matter removal pathway. Importantly, it should be noted that methane generation in wastewater treatment is not equal to methane emission. Collection of methane-rich biogas is a common practice in anaerobic wastewater or sludge treatment processes. Factors behind the methane generation (as described in Section 3.4) may not directly influence the methane emissions in a WWTP. In wastewater treatment plants employing anaerobic processes, most methane emissions could proceed from the release of dissolved methane in the effluent of those anaerobic processes (Figure 3.7). Section 3.5.1 presents the most commonly used anaerobic treatments and their potential methane emissions. On the other hand, in WWTPs without anaerobic treatment processes, methane is mainly emitted from sludge handling systems. Section 3.5.2 below will show CH_4 emissions related to the sludge handling processes.

3.5.1 Anaerobic wastewater treatments as sources of methane emissions

Anaerobic wastewater treatment processes have been regarded as sustainable technologies that enable energy recovery via the degradation of organic carbon to methane-rich biogas (Smith *et al.*, 2012). Compared with conventional aerobic treatment processes for organic carbon removal, anaerobic

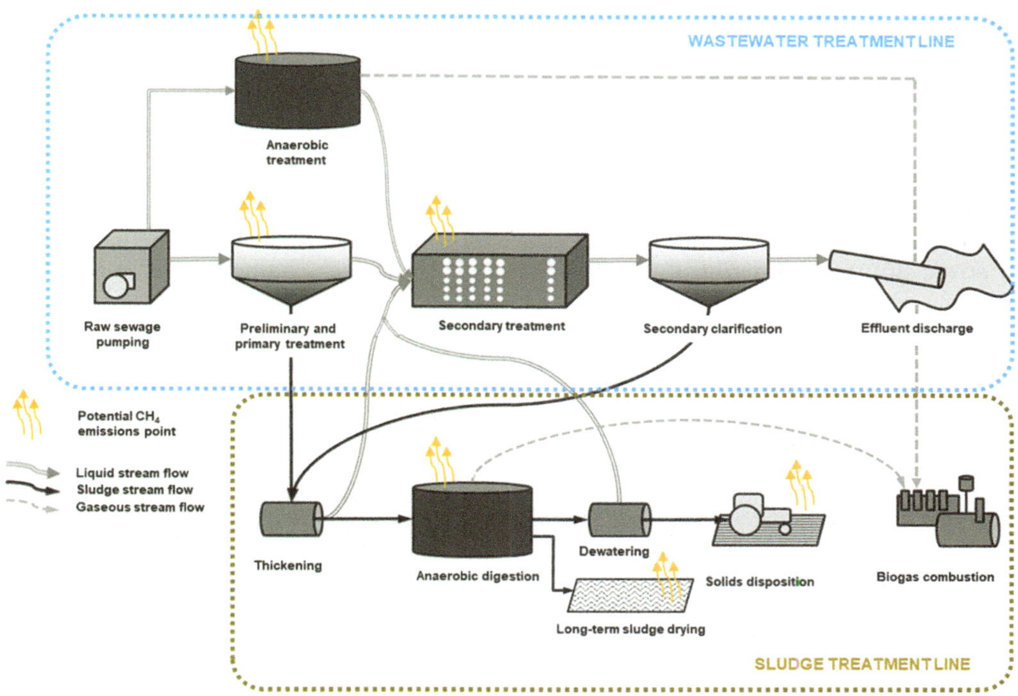

Figure 3.7 WWTP configuration where potential CH_4 emissions points, both in the wastewater and sludge treatment lines, are identified.

processes require less energy (no aeration demand) and produce less sludge (lower yield of anaerobic microorganisms) (Zeeman & Lettinga, 1999). Importantly, the generation of methane from anaerobic processes in WWTPs offers the unique opportunity to achieve energy neutral or even positive wastewater treatment. Due to the ease of operation, anaerobic wastewater treatment processes are often regarded as passive treatment processes, which are particularly suitable for developing countries and also for decentralized operation (Aiyuk *et al.*, 2006).

Anaerobic wastewater treatment involves a series of microbial processes such as hydrolysis, acidogenesis, acetogenesis and methanogenesis, as described in Section 3.2. Due to the slow growth of the anaerobic bacteria and the hydrolysis of solids under low temperatures, long solids retention time (SRT) is critical to retain the slow-growing anaerobic microbial populations in wastewater treatment systems. The conventional anaerobic treatment processes, such as the covered anaerobic lagoon process (as illustrated in Figure 3.8), have a relatively long HRT, as long as 3 months, to maintain a sufficiently long SRT (Khanal *et al.*, 2017). A typical covered anaerobic lagoon receiving municipal wastewater with an HRT of 6 days at subtropical climate, can achieve 60–80% BOD removal with biogas generation and collection (DeGarie *et al.*, 2000). To achieve a high rate of organic carbon removal, the HRT of the anaerobic process needs to be relatively low. The upflow anaerobic sludge blanket (UASB) reactor is one of the major successes in the development of high rate anaerobic wastewater technology that uncouples SRT and HRT. The UASB reactor maintains concentrated biomass by promoting sludge granulation. The sludge retention is based on the settling of sludge granules, irrespective of the short HRT of 4–8 h (Khanal *et al.*, 2017), as illustrated in Figure 3.8. The UASB reactor and its variants such as the expanded granular sludge bed (EGSB) reactor and internal circulation (IC) reactor represent 90% of all high rate anaerobic reactors currently in use (Crone *et al.*, 2016; Van Lier, 2008). The anaerobic membrane bioreactor (AnMBR) is a relatively new concept, which couples membrane filtration with anaerobic treatment, with the potential for a higher quality effluent (Ferrari *et al.*, 2019; Smith *et al.*, 2012).

During anaerobic wastewater treatment, methane is generated by methanogens and dissolves directly into the bulk liquid. Thermodynamic liquid-to-gas mass transfer, as described by Henry's Law, governs the emission of the dissolved methane into the reactor headspace. While the methane-rich biogas is captured for energy recovery, the methane dissolved in the effluent bulk liquid may be released into the atmosphere in the form of fugitive emissions. Studies have reported a large portion of methane remaining in the effluent of anaerobic treatment reactors, from 10 to 85% of total methane generation (Crone *et al.*, 2016). Considering the uncertainties of wastewater characteristics and operational

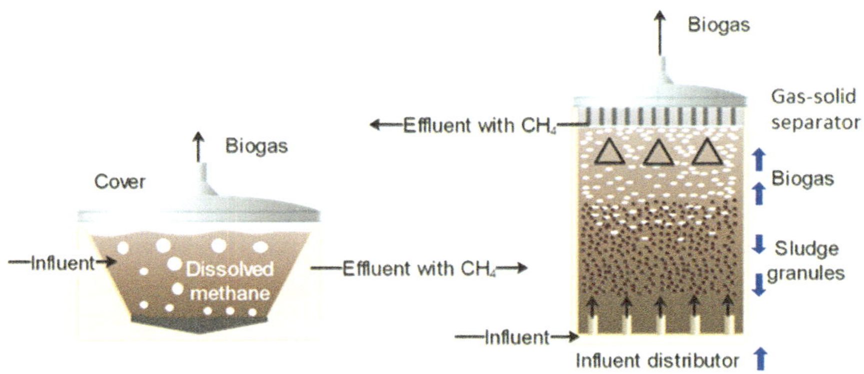

Figure 3.8 Illustration of the conventional covered anaerobic lagoon (left) and the commonly used high-rate anaerobic process, upflow anaerobic sludge blanket (UASB) reactor (right).

conditions using Monte Carlo simulation, Lobato *et al.* (2012) evaluated the dissolved methane losses in UASBs treating municipal wastewater and showed that 25 to 50% of the total methane produced could likely remain in the effluent. Studies have repeatedly shown that the anaerobic wastewater treatment process could produce significantly more GHG than its aerobic counterpart due to the release of dissolved methane in the effluent (Cakir & Stenstrom, 2005; Heffernan *et al.*, 2012). Detailed quantification of methane emissions from wastewater treatment systems is presented in Chapter 5.

Without proper post-treatment, dissolved methane in the anaerobic digestion (AD) effluent is largely released into the atmosphere. Post-treatment technologies have been developed to mitigate the emission of methane. The existing post-treatments can be classified mainly into two categories: (1) improving the gas-liquid transfer, including air stripping and membrane separation; and (2) encouraging the biological consumption of dissolved methane through aerobic methanotrophs. Air-stripping is a widely used technology to promote liquid-to-gas transfer efficiency and, for instance, it has been used to remove methane and ammonia from landfill leachate (Ferraz *et al.*, 2013). However, the use of air-stripping could dilute the methane in the biogas and prohibit the beneficial use of the biogas. Membrane-based technologies using microporous or non-porous membranes, have been developed to recover methane for subsequent reuse (Bandara *et al.*, 2011). The microporous membrane can be applied for AnMBR permeate due to its high substrate transfer efficiency, while the non-porous membrane is more suitable for the UASB effluent to avoid micropore wetting (Liu, 2020). Vacuum or sweeping gas could be introduced to create a methane gradient between the inside and outside of the membrane to promote methane liquid-to-gas transfer. A recovery efficiency of 72% was attained using nitrogen as the sweeping gas in the membrane. However, the use of sweeping gas reduced the purity of the methane in the biogas, compromising its reuse potential (Cookney *et al.*, 2012). In comparison, the vacuum operation could be more practical for subsequent utilization. A vacuum degasification membrane reactor achieved 97% methane recovery efficiency in a lab-scale UASB system. Moreover, the recovered biogas still maintained similar methane composition to that in the UASB headspace (Bandara *et al.*, 2011).

Aerobic methanotrophs can be applied as a biological approach to remove the dissolved methane due to their capability to oxidize methane to CO_2. Hatamoto *et al.* (2010) developed a down-flow hanging sponge (DHS) reactor to consume dissolved methane by aerobic methanotrophs. Sponge-cube carriers seeded with activated sludge are fixed in the reactor, and air is supplied to the reactor to provide oxygen. A maximal methane removal efficiency of 95% can be achieved in the DHS reactor with HRT of 2 h and air flux rate of 3.8 m^3 Air/m^3/d, at a methane removal rate of 0.2 kg CH_4/m^3/d. Furthermore, a two-stage DHS configuration was developed to recover and remove dissolved methane in two sequential DHS reactors, respectively (Matsuura *et al.*, 2015). Methane recovery was encouraged from the first DHS reactor through physical gasification due to low methane partial pressure in the influent air, resulting in an average of 76.8% of dissolved methane recovery efficiency. The remaining dissolved methane was then consumed in the second DHS reactor by aerobic methanotrophs, leading to a final methane concentration of 0.0025 mg CH_4/L in the effluent (99% removal efficiency) of the two-stage DHS (both DHS reactors were operated with 2 h HRT).

Effective recovery or removal of dissolved methane from anaerobically treated effluents could lead to a reduction in GHG emissions. However, while many technologies have been developed, their feasibility has not yet been fully evaluated in terms of practicability and economic viability. Without practical solutions, dissolved methane and the subsequent GHG emission is currently one of the main obstacles for broad application of anaerobic processes for wastewater treatment, particularly for low-strength wastewater (Liu *et al.*, 2014). A practical technology that is economically viable for effluent methane recovery could make anaerobic wastewater treatment even more sustainable and environmentally friendly for wastewater treatment.

3.5.2 Methane emissions from sludge handling processes

During wastewater treatment, the removal of biodegradable organic compounds, nutrients, and suspended solids generates a significant amount of biosolids that are commonly known as sewage sludge (Metcalf & Eddy, 2013). Before its final disposal, the generated sludge goes through a series of handling processes such as thickening, stabilization and dewatering. Sludge stabilization is practised in most WWTPs to reduce pathogens and offensive odours, and to inhibit, reduce, or eliminate the potential for putrefaction. The key point of achieving these objectives is to reduce the organic fraction of the sludge (i.e., volatile solids) or render them unsuitable for the survival of microorganisms (Duan *et al.*, 2019). This is because that growth of pathogens, odours emission, and putrefaction happen when microorganisms find suitable conditions to grow on the organic fraction of the sludge. Alkaline stabilization, anaerobic digestion, aerobic digestion and composting are some of the most common stabilization technologies developed. Among all sludge stabilization technologies, anaerobic digestion (AD) and aerobic digestion (AeD) are the two most used in WWTPs (Duan *et al.*, 2019; Meegoda *et al.*, 2018). Following sludge stabilization, the digested sludge is further dewatered to reduce the amount of sludge for final disposal. With regard to dewatering, different approaches are available including long-term sludge drying, sludge centrifugation, sludge filter press, and sludge belt press. During the sludge handling process, a significant amount of methane can be emitted from sludge anaerobic stabilization and long-term sludge drying processes.

Anaerobic digestion is widely applied in large WWTPs for sludge stabilization. A typical anaerobic high-rate digester is illustrated in Figure 3.9. While the generated methane in the headspace is captured in the sealed reactor, a significant amount of methane can be leaked from the reactor or the methane dissolved in the digestate (Daelman *et al.*, 2012; Schaum *et al.*, 2015, 2016; Tauber *et al.*, 2019). Methane leakage often occurs due to poor maintenance of anaerobic digesters. In an anaerobic digester, the gasholder (shown in Figure 3.9) prevents gas leakage and stores generated biogas. The pressure relief valves are designed for balancing the pressure of digester. However, the gasholder and valves could result in leakage if poorly maintained. Tauber *et al.* (2019) reported that methane leakage from AD manholes contributed to 0.4% of the total generated methane, due to poor maintenance. When the pressure-relief valves are insufficiently maintained or calibrated, they might release biogas to the atmosphere. The methane loss from pressure-relief valves reportedly contributes to 0.06–1.7% of the total methane generated in AD (Reinelt *et al.*, 2016).

As previously explained in Section 3.5.1, methane generated is dissolved directly into the liquid phase of AD reactors and will then be transferred into the gas phase to be captured. However, a

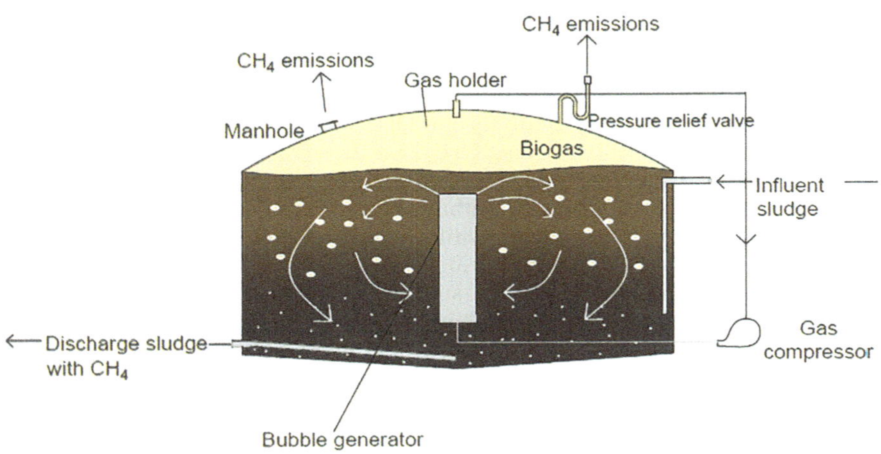

Figure 3.9 Schematic of the anaerobic high-rate digester and the methane emissions from an anaerobic digester. Adapted from Metcalf and Eddy (2013).

significant portion of dissolved methane remains in the liquid phase, which is discharged with the digested sludge. Consequently, dissolved methane will also be released to the atmosphere from the digested sludge. The reported range of methane emissions caused by dissolved CH_4 bubbles in digested sludge is 11–390 g CH_4/(PE·y), representing 0.4–1% of the total amount of methane generated in AD reactors (Schaum *et al.*, 2015, 2016; Tauber *et al.*, 2019).

Long-term sludge drying is a commonly applied technology in WWTPs in many countries, for example, Australia, for sludge dewatering due to its ease of operation. Long-term sludge drying can be carried out in sludge-drying lagoons, sludge-drying pans, or sludge-drying beds. Normally, the stabilized sludge will be stored for around a year for dewatering (by solar and wind) and further digestion. Figure 3.10 illustrates a typical sludge-drying lagoon. The air is continuously diffusing into the upper layer of the storage tank. The upper layer of the long-term sludge-drying unit is thus aerobic, in which COD is oxidized (Pan *et al.*, 2016). The bottom of the long-term sludge-drying unit is normally under anaerobic conditions. The operation and drying mechanisms of the sludge-drying pan and sludge-drying bed are similar to that of the drying lagoon (Pan *et al.*, 2016; Stickland *et al.*, 2013). As a significant portion of the stabilized sludge is still biodegradable, further anaerobic digestion could lead to methane generation during the long-term sludge drying process under anaerobic conditions. Methane is not normally captured in the long-term sludge drying process. Therefore, methane emissions from the long-term drying units could be significant. Pan *et al.* (2016) reported approximately 43% of the COD in sludge was converted to methane emissions in a sludge- drying lagoon. In the investigated WWTPs, the methane emissions from sludge-drying lagoons contributed to 24–65% of the total plant GHG emissions.

Overall, methane emissions from sludge handling processes could be significant, especially for WWTPs with anaerobic sludge digestion and/or long-term sludge-drying processes. Significant amounts of methane can be leaked from the anaerobic digestion process due to deficient maintenance and operation of AD reactors (Duan, 2019). In addition, the long-term sludge-drying process may emit the majority of GHG emissions from the sludge-handling process in large WWTPs. Regular maintenance of the anaerobic digester and effective removal of dissolved methane from digested sludge could mitigate the methane emissions. Further research is still required to understand and mitigate methane emissions from sludge-handling processes. This would help to develop effective CH_4 control strategies, still currently lacking due to the limited knowledge of the implications and contribution of these processes to the overall CH_4 and GHG emissions.

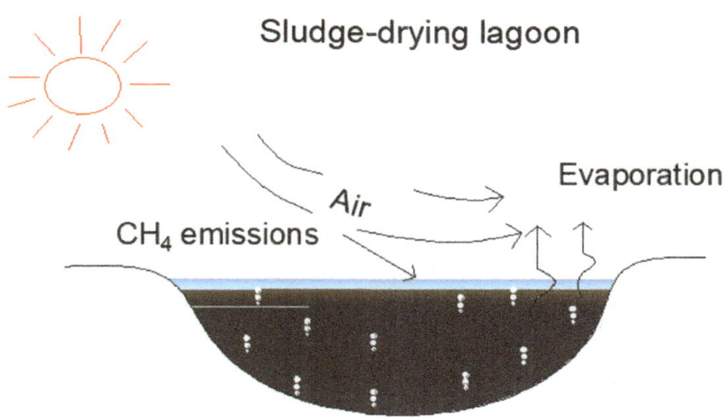

Figure 3.10 Schematic of methane emissions from a sludge-drying lagoon and sludge drying. Adapted from Pan *et al.* (2016).

3.6 CONCLUDING REMARKS

Wastewater treatment is a significant source of methane, contributing to approximately 5% of worldwide methane emissions. The mechanisms, sources, and factors that affect methane emissions in urban wastewater systems are discussed in this chapter. As presented above, methane production can occur in different sections of the UWS once anaerobic conditions prevail. Due to the natural presence of organics and the lack of oxygen in sewage, methanogenic archaea can find conditions to carry out their metabolism and produce methane.

The main biological processes involved have been described and, for sewer systems, linked to their interactions with the sulfate-reducing bacteria, often coexisting with methanogens thanks to the spatial distribution within thick sewer biofilms and sediments. It also has been stated that in sewer systems, methane production generally results in direct fugitive emissions. This is because of the large extensions and multiple CH_4-generating points in sewers which make it challenging not only to monitor each emission point but also to capture the emissions from those hotspots.

On the other hand, methane emissions in WWTPs are mainly produced from undesired leaks from anaerobic treatment technologies. For anaerobic wastewater treatment plants, the majority of methane is released from the dissolved methane in anaerobically treated wastewater. This significant methane loss (emission) is currently one of the main obstacles for the broad application of anaerobic wastewater treatment technologies. In WWTPs with aerobic wastewater treatment processes, methane emissions mainly occur via leakage from anaerobic sludge-handling processes. Anaerobic sludge digestion is a commonly practised technology for sludge stabilization, where methane is produced and collected. Nevertheless, a portion of the generated methane leaks from the facilities. Following sludge stabilization, an anaerobic sludge-drying process in long-term drying lagoons is also a common practice where optimal conditions for methane formation exist, thus producing substantial methane emissions. Overall, methane-emission points from UWSs are well identified. However, the lack of suitable instruments/technologies for the dynamic quantification of methane emission prevents a full picture of its contribution to the complete GHG emission of urban wastewater systems being obtained.

ACKNOWLEDGEMENTS

Oriol Gutierrez acknowledges the support from the Economy and Knowledge Department of the Catalan Government through a Consolidated Research Group (ICRA-TECH – 2017 SGR 1318) – Catalan Institute for Water Research. Haoran Duan and Ziping Wu acknowledge the support of the Australian Research Council (ARC) through project DP180103369.

REFERENCES

Aiyuk S., Forrez I., Lieven D. K., van Haandel A. and Verstraete W. (2006). Anaerobic and complementary treatment of domestic sewage in regions with hot climates–a review. *Bioresource Technology*, **97**(17), 2225–2241. https://doi.org/10.1016/j.biortech.2005.05.015

Ashley R. M. and Verbanck M. A. (1996). Mechanics of sewer sediment erosion and transport. *J. Hydraul. Res.* **34**(6), 753–770. https://doi.org/10.1080/00221689609498448

Ashley R.M., Bertrand-Krajewski J. L., Hvitved-Jacobsen T. and Verbanck M. (2005). Solids in Sewers. *Water Science and Technologies*, **52**(3), 73–84. https://doi.org/10.2166/wst.2005.0063

Buendia E. C., Guendehou S., Limmeechokchai B., Pipatti R., Rojas Y., Sturgiss R., Tanabe K., Wirth T., Romano D., Witi J., Garg A., Weitz M. M., Cai B., Ottinger D. A., Dong H., MacDonald J. D., Ogle S. M., Rocha M. T., Sanz Sanchez M. J., … Irving W. (2019). 2019 Refinement to the 2006 IPCC Guidelines for National Greenhouse Gas Inventories.

Cakir F. Y. and Stenstrom M. K. (2005). Greenhouse gas production: a comparison between aerobic and anaerobic wastewater treatment technology. *Water Research*, **39**(17), 4197–4203. https://doi.org/10.1016/j.watres.2005.07.042

Chaosakul T., Koottatep T. and Polprasert C. (2014). A model for methane production in sewers. *Journal of Environmental Science and Health, Part A*, **49**(11), 1316–1321. https://doi.org/10.1080/10934529.2014.910071

Cookney J., Cartmell E., Jefferson B. and McAdam E. J. (2012). Recovery of methane from anaerobic process effluent using poly-di-methyl-siloxane membrane contactors. *Water Science and Technology*, **65**(4), 604–610. https://doi.org/10.2166/wst.2012.897

Crone B. C., Garland J. L., Sorial G. A. and Vane L. M. (2016). Significance of dissolved methane in effluents of anaerobically treated low strength wastewater and potential for recovery as an energy product: a review. *Water Research*, **104**, 520–531. https://doi.org/10.1016/j.watres.2016.08.019

Daelman M. R. J., van Voorthuizen E. M., van Dongen U. G. J. M., Volcke E. I. P. and van Loosdrecht M. C. M. (2012). Methane emission during municipal wastewater treatment. *Water Research*, **46**(11), 3657–3670. https://doi.org/10.1016/j.watres.2012.04.024

Dar S. A., Kleerebezem R., Stams A. J. M., Kuenen J. G. and Muyzer G. (2008). Competition and coexistence of sulfate-reducing bacteria, acetogens and methanogens in a lab-scale anaerobic bioreactor as affected by changing substrate to sulfate ratio. *Applied Microbiology and Biotechnology*, **78**(6), 1045–1055. https://doi.org/10.1007/s00253-008-1391-8

DeGarie C. J., Crapper T., Howe B. M., Burke B. F. and McCarthy P. J. (2000). Floating geomembrane covers for odour control and biogas collection and utilization in municipal lagoons. *Water Science and Technology*, **42**(10–11), 291–298. https://doi.org/10.2166/wst.2000.0664

Duan H. (2019). Controlling Nitrite Oxidizing Bacteria Using Nitrogen Released from Sludge Digestion.

Duan H., Ye L., Lu X., Batstone D. J. and Yuan Z. (2019). Self-sustained nitrite accumulation at low pH greatly enhances volatile solids destruction and nitrogen removal in aerobic sludge digestion. *Environmental Science & Technology*, **53**(3), 1225–1234. https://doi.org/10.1021/acs.est.8b04498

Ferrari F., Balcazar J. L., Rodriguez-Roda I. and Pijuan M. (2019). Anaerobic membrane bioreactor for biogas production from concentrated sewage produced during sewer mining. *The Science of the Total Environment*, **670**, 993–1000. https://doi.org/10.1016/j.scitotenv.2019.03.218

Ferraz F. M., Povinelli J. and Vieira E. M. (2013). Ammonia removal from landfill leachate by air stripping and absorption. *Environmental Technology*, **34**(15), 2317–2326. https://doi.org/10.1080/09593330.2013.767283

Foley J., Yuan Z. and Lant P. (2009). Dissolved methane in rising main sewer systems: field measurements and simple model development for estimating greenhouse gas emissions. *Water Science and Technology*, **60**(11), 2963–2971. https://doi.org/10.2166/wst.2009.718

Foley J., Yuan Z., Keller J., Senante E., Chandran K., Willis J., Shah A., van Loosdrecht M. and van Voorthuizen E. (2011a). N2O and CH4 Emission from Wastewater Collection and Treatment Systems: State of the Science Report. http://espace.library.uq.edu.au/view/UQ:267322#.VTY1BQDSue4.mendeley

Foley J., Yuan Z., Senante E., Chandran K., Willis J., van Loosdrecht M. and van Voorthuizen E. (2011b). *Global Water Research Coalition*.

Guisasola A., de Haas D., Keller J. and Yuan Z. (2008). Methane formation in sewer systems. *Water Research*, **42**(6–7), 1421–1430. http://www.sciencedirect.com/science/article/B6V73-4PX046M-4/2/5ab87c75f8b25b5a369e462a29ef50fa

Guisasola A., Sharma K. R., Keller J., Yuan Z. and Jiang G. (2009). Development of a model for assessing methane formation in rising main sewers. *Water Research*, **43**(11), 2874–2884. http://www.scopus.com/inward/record.url?eid=2-s2.0-70149111960&partnerID=40&md5=a7dc45466e5ce0232a4e9d25adff8df5

Gutierrez O., Mohanakrishnan J., Sharma K. R., Meyer R. L., Keller J. and Yuan Z. (2008). Evaluation of oxygen injection as a means of controlling sulfide production in a sewer system. *Water Research*, **42**(17), 4549–4561. http://www.sciencedirect.com/science/article/B6V73-4T5TPH6-5/2/b08ec8aa7ac6ce029b65e5e3cb1555f9

Hatamoto M., Yamamoto H., Kindaichi T., Ozaki N. and Ohashi A. (2010). Biological oxidation of dissolved methane in effluents from anaerobic reactors using a down-flow hanging sponge reactor. *Water Research*, **44**(5), 1409–1418. https://doi.org/10.1016/j.watres.2009.11.021

Heffernan B., Blanc J. and Spanjers H. (2012). Evaluation of greenhouse gas emissions from municipal UASB sewage treatment plants. *Journal of Integrative Environmental Sciences*, **9**(sup1), 127–137. https://doi.org/10.1080/1943815X.2012.696546

Hvitved-Jacobsen T. (2002). Sewers Systems and Process. In: Sewer Process – Microbial and Chemical Process Engineering of Sewer Network. Published April 16, 2013 by CRC Press. ISBN 9781439881774.

IPCC (2006). Guidelines for National Greenhouse Gas Inventories. volume 5–waste. In: The National Greenhouse Gas Inventories Programme, The Intergovernmental Panel on Climate Change, H. Eggleston, L. Buendia, K. Miwa, T. Ngara and K. Tanabe (eds.), pp. 6.1–6.28. Ed. Intergovernmental Panel on Climate Change (IPCC) of the United Nations (UN). https://www.ipcc-nggip.iges.or.jp/public/2006gl/

IPCC (2019). Refinement to the 2006 IPCC Guidelines for National Greenhouse Gas Inventories. Task Force Report on National Greenhouse Gas Inventories in accordance with the decision taken at the 44th Session of IPCC in Bangkok, Thailand, in October 2016. Ed. Intergovernmental Panel on Climate Change (IPCC) of the United Nations (UN). https://www.ipcc.ch/report/2019-refinement-to-the-2006-ipcc-guidelines-for-national-greenhouse-gas-inventories/

Kehrein P., van Loosdrecht M., Osseweijer P., Garfí M., Dewulf J. and Posada J. (2020). A critical review of resource recovery from municipal wastewater treatment plants – market supply potentials, technologies and bottlenecks. *Environmental Science: Water Research & Technology*, **6**(4), 877–910. https://doi.org/10.1039/C9EW00905A

Khanal S. K., Giri B., Nitayavardhana S. and Gadhamshetty V. (2017). 10 – anaerobic bioreactors/digesters: design and development. In: Current Developments in Biotechnology and Bioengineering, pp. 261–279. https://doi.org/10.1016/ B978-0-444-63665-2.00010-2

Liu T. (2020). Exploring Versatile Applications of Nitrite/Nitrate-Dependent Anaerobic Methane Oxidation Through Modelling and Experimental Investigations.

Liu Z., Yin H., Dang Z. and Liu Y. (2014). Dissolved methane: a hurdle for anaerobic treatment of municipal wastewater. *Environmental Science & Technology*, **48**(2), 889–890. https://doi.org/10.1021/es405553j

Liu Y., Ni B. J., Sharma K. R. and Yuan Z. (2015a). Methane emission from sewers. *Science of the Total Environment*, **524–525**, 40–51. https://doi.org/10.1016/j.scitotenv.2015.04.029

Liu Y., Sharma K. R., Fluggen M., O'Halloran K., Murthy S. and Yuan Z. (2015b). Online dissolved methane and total dissolved sulfide measurement in sewers. *Water Research*, **68**, 109–118. https://doi.org/10.1016/j.watres.2014.09.047

Liu Y., Tugtas A. E., Sharma K. R., Ni B. J. and Yuan Z. (2016). Sulfide and methane production in sewer sediments: field survey and model evaluation. *Water Research*, **89**, 142–150. https://doi.org/10.1016/j.watres.2015.11.050

Lobato L. C. S., Chernicharo C. A. L. and Souza C. L. (2012). Estimates of methane loss and energy recovery potential in anaerobic reactors treating domestic wastewater. *Water Science and Technology*, **66**(12), 2745–2753. https://doi.org/10.2166/wst.2012.514

Mannina G., Butler D., Benedetti L., Deletic A., Fowdar H., Fu G., Kleidorfer M., McCarthy D., Steen Mikkelsen P., Rauch W., Sweetapple C., Vezzaro L., Yuan Z. and Willems P. (2018). Greenhouse gas emissions from integrated urban drainage systems: where do we stand? *Journal of Hydrology*, **559**, 307–314. https://doi.org/10.1016/j.jhydrol.2018.02.058

Matsuura N., Hatamoto M., Sumino H., Syutsubo K., Yamaguchi T. and Ohashi A. (2015). Recovery and biological oxidation of dissolved methane in effluent from UASB treatment of municipal sewage using a two-stage closed downflow hanging sponge system. *Journal of Environmental Management*, **151**, 200–209. https://doi.org/10.1016/j.jenvman.2014.12.026

Meegoda J. N., Li B., Patel K. and Wang L. B. (2018). A review of the processes, parameters, and optimization of anaerobic digestion. *International Journal of Environmental Research and Public Health*, **15**(10), 2224, https://doi.org/10.3390/ijerph15102224

Metcalf and Eddy I. (2003). 4th edn. Wastewater Engineering, Treatment and Reuse. In: MacGraw-Hill (ed.), Series in civil and environmental engineering.

Metcalf A. and Eddy I. (2013). Wastewater Engineering: Treatment and Resource Recovery. Ed McGraw-Hill Higher Education. ISBN-13: 978-0073401188

Mohanakrishnan J., Gutierrez O., Sharma K. R., Guisasola A., Werner U., Meyer R. L., Keller J. and Yuan Z. (2009). Impact of nitrate addition on biofilm properties and activities in rising main sewers. *Water Research*, **43**(17), 4225–4237. http://www.sciencedirect.com/science/article/B6V73-4WJBBRF-1/2/8a5436c0c1339d83010f23a6c7616064

Muyzer G. and Stams A. J. M. (2008). The ecology and biotechnology of sulphate-reducing bacteria. *Nature Reviews. Microbiology*, **6**(6), 441–454. https://doi.org/10.1038/nrmicro1892

Pan Y., Ye L., van den Akker B., Ganigue Pages R., Musenze R. S. and Yuan Z. (2016). Sludge-drying lagoons: a potential significant methane source in wastewater treatment plants. *Environmental Science & Technology*, **50**(3), 1368–1375. https://doi.org/10.1021/acs.est.5b04844

Pikaar I., Sharma K. R., Hu S., Gernjak W., Keller J. and Yuan Z. (2014). Reducing sewer corrosion through integrated urban water management. *Science*, **345**(6198), 812–814. https://doi.org/10.1126/science.1251418

Puyol D., Batstone D. J., Hülsen T., Astals S., Peces M. and Krömer J. O. (2017). Resource recovery from wastewater by biological technologies: opportunities, challenges, and prospects. *Frontiers in Microbiology* **7**, 2106. https://www.frontiersin.org/article/10.3389/fmicb.2016.02106

Reinelt T., Liebetrau J. and Nelles M. (2016). Analysis of operational methane emissions from pressure relief valves from biogas storages of biogas plants. *Bioresource Technology*, **217**, 257–264. https://doi.org/10.1016/j.biortech.2016.02.073

Schaum C., Lensch D., Bolle P.-Y. and Cornel P. (2015). Sewage sludge treatment: evaluation of the energy potential and methane emissions with COD balancing. *Journal of Water Reuse and Desalination*, **5**(4), 437–445. https://doi.org/10.2166/wrd.2015.129

Schaum C., Fundneider T. and Cornel P. (2016). Analysis of methane emissions from digested sludge. *Water Science and Technology*, **73**(7), 1599–1607. https://doi.org/10.2166/wst.2015.644

Sharma K., de Haas D., Corrie S., O'Halloran K., Keller J. and Yuan Z. (2008). Predicting hydrogen sulfide formation in sewers: a new model. *Water*, **35**(2), 132–137. http://www.scopus.com/inward/record.url?eid=2-s2.0-77953083359&partnerID=40&md5=62d8ae99a1a1bac8892a44efe6c38403

Sharma K., Ganigue R. and Yuan Z. (2013). pH dynamics in sewers and its modeling. *Water Research*, **47**(16), 6086–6096. https://doi.org/10.1016/j.watres.2013.07.027

Smith A. L., Stadler L. B., Love N. G., Skerlos S. J. and Raskin L. (2012). Perspectives on anaerobic membrane bioreactor treatment of domestic wastewater: a critical review. *Bioresource Technology*, **122**, 149–159. https://doi.org/10.1016/j.biortech.2012.04.055

Stickland A. D., Rees C. A., Mosse K. P. M., Dixon D. R. and Scales P. J. (2013). Dry stacking of wastewater treatment sludges. *Water Research*, **47**(10), 3534–3542. https://doi.org/10.1016/j.watres.2013.04.002

Sudarjanto G. and Yuan Z. (2011). A laboratory assessment of the impact of brewery wastewater discharge on sulfide and methane production in a sewer. *Water Science and Technology*, **64**(8), 1614–1619. https://doi.org/10.2166/wst.2011.733

Sun J., Hu S., Sharma K. R., Ni B.-J. and Yuan Z. (2014). Stratified microbial structure and activity in sulfide- and methane- producing anaerobic sewer biofilms. *Applied and Environmental Microbiology*, **80**(22), 7042–7052. https://doi.org/10.1128/AEM.02146-14

Tauber J., Parravicini V., Svardal K. and Krampe J. (2019). Quantifying methane emissions from anaerobic digesters. *Water Science and Technology*, **80**(9), 1654–1661. https://doi.org/10.2166/wst.2019.415

Van Lier J. B. (2008). High-rate anaerobic wastewater treatment: diversifying from end-of-the-pipe treatment to resource-oriented conversion techniques. *Water Science and Technology*, **57**(8), 1137–1148. https://doi.org/10.2166/wst.2008.040

Willis J., Brower B., Graf W., Murthy S., Peot C., Regmi P., Sharma K. and Yuan Z. (2016). New GHG methodology to quantify sewer methane. 91st Annual Water Environment Federation Technical Exhibition and Conference, WEFTEC 2016, January, 4745–4752. https://doi.org/10.2175/193864718825139465

Zeeman G. and Lettinga G. (1999). The role of anaerobic digestion of domestic sewage in closing the water and nutrient cycle at community level. *Water Science and Technology*, **39**(5), 187–194. https://doi.org/10.1016/S0273-1223(99)00101-8

NOMENCLATURE

AeD	Aerobic digestion
AD	Anaerobic digestion
AnMBR	Anaerobic membrane bioreactor
A/V	Area to volume ratio
CH_4	Methane
COD	Chemical oxygen demand
DO	Dissolved oxygen
DHS	Down-flow hanging sponge reactor
EGSB	Expanded granular sludge bed
FISH	Fluorescence in situ hybridization

GHG	Greenhouse gas
GWRC	Global Water Research Coalition
HRT	Hydraulic retention time
H_2S	Hydrogen sulfide
IC	Internal circulation
MA	Methanogenic archaea
SRB	Sulfate reducing bacteria
SRT	Solids retention time
UASB	Upflow anaerobic sludge blanket
UWS	Urban wastewater systems
VFA	Volatile fatty acids
WRRF	Water resource recovery facilities
WWTP	Wastewater treatment plant

doi: 10.2166/9781789060461_063

Chapter 4

Reporting guidelines

Ariane Coelho Brotto[1] and Amanda Lake[2]

[1]Jacobs, Cottons Lane, London SE1 2QG, UK. E-mail: ariane.brotto@jacobs.com
[2]Jacobs, 160 Dundee Street, Edinburgh EH11 1DQ, Scotland. E-mail: amanda.lake@jacobs.com

SUMMARY

The *Reporting Guidelines* chapter focuses on the accounting methodologies and protocols supporting top-down greenhouse gas (GHG) emissions assessment and reporting of relevance to the urban water system in wastewater treatment of domestic and industrial wastewaters. It summarizes the basis for existing methane (CH_4) and nitrous oxide (N_2O) emission factors, the three-tier approach set out in the internationally accepted Intergovernmental Panel on Climate Change (IPCC) methodology and areas where further work is required. This chapter also summarizes the implications of the 2019 IPCC Refinement on the magnitude of N_2O emissions from secondary treatment, as well as country-specific emission factors developed through national bottom-up monitoring. Finally, this chapter highlights the importance of bottom-up approaches in understanding the opportunities to optimize treatment processes and conditions that minimize direct GHG emissions and help move the water industry towards net zero GHG emissions.

Keywords: Bottom-up, emission factor, emission inventory, IPCC, methodology, top-down

TERMINOLOGY

Term	Definition
Bottom-up	Estimation of emissions based on direct on-site measurement of concentration and emission fluxes, typically at the facility level.
Fugitive emissions	Intentional or unintentional emissions of greenhouse gas that are not produced intentionally by a stack or vent, which may include leaks from process units and pipelines.
Greenhouse gas	Gas that absorbs and emits radiant energy within the thermal infrared range.
Greenhouse gas inventory	Accounting of all greenhouse gas emissions and removals from given sources and sinks from a defined region in a specific period of time.
Methodology	A specific accounting guideline with a foundational set of equations and emission factors based on scientific and applied research to estimate emissions, typically at organizational or regional level.

Protocol	A standardized framework for measuring and reporting GHG emissions. These are usually based on the guiding principles of relevance, completeness, consistency, accuracy and transparency.
Tier 1 method	The IPCC Tier 1 method applies default values for an emission factor and activity parameters. It is considered good practice for countries with limited data.
Tier 2 method	The IPCC Tier 2 method follows the same method as Tier 1 but allows for incorporation of a country-specific emission factor and country-specific activity data, which could include country-specific factors and/or field measurement data from the reporting country.
Tier 3 method	The IPCC Tier 3 method is applied for a country with good data and advanced methodologies. It applies country-specific factors and field measurement data at a country and/or facility level.
Top-down	Estimation of GHG emissions based on generalized equations and emissions factors applied to activity data.

4.1 INTRODUCTION

The influence of human activities on the world's climate system is unequivocal – the unparalleled levels of greenhouse gas (GHG) emissions since pre-industrial times have already caused an estimated 1.09°C (range 0.95–1.2°C) of global warming above pre-industrial levels (IPCC, 2021). Human-induced climate change is already affecting many weather and climate extremes in every region across the globe, resulting in significant and increasingly catastrophic impacts on communities, as well as on ecosystems and natural resources. Under all emissions scenarios, global warming is likely (ranging from *very likely* to *more likely than not*) to exceed 1.5°C between 2021 and 2040 (IPCC, 2021). Limiting global warming to 1.5°C to meet the Paris Agreement will require sharp GHG reduction to net zero emissions by 2050 (Rogelj *et al.*, 2018).

The management of domestic and industrial wastewaters causes anthropogenic GHG emissions throughout the urban water cycle. These GHG emissions are related to fossil derived energy (electricity and heat) use for water abstraction, treatment and conveyance, and for wastewater collection and treatment, as well for direct GHG emissions from the treatment processes. Emissions also occur when wastewaters are discharged, treated or untreated, to the environment from centralized and decentralized systems. This includes the discharge of sewage effluent to the environment and, where applicable, the application of sludge residuals, or biosolids, to land. This chapter considers emissions from sewage conveyance and at wastewater treatment plants (WWTPs) but not emissions from natural treatment systems (e.g., wetlands), and from the release of final effluent or sludge residuals to the environment.

Methane (CH_4) and nitrous oxide (N_2O) are the main GHGs emitted during the collection and treatment of wastewater and in the on-site treatment and management of sludge residuals. These direct process emissions are required to be reported under international agreements and are gaining increased attention with the continued reduction of indirect energy-related GHG due to decarbonization of electricity grids with renewable energy. Of particular importance to the water sector, and as discussed in Section 4.2, the relative warming impact (global warming potential, GWP) of CH_4 and N_2O are substantially greater than that of carbon dioxide (CO_2). As a result, these process emissions may form a very substantial part of a facility's operational carbon emissions.

The United Nationals Framework Convention on Climate Change (UNFCCC), an international treaty which came into force in 1994 and seeks to reduce emissions of GHGs, requires Parties to develop, update and publish national emissions inventories. National GHG inventories are essential tools for transparent reporting of anthropogenic emissions and removal of GHGs. The Intergovernmental Panel on Climate Change (IPCC) provides global guidelines and methodologies for quantifying GHG emissions for these national GHG inventories, including for CH_4 and N_2O from wastewater treatment. Guidelines provide a basis for the mutual trust and confidence that are needed for effective implementation of international agreements to address climate change and provide an essential tool for developing policies and monitoring impact (Bartram *et al.*, 2019).

In 2015, an historic agreement was reached in Paris between 196 Parties of the UNFCCC. The Paris Agreement seeks to limit global temperature increase to well below 2°C and requires efforts to limit global temperature increase to 1.5°C above pre-industrial levels. It requires each Party to prepare, communicate and maintain successive nationally determined contributions (NDCs) for GHG emissions that it intends to achieve. NDCs must be reported every five years, using international guidance provided by the IPCC. The Paris Agreement allows for a first 'global stocktake' of emissions in 2023 and a review every five years after, with the aim that Parties increase their mitigation efforts and ambition through successive reviews (IPCC, 2020b).

In alignment with the Paris Agreement, countries, local governments, and economic sectors around the world are pledging to achieve net zero within the decades leading up to the recognized requirement for net zero by 2050 to minimize global heating. Reliable accounting of GHG inventories, aligned with international guidelines, are essential within all these spheres of influence to evaluate the magnitude of emissions as accurately as possible and to assess the efforts required to achieve target GHG reductions.

This chapter focuses on the accounting methodologies and protocols supporting top-down GHG emissions assessment and reporting of relevance to wastewater collection and treatment and to water utilities and water industry sectors. It summarizes the basis for existing N_2O and CH_4 emission factors from wastewater collection and treatment, the three-tier approach set out in the internationally accepted IPCC methodology and considerations of how this is being applied, including ongoing work and challenges in the development of country-specific emission factors (EFs) through national monitoring. It explains the implications of the 2019 IPCC Refinement on the magnitude of N_2O emissions from secondary treatment and the development of a revised EF for N_2O. It also considers uncertainty in accounting methodologies and protocols as defined by the IPCC. The following chapters of this book address issues of uncertainty in emissions and emission factors across sites and process types and uncertainty around measurement and analysis methods.

Finally, this chapter highlights the importance of bottom-up approaches in understanding the opportunities to optimize treatment processes and conditions that minimize direct GHG emissions and help move the water industry towards net zero GHG emissions.

4.2 ACCOUNTING CONSIDERATIONS

This section provides an overview of accounting considerations in applying GHG protocols and international best practice. It first considers how wastewater treatment emissions are defined in carbon accounting practice and then the basis for accounting methodologies – defining the concepts of top-down and bottom-up accounting.

4.2.1 Reporting scope considerations for the water industry

GHG emissions can be quantified and reported, whether by country, company or other organization/individual, by Scope. The commonly accepted definitions of emissions Scope 1, 2 and 3 were introduced by the GHG Protocol of the World Business Council for Sustainable Development (WBCSD)/World Resources Institute (WRI) to categorize emissions by ownership levels, that is direct (Scope 1) and indirect (Scope 2 and 3) emissions (Greenhouse Gas Protocol, 2020a, 2020b). These are shown below in Figure 4.1.

The GHG Protocol establishes comprehensive global standardized frameworks to measure and manage greenhouse gas (GHG) emissions from private and public sector operations, value chains and mitigation actions – including for the water sector. Within the water sector, ownership and reporting of emissions as Scope 1 or 3 may be differentiated depending on the type of organization and defined reporting boundaries.

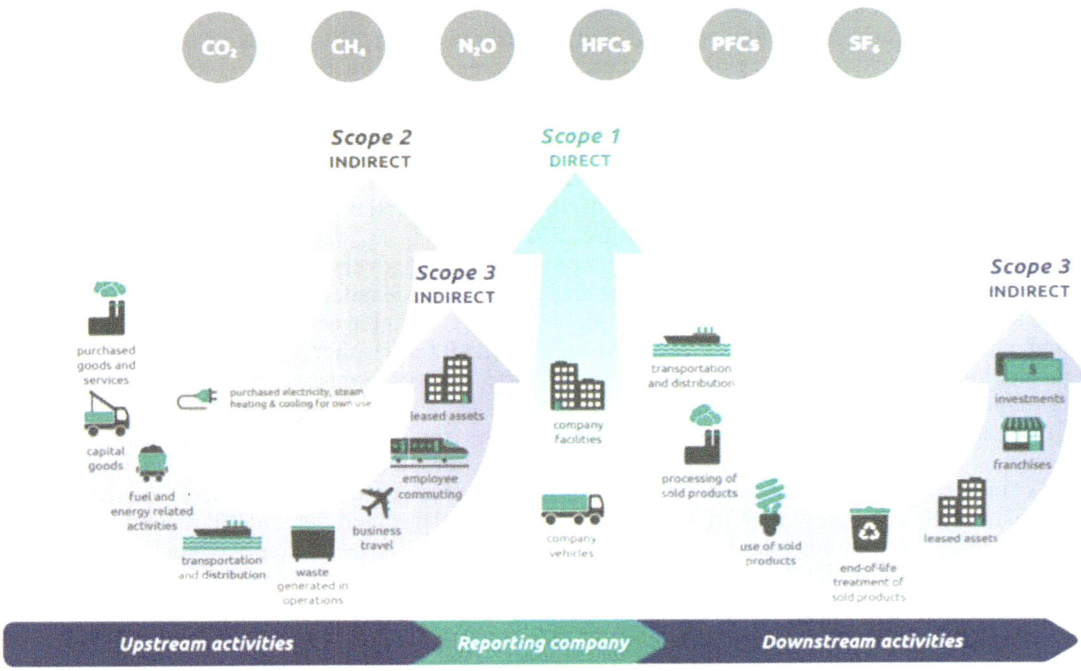

Figure 4.1 Sources and categories of GHG emissions for corporate reporting (Greenhouse Gas Protocol, 2020a, 2020b). HFCs, hydrofluorocarbons; PFCs, perfluorocarbons. With thanks to the World Resources Institute - licensed under a creative commons licence (http://creativecommons.org/licenses/by-nc-nd/3.0)

Reference is also made to the International Standards Organization ISO 14064-1:2018 Greenhouse Gases: Part 1: Specification with guidance at the organization level for quantification and reporting of greenhouse gas emissions and removals. This provides definitions for six categories of GHG emissions and removals for an organization to consider in reporting.

Regional water companies, which may be public or privately owned and provide water services over a regional geography such as city, state, country or other geographic area, usually adopt what is called the control approach (relating to either financial or operational control). Under this approach, these companies account for 100% of all the emissions from operations over which they have control. The relevant international best practice for reporting their emissions is the GHG Protocol Corporate Standard (Greenhouse Gas Protocol, 2015, 2020a, 2020b).

This defines scopes and examples for the waste sector (with respect to wastewater treatment) as follows (Greenhouse Gas Protocol, 2015):

- Scope 1: direct GHG emissions from sources owned or controlled by the company from stationary combustion (incinerators, boilers, flaring), process emissions from the transformation of raw materials (e.g., N_2O emissions from the oxidation of ammoniacal nitrogen in sewage treatment), and CH_4 emissions from the anaerobic treatment of wastewater and/or sludges. Direct GHG emissions from the water sector also include CO_2 emissions from wastewater treatment and emissions from mobile combustion (e.g., from gas boilers or owned or leased cars, vans and lorries for transportation of waste/products).

- Scope 2: indirect emissions from the generation of purchased electricity, heat or steam that is consumed in its owned or controlled equipment or operations.
- Scope 3: indirect GHG emissions which, based on the selected consolidation approach (e.g. control) used in setting its organizational boundaries, are not owned or controlled by the company. There are 15 Scope 3 categories shown in the GHG Protocol. With respect to the water sector, upstream Scope 3 emissions would include materials and consumables for the treatment of water and wastewater – for example chemicals manufacture and transport and the emissions associated with purchased goods and services, including those for capital infrastructure works, and waste generated by company operations, as well as employee travel and commuting. Examples of downstream Scope 3 emissions for the water sector include emissions associated with the use of treated water or wastewater, use of products sold, transportation and distribution of drinking water, biosolids recycled to land or sludge products used as fuel at off-site processes.

It is noted that CO_2 produced during biological wastewater treatment through biological processes is considered biogenic and not included in reporting. However, CO_2 emissions which occur as a result of fossil carbon in feedstocks used to manufacture a wide range of personal care and/or cleaning products which find their way into sewer systems and onto treatment facilities should be considered for inclusion.

Municipal water companies which are publicly owned and affiliated with a city may adopt a geographic boundary approach, differentiating emissions occurring physically within and outside the city boundary. Global best practice in this case would follow the GHG Global Protocol for Cities – for example as applied by municipal water companies for cities under the C40 Cities initiative (Greenhouse Gas Protocol, 2014) (C40, 2020). This would consider the defined city boundary for emissions reporting and key sources of emissions as per Figure 4.2 below. In the case of wastewater management and treatment the GHG protocol provides scope definitions as (Greenhouse Gas Protocol, 2014):

- Scope 1: GHG emissions from treatment and disposal of waste within the city boundary regardless of whether the waste is generated within or outside the city boundary.
- Scope 2: not applicable to wastewater treatment – all emissions from the use of grid-supplied electricity in waste treatment facilities within the city boundary are typically reported separately and not by the water sector.
- Scope 3: GHG emissions from treatment of waste generated by the city and activities associated with waste treatment (chemical supply, consumables, employee travel) which are treated outside the city boundary or imported from outside the city boundary.

Within the GHG Protocol and, generally, for GHG inventory guidance for organizations reporting and disclosing GHG emissions, quantification of Scope 1 and 2 emissions is mandatory, while Scope 3 emissions quantification can be voluntary in some cases. It is important to consider that there could be significant variation on the overall GHG estimate depending on the boundary definition and reporting requirements.

Table 4.1 provides further examples of GHG emissions relevant to the urban water cycle based on these definitions. For any given case, establishing the basis for defining boundaries and reporting emissions is important, with reference to guiding global best practice set out in the GHG Protocol and/or other relevant guidelines or policy. Reference is again made to ISO 14064-1:2018 which provides useful categories for understanding and reporting of GHG emissions and removals and which may be of benefit to companies or utilities in the water sector.

This chapter covers GHG emissions from sewerage and on-site centralized wastewater treatment processes, with a focus on emissions of CH_4 and N_2O from WWTPs. It does not include discussion of emissions from natural treatment systems (e.g. wetlands) and it does not include emissions from the discharge of treated effluent to aquatic environments – for example rivers and oceans or disposal of used water to land (e.g. for irrigation). However, wetlands for wastewater treatment are included in Chapter 6 of the 2014 Supplement to the 2006 IPCC Guidelines (IPCC, 2014b).

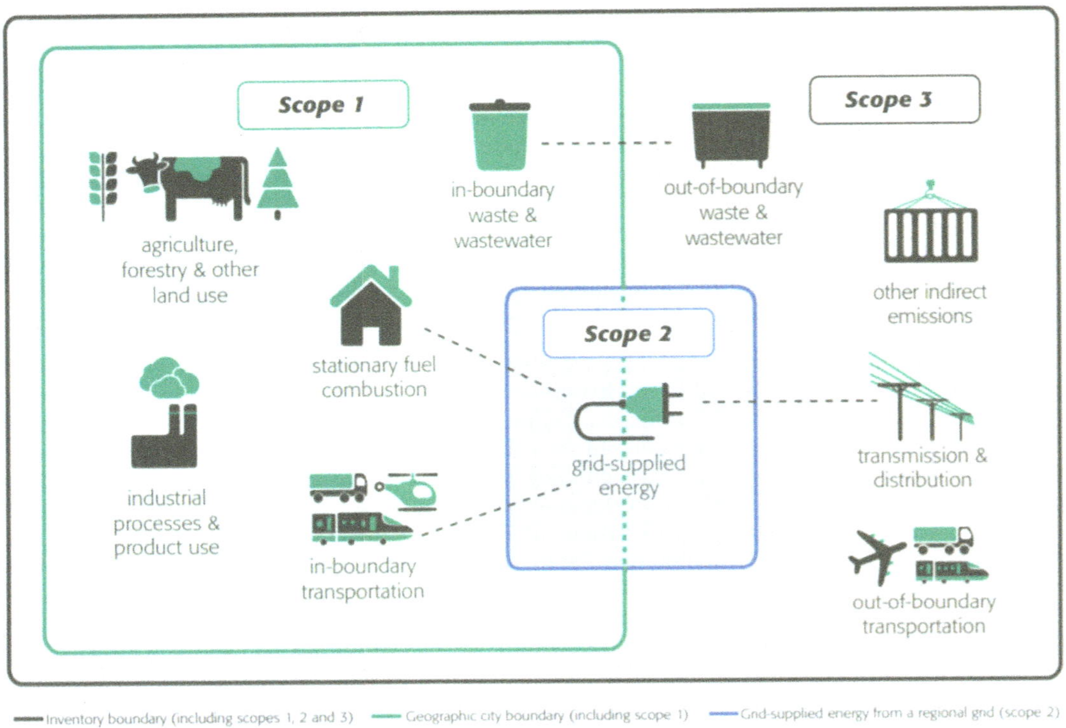

Figure 4.2 Sources of city GHG emissions (Greenhouse Gas Protocol, 2020a, 2020b). With thanks to the World Resources Institute - licensed under a creative commons licence (http://creativecommons.org/licenses/by-nc-nd/3.0)

Also, it does not cover emissions of GHG from the use or management of wastes or emissions from off-site use of resources recovered from wastewater. Nevertheless, it is important to understand that the CH_4 and N_2O emissions associated with these off-site activities to manage (whether use or disposal of) products can be significant relative to on-site activities, particularly the recycling of sewage sludges to land or their disposal to landfill. Depending on the approach as exemplified above, these emissions can be considered either Scope 1 or Scope 3.

It is important to recognize that off-site emissions may be accounted for differently, leading to discrepancies in accounting and reporting of Scope 1 or Scope 3 emissions. Clarity of boundaries for water sector emissions relative to nationally reported emissions inventories is important to provide reporting consistency and relevant baselining for the ambitious GHG emissions required under the Paris Agreement.

As water utilities and companies move towards net zero and Paris-aligned GHG emissions, a more holistic carbon management approach is being adopted to account for all direct and indirect emissions, regardless of the control they have over downstream emissions – adopting methods for life cycle analysis of carbon and other non-economic impacts to enable decision making.

A relevant example where water companies are reporting aligned with corporate Scope 1, 2 and 3 emissions is with respect to end use of treated effluent – for example return to the natural environment – and the disposal of biosolids to land. Whilst they may not be required to report these emissions under existing corporate reporting standards or to support National Inventories, water companies have the potential to substantially influence downstream emissions; for example, the residual N_2O emissions

Table 4.1 Examples of emissions scopes relevant to the urban water cycle.

Scope	Private and/or regional water company serving a defined geographic area (city, town, state, country level)	Municipal water company serving a defined city geographic area[a]
Scope 1	Direct GHG emissions occurring from sources that are owned or controlled by the company	GHG emissions from sources located within the city boundaries
Example:	*Stationary and mobile fuel combustion (on-site use of natural gas and other fuels), process emissions from water and wastewater treatment (N_2O emissions from biological wastewater treatment), fugitive CH_4 emissions during anaerobic treatment and sludge management, and from owned or controlled sewerage networks emissions*	*Stationary and mobile fuel combustion (on-site use of natural gas and other fuels), in-boundary process emissions from water and wastewater treatment (N_2O emissions from biological wastewater treatment), fugitive CH_4 emissions during anaerobic treatment and sludge management, and from owned or controlled sewerage networks, emissions from discharge of treated effluent into aquatic environments if these are within the city boundary (N_2O emissions from receiving water body)*
Scope 2	Indirect GHG emissions from the generation of purchased electricity, heat or steam consumed by the company in its owned or controlled equipment or operations	GHG emissions occurring as a consequence of the use of grid-supplied electricity, heat, steam within the city boundary would be reported in City accounting and not by the water company
Example:	*Purchased electricity, heat and steam*	*None reported by water company/authority*
Scope 3	Indirect emissions as a consequence of the activities of the company, but occurring from sources not owned or controlled by the company	All other GHG emissions that occur outside the city boundary as a result of activities taking place within the city boundary
Example:	*Employee business travel, emissions from waste disposal of effluent and residual streams including N_2O from discharge of treated effluent and N_2O and CH_4 from the storage and recycling of effluent or biosolids to land. Transmission & distribution of electricity, production and distribution of chemicals or other materials*	*Out-of-boundary process emissions from water and wastewater treatment, out-of-boundary transportation, out-of-boundary waste disposal, transmission and distribution of electricity, production and distribution of chemicals*

[a]For municipal water companies, example of emissions may change depending on what is included and excluded from the city boundary.

from treated effluent or biosolids recycled to land. Further, water companies with targets aligned to the Science Based Targets initiative (SBTi) will be required to report on these emissions, where they are significant Scope 3 emissions.

4.2.2 Top-down and bottom-up approach considerations for the water sector

Top-down and bottom-up are the two main GHG accounting inventory methodologies, distinguished based on how the data is obtained and the level of confidence. A top-down approach refers to GHG emissions estimated based on equations with factors and constants which are defined at global level. These are developed from data collected from research or accepted industry practice or based on general assumptions. They provide a methodology for estimation of GHG emissions, significantly relying on default factors.

A bottom-up approach consists of measurements of the actual GHG emissions at the facility level, based on a defined methodology. This could include averaged EFs from facilities to provide a national dataset or specific WWTP data from the measurement of emissions for each facility. A bottom-up approach is preferable and results in an improved methodology for a more accurate GHG inventory. A bottom-up assessment of GHG is possible where high quality data and advanced methodologies exist at a country level.

The following sections provide a description of how top-down and bottom-up emissions are calculated based on best global practice.

4.2.2.1 Top-down methodologies

The top-down estimation of GHG emissions for inventory of emissions can be exemplified in the generalized Equation (4.1):

$$\text{Emission Rate (ER)} = \text{Emission Factor (EF)} \times \text{Activity Data (AD)} \tag{4.1}$$

where the emission rate, usually in mass per a period of time (e.g. kg N_2O/year) is a factor of the human activity by the emitting activity based on site measurements or lookup factors for specific countries, and on appropriate EFs for different emitting sources. For a top-down approach, the EF and activity data will be derived from higher level (e.g. international literature) data compared with a bottom-up estimate, which will use in-country or facility-level datasets.

GHG emissions from chemical and biological processes in the water sector are not as straightforward to estimate as GHG emissions from the power sector, such as quantifying emissions from the burning of fuels. In the case of burning of fuels, the amount of GHG produced is a function of the carbon content of the fuel, thus a direct stoichiometric correlation. Biological processes, conversely, are highly complex and emissions are dependent on the environmental and operational conditions in which the treatment is carried out. As has been discussed, in Chapter 2, N_2O is produced as a by-product or intermediate during biological wastewater treatment of nitrogen-containing resource streams under aerobic and anoxic conditions, and CH_4 is produced during anaerobic treatment of resource streams.

Research to develop the fundamental understanding of GHG production and emissions from biological wastewater treatment processes, in particular N_2O, has been an area of continued progress for almost three decades. Since the first publication on N_2O emission from a small activated sludge treatment works in New Hampshire, USA (Czepiel *et al.*, 1995), significant research has been conducted around the world both at lab- and full-scale to determine the microbial pathways, mechanisms and factors leading to N_2O production and emission from different configurations of WWTPs. Although a general consensus exists, there are still gaps and it remains an area of multi-layered research.

CH_4 emissions generally are due to leakages of CH_4 produced during anaerobic processes used for wastewater and sludge treatment and, whilst often captured for beneficial use as biogas at large centralized facilities, may be emitted unintentionally from tanks, pipework and fittings. CH_4 may also be produced and emitted in sewerage systems. The extent of CH_4 emissions for a site are likely to be highly dependent on on-site operations and gas management controls as well as the nature of processes employed (e.g. enclosed anaerobic digestion versus open secondary digesters or sludge lagoons).

By applying an averaged global EF from a top-down approach to these N_2O and CH_4 emissions from WWTPs, a higher degree of uncertainty is inferred, leading to a lower level of confidence in the GHG estimations for the water sector. However, it is noted that bottom-up emissions of N_2O and CH_4 from different treatment processes when measured at site level (e.g. bottom-up) have been shown to vary significantly, even for the same type of treatment process but with different operational conditions. Bottom-up methodologies based on current reporting protocols are discussed subsequently in Section 4.2.2.2; Chapter 5 discusses full scale quantification of emissions and site level approaches.

4.2.2.2 Bottom-up methodologies

A bottom-up approach follows the same general approach as above in Equation (4.1) but data sources differ – for example, this may consider EFs derived from a national dataset developed from a sub-set of in-country facilities or from data specifically at individual facility (WWTP) level. Increasingly, country level work seeks to develop methodologies which include a country-specific EF for N_2O or CH_4, based on the measurement of emissions across a number of in-country facilities or based on the adoption of a published dataset applicable to the in-country facilities.

A bottom-up approach with a nationally-derived dataset offers an improvement on a top-down approach but does not offer facility-level understanding. Development of advanced methodologies to measure emissions accurately and effectively is important, particularly given the variations exhibited in GHG emissions due to temporal and operational conditions. Hence, a bottom-up approach may also consider a methodology which measures actual GHG emissions from a treatment facility and uses these to develop the GHG inventory for each facility. In this case, compiling the GHG emissions inventory may not require the development and application of an EF and application of Equation (4.1), but instead, may be able to directly report measured emissions. Alternatively, long-term facility monitoring may be used to develop a facility-level EF which, when used in Equation (4.1) with appropriate facility or geographical activity data, can provide GHG inventory emissions estimation. Chapter 5 discusses site-level full-scale quantification of GHG emissions further.

When comparing top-down and bottom-up approaches and considering nationally-derived datasets versus a facility-level emissions quantification, it is important to recognize that an approach which considers globally- or nationally-derived factors does not give insights into the conditions leading to these GHG emissions from a specific treatment process. Given the potential for significant variability in emissions of N_2O and CH_4 between facilities, this limits the ability of a treatment process to be investigated and understood such that mitigations can be applied resulting in low reduction potential value. Conversely, on-site monitoring and development of mitigation and control strategies have a high potential to lead to sustained reduction in process emissions. Moreover, the outcomes of monitoring and resulting mitigations are likely to offer wider benefits to the sector in terms of process safety, stability, performance, operational cost and proactive maintenance.

A top-down approach is considered *good practice*, as explained in the following sections, for situations where data, methodologies and resources are not available at country-level to develop a bottom-up approach to full-scale quantification of GHG emissions. Where resources do exist to develop and apply advanced methodologies, the aim should be to apply advanced methodologies for quantification of GHG emissions at country-level and ideally based on full-scale quantification of GHG emissions. These should be facility-level bottom up approaches, as covered in Chapter 5.

4.3 INTERNATIONAL METHODOLOGIES

This section sets out global practice in GHG accounting methodology, as defined by the IPCC in the IPCC Guidelines (Bartram *et al.*, 2019; IPCC, 2006). The IPCC methodology provides a globally consistent approach for high-level government GHG emissions accounting and reporting. It is also the foundation of most protocols for GHG accounting, such as the most widely used protocol for GHG accounting for businesses and government leaders, the World Resource Institute (WRI) GHG Protocol (Greenhouse Gas Protocol, 2020a, 2020b). Therefore, this chapter focuses mainly on the IPCC methodology while also bringing the perspective of other national and sector-specific methodologies.

4.3.1 The intergovernmental panel on climate change

The IPCC is the international body for assessing the science related to climate change. It was created by the World Meteorological Organization (WMO) and United Nations Environment Programme (UNEP) in 1988 to provide policymakers with regular assessments of the scientific basis of climate change, its impacts and future risks, and options for adaptation and mitigation (IPCC, 2018a).

Through the Task Force on National Greenhouse Gas Inventories, the IPCC provides internationally agreed methodologies for measuring national GHG emissions from the different sectors of the economy based on published research conducted around the world. The methodology is used by

Parties reporting through the National Inventory Reports (NIRs) under the UNFCCC, in compliance with the Kyoto Protocol. The signatory Parties of the 1992 UNFCCC are required to monitor and report annually, at a national scale, their emissions of the six key GHGs, namely: carbon dioxide (CO_2), CH_4, N_2O, hydrofluorocarbons (HFCs), perfluorocarbons (PFCs), and sulfur hexafluoride (SF_6) (United Nations, 2020).

As each GHG has its unique radiative effects over a given period, the GHG emission calculations convert each GHG into one ton of CO_2, known as CO_2 equivalent (CO_2e) based on their global warming potential (GWP) (Equation (4.2)).

$$CO_2e(\text{tonnes/yr}) = GHG(\text{tonnes/yr}) \times GWP_{100} \tag{4.2}$$

The 100-year GWP (GWP_{100}) was adopted by the UNFCCC and its Kyoto Protocol and is now used widely as the default metric (IPCC, 2014a, 2014b). The IPCC is responsible for updating the GWP values as scientific understanding develops. The most recent values are presented in Table 5.1 of the fifth assessment report (AR5) published in 2014, reproduced in Table 4.2 below. The IPCC is currently in its sixth assessment cycle (AR6), which will be finalized in 2022 (IPCC, 2020a, 2020b).

The most recent guidelines for National GHG Inventories are provided in the 2019 Refinement to the 2006 IPCC Guidelines for National Greenhouse Gas Inventories (the 2019 Refinement). The 2019 Refinement was adopted by the IPCC at its 49th Session in May 2019 in Kyoto, Japan (Bartram *et al.*, 2019; Federici, 2019). It includes 5 volumes, comprising: Volume 1 – General Guidance and Reporting, Volume 2 – Energy, Volume 3 – Industrial Processes and Product Use, Volume 4 – Agriculture, Forestry and Other Land Use, and Volume 5 – Waste.

4.3.2 IPCC methodologies for the water sector

Within the water sector, for water companies or water utilities treating wastewater the relevant IPCC methodology is found in the 2019 Refinement Volume 5: Waste, Chapter 6: Wastewater Treatment and Discharge. Volume 1: General Guidance for Reporting provides an overview of GHG inventories and includes chapters on uncertainties, consistency, quality assurance and quality control and verification in the protocol.

The IPCC Guidelines follow the top-down approach described in Equation (4.1), comprising EFs and activity constants for estimation of CH_4 and N_2O from wastewater treatment and discharge. The IPCC provides a three-tier methodology to select the EFs and activity data, as discussed in more detail in subsequent sections.

- Tier 1 (*good practice*) method: uses default values for the EF and activity parameters. It is considered good practice for countries with limited data.
- Tier 2 (*good practice*) method: uses a country-specific EF based on field measurements and country-specific activity data.
- Tier 3 (*advanced*) method: uses a country-specific method – for example, based on plant-specific emissions from large WWTPs. It is for countries with good data and advanced methodologies, where direct measurement methods provide a more accurate measurement from each facility.

Table 4.2 Global warming potential (IPCC, 2014a, 2014b).

GHG	GWP_{100}
CO_2	1
CH_4	28
N_2O	265

The three-tier method represents the level of methodological complexity and data requirements. A progression from Tier 1 to Tier 3 represents an increase in confidence in the GHG estimates, and generally requires more extensive resources for site measurement and data collection. Thus, it may not be feasible to use the higher tier methods (Tier 2 and 3), which are generally considered to be more accurate, for every category of the emissions inventory. The IPCC guidance provides considerations on quantitative and qualitative approaches to identify categories that are key sources of emissions to help manage overall inventory uncertainty (IPCC, 2006).

The following sections set out the 2019 Refinement methodology for estimation of CH_4 and N_2O emissions from wastewater treatment. It is noted that the 2019 Refinement did not revise Chapter 5 of the guidelines, which covers solid waste and also covers emissions calculation for the anaerobic treatment of wastewater treatment sludges. Whilst these are reported under the Wastewater Treatment and Discharge category described in Chapter 6 of the guidelines, in the 2019 Refinement Chapter 6 refers to the previous 2006 Guidelines for the relevant methodology for estimation of emissions for the anaerobic treatment of sewage sludges.

4.3.2.1 Methane

The methodology to estimate CH_4 emissions from wastewater treatment and discharge was first introduced by the IPCC in the 1995 IPCC Guidelines for National Greenhouse Gas Inventories, later replaced by the 1996 Revised Guidelines and further revised in the 2006 Guidelines and 2019 Refinement (Volume 5: Waste, Chapter 6: Wastewater Treatment and Discharge) (Bartram *et al.*, 2019).

Chapter 6 (including Table 6.1) in the 2019 Refinement provides methane emissions potential from wastewater and sludge treatment systems and discharge systems. Sewers, and aerobic, anaerobic and sludge treatment systems are considered; composting and incineration are considered elsewhere in the IPCC Guidelines.

CH_4 emissions from decentralized and industrial wastewater treatment and the anaerobic digestion of non-domestic sludges are considered and reported separately from wastewater treatment emissions under the IPCC Guidelines. For this chapter of the Refinement, only CH_4 emissions from centralized treatment plants are considered. As according to Figure 6.1 in Chapter 6 of the 2019 Refinement, on-site sludge treatment, emissions should be reported under the Wastewater Treatment and Discharge category. Conversely, emissions from the off-site treatment of sludges, from composting, incineration, and landfilling of sludges, and for sludge application to land are described in other IPCC chapters and should not be reported within the Wastewater Treatment and Discharge category in National GHG Inventory assessments (Bartram *et al.*, 2019).

The 2019 Refinement acknowledges that in addition to sludge treatment and management, CH_4 emissions may also occur from settling basins and other anaerobic pockets, and especially resulting from sewer networks and being stripped out in turbulent aerobic zones of secondary treatment (Bartram *et al.*, 2019). The potential EF for CH_4 emissions from upstream sewer networks is currently not included in IPCC Guidelines, although there has been significant discourse on the subject. Chapter 3 and Chapter 8 of this book provide further discussion on CH_4 emissions and modeling approaches. The lack of methods to estimate CH_4 emissions from sewers that can easily be implemented by a water company is the key challenge in the assessment of their potential contribution to their CH_4 emissions.

4.3.2.1.1 Methane emissions from wastewater treatment

For CH_4 from wastewater treatment, the IPCC methodology provides a three-tiered approach. For each of these tiered approaches, methane emissions are calculated as the sum of the emissions from each treatment unit and the CH_4 recovered or flared, as detailed in Equation (4.3) (see Section 6.2.2 and Equations 6.1 and 6.1A in IPCC 2019 Refinement), in kg CH_4/year, and Equation (4.4) for calculation of the EF:

$$CH_{4\,\mathrm{emissions}} = ([\mathrm{TOW}_j - S_j] \times \mathrm{EF}_j - R_j) \times [10^{-6}] \tag{4.3}$$

$$EF_j = B_o \times MCF_j \tag{4.4}$$

where TOW_j is the organics in wastewater treatment/discharge pathway or system, j, in inventory year (kg BOD/year), S_j is the organic component removed from wastewater in the form of sludge (kg BOD/year), j is each type of treatment, EF_j is the EF for treatment/discharge pathway or system, j (kg CH_4/kg BOD), R_j is the amount of CH_4 recovered, for example through anaerobic digestion (AD), or flared from treatment/discharge pathway or system, j, in inventory year (kg CH_4/year), and 10^{-6} is the conversion of kg to Gg. The EF for CH_4 from wastewater treatment is a function of the maximum CH_4 producing potential (B_o) and the CH_4 correction factor (MCF) for the wastewater treatment and discharge system.

The B_o indicates the maximum amount of CH_4 that can be produced for a given amount of biochemical oxygen demand (BOD), while the MCF is the extent to which the CH_4 producing capacity (B_o) is realized. It is *good practice* as set out in the IPCC Guidelines to use country-specific data (Tier 2 or 3) for EFs, which are made up of B_o and MCF values (Bartram *et al.*, 2019).

The default B_o derived based on expert judgment by the lead authors and based on Doorn *et al.* (1997) is 0.6 kg CH_4/kg BOD with the BOD to be estimated based on incoming population per capita figures. MCFs are provided in Table 6.3 of the 2019 Refinement (Bartram *et al.*, 2019).

The MCF recommended for a centralized aerobic treatment plant is 0.03 (0.003–0.09) or 3% CH_4 emission from total influent BOD and was calculated based on on-site measurements from 14 full-scale USA WWTPs (Bellucci *et al.*, 2010; Czepiel *et al.*, 1995; Daelman *et al.*, 2013; Delre *et al.*, 2017; Kozak *et al.*, 2009; Kyung *et al.*, 2015; Wang *et al.*, 2015). These studies include both activated sludge (assumed nitrifying), biological nutrient removal, a sequencing batch reactor (SBR) and an anaerobic/anoxic/oxic (A^2O) process. The highest and 2nd lowest MCFs are reportedly from activated sludge. The data shows significant variability and the IPCC recommend more extensive monitoring and collection of site data to allow better understanding between treatment process types.

Based on the above, and on the default B_o and MCF values (0.6 kg CH_4/kg BOD and 0.03 kg CH_4/kgCH_4), the default EF for CH_4 from wastewater treatment is effectively 0.018 kg CH_4/kg BOD influent. If a country chooses to introduce country-specific data for B_o based on measured composition of wastewater, the MCF must also be updated as MCFs are developed using the default B_o values.

4.3.2.1.2 Methane emissions from sludge treatment

The methodology for CH_4 emissions from on-site sludge treatment was not updated in the IPCC 2019 Refinement and is covered by the 2006 IPCC Guidelines under Volume 5: Waste, Chapter 4: Biological Treatment of Solid Waste for emissions from AD of organic waste (Doorn *et al.*, 2006). This follows a *similar* tiered approach to that described in Section 4.3.2. Note the units and notation used for equation components which are taken directly from the 2006 IPCC Guidelines differ slightly to those used in Section 4.3.2.1.1 above, as provided in the IPCC 2019 Refinement.

For CH_4 from sludge treatment, the default methodology approach to calculate CH_4 emissions from sludge treatment for any tier is the difference between the methane emissions produced from the mass of organic waste (sewage sludge) treated by biological treatment type i (either composting or anaerobic digestion) and the emission factor for each treatment type as g CH_4/kg waste treated and the CH_4 recovered or flared, as detailed in Equation (4.5), in kg CH_4/year, and Equation (4.6) for calculation of the EF:

$$CH_{4\,emissions} = M_i \times EF_i \times [10^{-3}] - R \tag{4.5}$$

$$EF_i = B_o \times MCI \tag{4.6}$$

where M_i is the mass of organic waste treated by biological treatment (Gg) in each type of treatment, EF_{ji} is the EF for treatment (g CH_4/kg waste treated), R is the amount of CH_4 recovered and utilized (e.g. combusted to biogenic carbon dioxide on site or valorized to biomethane for use off site) or flared in inventory year (Gg), and 10^{-3} is the conversion of g to kg. The EF for CH_4 from biological treatment of sludge, *i*, is provided by Tier 1, 2 or 3 measurement, with Tier 1 values provided in Table 4.1 of Chapter 5 of the 2006 IPCC Guidelines.

As the IPCC Guidelines note, consistency between CH_4 (and N_2O) emissions from composting or anaerobic treatment of sludge and emissions from treatment of sludge reported in the Wastewater Treatment and Discharge category should be checked. Further, if emissions from anaerobic digestion are reported under Biological Treatment of Solid Waste, practitioners should check that these emissions are not also included under the Energy Sector.

Estimation of CH_4 emissions from sludge that is managed off-site from the WWTP using landfill (Volume 5, Chapter 3: Solid Waste Disposal), incineration (Volume 5, Chapter 5: Incineration and Open Burning of Waste), composting (Volume 5, Chapter 4: Biological Treatment of Solid Waste) and for land application (Volume 4, Chapter 11: Agriculture, Forestry and Other Land Use) are reported separately. Emissions from on-site wastewater and sludge treatment processes, such as AD, commonly situated within the WWTP boundary are considered further in this coverage of accounting methodologies. Other sludge treatment processes such as composting, incineration and the application of biosolids to land are not considered further in this discussion of accounting methodologies.

For CH_4 emissions from AD due to unintentional leakages as a result of pipework, valve and tank roof leaks, where biogas is lost and not recovered for valorization and/or is not combusted, the IPCC states in the 2006 Guidelines, Volume 5, Chapter 4, that emissions will generally be between 0% and 10% of the amount of CH_4 generated and that 5% should be used, in the absence of other information. It does not include an emission factor for these unintentional leakages. It also suggests that where technical standards for biogas plants ensure that unintentional emissions are flared, emissions are likely to be close to zero. Emissions from flaring are not considered significant, as the majority of CO_2 emissions are of biogenic origin, and the CH_4 and N_2O emissions are very small. Therefore, *good practice*, according to the 2019 Refinement does not require estimation of emissions from flaring of biogas. However, if estimating these emissions is desired, they should be reported under the Waste sector and the IPCC Guidelines refer to Volume 2 (Energy), Chapter 4 for guidance on flaring (Bartram *et al.*, 2019). This guidance assumes a conversion of 98% efficiency for combustion of methane in Equation (4.2.4). This figure is also implemented in the Water and Wastewater Companies for Climate Mitigation (WaCCliM) tool, ECAM, which uses a 2% methane loss due to incomplete combustion due to flaring, based on technical judgment and Volume 2 of the IPCC

There are no further considerations for CH_4 emissions in either the 2019 Refinement or 2006 Guidelines. A number of other fugitive sources of CH_4 emissions from sludge treatment which are not currently considered under the IPCC guidelines could be significant. These are likely to be very site specific and limited industry level data exists to support emission factors. A number of these have been subject to published research, as highlighted below and as recently summarized by UKWIR (United Kingdom Water Industry Research, 2020):

- Sludge storage and processing: Storage, thickening and dewatering of sludges with or without anaerobic digestion facilities, could be a significant source of fugitive CH_4 emissions from WWTPs. Type of sludges, their extent of treatment, the nature of storage and venting/extraction, and general asset operation are all likely to influence fugitive emissions.
- Gas line: Leakages in the gas line are highly dependent on on-site maintenance and gas measurement control and could be a potentially important source of fugitive emissions from WWTPs.
- Combined Heat and Power (CHP): CHP engines can reach an efficiency of up to approximately 42% conversion of biogas into electricity by burning CH_4, with high- and low-grade heat making up most of the leftover energy in the biogas (9% is parasitic energy). The percentage of unburned CH_4 measured in combustion that will be released as emissions has been found to be between 1.5% (Daelman *et al.*, 2012) and 1.8% (Woess-Gallash *et al.*, 2010).
- Dewatering digestate: The dissolved CH_4 remaining in digestate will be released to a great extent in the dewatering process. Digestate is blended with polymer for flocculation for fast

gravity water release. Here some of the CH_4 will stay in the centrate (the water released) or stay in the flocculated sludge, with very little going into the atmosphere. It is in the next step, dewatering, where pressure is applied, either by belts pressing against each other (e.g., belt presses) or by centrifugal forces (e.g., centrifuges), where more dissolved CH_4 is likely to be released. Centrifuges are believed to release more CH_4 from digestate due to the high forces being applied. It has been reported that dewatering and storing digestate might lead to emissions of 2–4.5% the total CH_4 production (Daelman *et al.*, 2012; Schaum *et al.*, 2015).

- Maturation pads and digestate storage: Further to the above, any storage of digestate prior to land spreading or landfill will have potential for CH_4 emissions – this may be in addition to secondary digestion emissions and could add additional emissions to existing conventional AD sites with the common storage of sludge cake. There has been limited research, but this could be considered a potentially significant source of CH_4 (Daelman *et al.*, 2012).
- Biogas upgrading: There is increasing interest in the potential economic benefits of injecting biogas into existing gas networks. There are a number of biogas-to-grid projects being implemented in several countries. In order to inject biogas into the grid, the biogas needs to be cleaned up and processing technologies, for example membrane or water wash based removal of impurities in biomethane, will have a potential fugitive emission consideration. The potential slippage is subject to ongoing investigation but may be in the order of 1–3% of produced biogas for grid injection.

4.3.2.2 Nitrous oxide

4.3.2.2.1 Nitrous oxide emissions from wastewater treatment

The methodology to estimate N_2O emissions from wastewater treatment was first introduced by the IPCC in the 2006 Guidelines Refinement, Volume 5: Waste, Chapter 6: Wastewater Treatment and Discharge. Based on limited data quantification of N_2O emissions from full-scale WWTPs, the 2006 Guidelines recommended a default EF of 3.2 g N_2O/person/year from a single study carried out at a small domestic activated sludge WWTP (4 Ml/day) in the University of New Hampshire, USA (Czepiel *et al.*, 1995). It stated that direct N_2O emissions from WWTPs were considered as a minor source in comparison with the indirect N_2O emissions from effluent discharges. Since then, this EF has been used to estimate direct N_2O emissions from different treatment processes around the world for countries where emissions from WWTPs are considered.

The 2006 Guidelines methodology did not provide any higher tier guidance. It noted that it was considered *good practice* to estimate total N_2O emission from domestic wastewater treatment only for countries that have predominantly advanced, centralized, WWTPs with nitrification and denitrification steps. For these, it suggested, using the following equation, in kg N_2O/year:

$$N_2O_{WWTP} = P \times T_{WWTP} \times F_{IND\text{-}COM} \times EF_{WWTP} \tag{4.7}$$

where P is the human population; T_{WWTP} is the degree of utilization of modern, centralized WWTPs (%, country-specific), $F_{IND\text{-}COM}$ is the fraction of industrial and commercial co-discharged protein (default = 1.25, based on data in Metcalf and Eddy (2003) and expert judgment), and EF_{WWTP} is 3.2 (2–8) g N_2O/person/year.

After several years of research and national monitoring campaigns at full-scale WWTPs employing different treatment processes in various countries, there has been a consensus that the 2006 Guidelines methodology with the applicability of a single EF to represent N_2O emissions from different processes presented several limitations, including:

- N_2O emission was attributed to denitrification as the dominant source of N_2O, with nitrification being considered a minor contributor. Research has shown that nitrification is a significant contributor to N_2O emissions from aerobic zones, and that the importance of one pathway over another will depend on the environmental and operational conditions of the WWTP (Ahn *et al.*, 2010; Pan *et al.*, 2016).

- *It did not consider spatial and diurnal variability in N_2O emissions.* Significantly high spatial and diurnal variability is observed in all studies, by monitoring N_2O emissions at across the secondary treatment tanks and for longer periods of time (Ni *et al.*, 2015; Pan *et al.*, 2016; Vasilaki *et al.*, 2019).
- It did not make a distinction between WWTPs with different treatment types or different operational conditions and the associated effect on N_2O production. The production pathways of N_2O in wastewater treatment are highly complex and vary depending on the type of treatment but also on the operational conditions – even within the same treatment processes. For instance, SBRs and step-feed plug flow reactors are generally associated with higher N_2O emissions compared to other process configurations due to sudden operational changes (Pan *et al.*, 2016; Pijuan *et al.*, 2014; Vasilaki *et al.*, 2019).
- *It did not consider WWTPs that are located in different climate zones.* It has been shown that N_2O emissions from tropical climate zones are higher than from temperate zones, as a factor of temperature and bacterial activity (Brotto *et al.*, 2015b).

In 2018/2019, the IPCC Guidelines went through a peer-review of the science developed since the 2006 Guidelines, providing significantly more guidance, in particular with respect to N_2O from wastewater treatment (Bartram *et al.*, 2019). In recognition of the high variability of N_2O production and its dependency on the treatment design and operations, a new significantly higher EF of 0.016 (with range minimum 0.00016 – maximum 0.045) kg N_2O-N/kg N load is recommended in the 2019 Refinement (Bartram *et al.*, 2019). This represents a change of two orders of magnitude in the default EF for the Tier 1 method application, from 0.00035 kg N_2O-N/kg N load (conversion of 3.2 g N_2O/person/year based on the IPCC protein/N conversion of 0.16 g N/g protein) to 0.016 kg N_2O-N/kg N load.

The new EF is derived from linear regression of 29 full-scale monitoring data on N_2O emissions and influent nitrogen load from a variety of the most common suspended growth (e.g. activated sludge, including continuous and batch operating modes) treatment processes in Australia, Brazil, China, Japan, the Netherlands, Spain, and the USA – regarding these as *the most typically and widely used treatment processes globally* (Bartram *et al.*, 2019). A summary of the studies included is provided in Table 4.3 however there is some evidence that the originally cited references may require review due to some apparent variation in conversion of units and other issues – these are shown as footnote comments. Recent analysis also includes discussion of data points and the IPCC 2019 Refinement method and considers the extent to which larger treatment plants (treating greater than 300 000 population equivalent, e.g.,) drive the linear regression and resultant emission factor derivation as well as the recognition that treatment performance, in particular in terms of nitrogen removal, likely requires further consideration (de Haas & Ye, 2021). Work by others (e.g., Valkova *et al.*, 2020) also discusses the importance of considering nitrous oxide emissions with respect to total nitrogen removed though for higher levels of total nitrogen removed (~50–92%). Conversely, sector-level work in Denmark does not exhibit a similar correlation with total nitrogen removal and highlights significant variability for a similar (~60–95%) degree of total nitrogen removal (Lake *et al.*, 2021 (unpublished)). Chapter 11 provides further discussion of this and other emerging issues.

Chapter 5 covers more details on the different types of monitoring methodologies (e.g. online off-gas, grab sampling) as specified in Table 4.3 and Chapter 11 provides further discussion on the derivation of emission factors from site measurement campaigns with emerging research and practical application of methods and consideration of best practice.

The 2019 Refinement provides a three-tier methodology for assessment of greenhouse gas emission factors and a decision tree to help identify which tier should be applied depending on the available data and activity parameters. Historically, the three-tier methodology was not applied to N_2O emissions assessment, but the 2019 Refinement provides for the following methods which are subsequently discussed:

Table 4.3 Summary of N_2O studies considered in the 2019 IPCC refinement (adapted from Bartram *et al.*, 2019).

Treatment Type[a]	Category	EF (kg N_2O-N/kg $N_{influent}$)	Country	Frequency/sampling method	Reference
AO	BNR	0.028	Netherlands	16-months on-line monitoring period (1-month interruption due to technical failure)	Daelman *et al.* (2015)
AO	BNR	0.021	Australia	2 days, twice a day off-line grab sample	Foley *et al.* (2010)
AO	BNR	0.045	Australia	2 days, twice a day off-line grab sample	Foley *et al.* (2010)
A2O	BNR	0.013	Australia	2 days, twice a day off-line grab sample	Foley *et al.* (2010)
SBR	BNR	0.023	Australia	2 days, twice a day off-line grab sample	Foley *et al.* (2010)
OD	BNR	0.008	Australia	2 days, twice a day off-line grab sample	Foley *et al.* (2010)
IA	BNR	0.0005	Japan	Short-term, off-line grab samples	Kimochi *et al.* (1998)
EA	BNR	0.015	Australia	2 days, twice a day off-line grab sample	Foley *et al.* (2010)
A2O	BNR	0.013	China	24-h monthly, 1 year, online off-gas monitoring	Wang *et al.* (2016)
CAS	BNR	0.00036	UK	8 weeks, online off-gas measurement	Aboobakar *et al.* (2013)
AO[b]	BNR	0.12[c]	Spain	10 weeks, 2–3 days a week, online off-gas measurement	Rodriguez-Caballero *et al.* (2014)
OD	BNR	0.00016	Japan	4 times throughout the year, off-gas grab samples	Masuda *et al.* (2018)[d]
AO	BNR	0.0013	Japan	4 times throughout the year, off-gas grab samples	Masuda *et al.* (2018)
AO	BNR	0.0049	Japan	4 times throughout the year, off-gas grab samples	Masuda *et al.* (2018)
Separate-stage	BNR	0.00019	USA	5 days, summer and winter, online off-gas measurement	Ahn *et al.* (2010)
Bardenpho	BNR	0.0036	USA	5 days, summer and winter, online off-gas measurement	Ahn *et al.* (2010)

Step-feed	BNR	0.011	USA	5 days, summer and winter, online off-gas measurement	Ahn *et al.* (2010)
MLE	BNR	0.0007	USA	5 days, summer and winter, online off-gas measurement	Ahn *et al.* (2010)
MLE	BNR	0.0006	USA	5 days, summer and winter, online off-gas measurement	Ahn *et al.* (2010)
OD	BNR	0.0003	USA	5 days, summer and winter, online off-gas measurement	Ahn *et al.* (2010)
Step-feed	BNR	0.015	USA	5 days, summer and winter, online off-gas measurement	Ahn *et al.* (2010)
Step-feed, plug flow	BNR	0.019	Australia	7 weeks, online off-gas measurement	Ni *et al.* (2015); Pan *et al.* (2016)
SBR	BNR	0.029[e]	China	Short-term (undefined), off-gas grab samples	Bao *et al.* (2016)
SBR	BNR	0.038	Spain	33 days, online off-gas measurement	Rodriguez-Caballero *et al.* (2015)
Plug flow	non-BNR	0.004	USA	5 days, summer and winter, online off-gas measurement	Ahn *et al.* (2010)
Plug flow	non-BNR	0.0062	USA	5 days, summer and winter, online off-gas measurement	Ahn *et al.* (2010)
Step-feed	non-BNR	0.0018	USA	5 days, summer and winter, online off-gas measurement	Ahn *et al.* (2010)
Plug flow	non-BNR	0.023[f]	Japan	5 times throughout the year, off-gas grab samples	Masuda *et al.* (2015)
AO	non-BNR	0.013[f]	China	Short-term (undefined), off-gas grab samples	Bao *et al.* (2016)
IA	non-BNR	0.0016	Brazil	1-day, off-line grab samples	de Mello *et al.* (2013)

[a]BNR: biological nutrient removal, AO: anaerobic-oxic activated sludge process, A2O: anaerobic-anoxic-oxic activated sludge process, SBR: sequencing batch reactor, OD: oxidation ditch, IA: intermittent aeration process, EA: extended aeration process, CAS: conventional activated sludge process, MLE: modified Ludzack-Ettinger.
[b]Reference not considered for the derivation of the new EF but considered as an outlier.
[c]It appears this value was incorrectly cited and should be 0.0012 kg N_2O-N/kg $N_{influent}$ based on influent TN.
[d]It appears Masuda *et al.* (2018) considered TN removed (and TN still in final effluent) in their analysis, thus factors from Masuda *et al.* (2018) should be considered further.
[e]It appears this value was incorrectly cited and should be 0.019 kg N_2O-N/kg $N_{influent}$.
[f]Review (unpublished) suggests that the derivation and/or citation of these factors from original studies may require further review.

4.3.2.2.1.1 Tier 3 – asset-specific EFs

Asset-specific EFs are emissions estimated using bottom-up, on-site measurements at the facility-level, are recognized as a 'Tier 3' methodology and are advocated as most preferable in the IPCC Guidelines. Although direct monitoring is now recognized by the IPCC methodology as the preferable option, few water utilities have undertaken direct monitoring historically. Most of the available data are the result of research to investigate the pathways and process conditions leading to N_2O emissions, and not to derive an EF. The 2019 Refinement does not provide a methodology to develop site-level emission monitoring campaigns and guidance for deriving emission factors from these which is a recognized area for ongoing work. Given the substantial variation in EFs calculated across WWTPs, even where these are very similar treatment process types, a focus on Tier-3-level assessment and long-term study to develop robust WWTP-level EFs is likely to remain a key area of focus for the water sector – with key recent discussions considering that the use of emission factors relative to the extent of total nitrogen removal may be most applicable (de Haas & Ye, 2021; Valkova *et al.*, 2020).

4.3.2.2.1.2 Tier 2 – country-specific EFs

The 2019 Refinement guidelines suggest that if asset-specific EFs are not available, country-specific EFs are most preferable (i.e., Tier 2). Few countries have currently taken this route, as further discussed in Section 4.4. Given the high number of WWTPs, the variety of treatment processes and variability of N_2O emissions, deriving a single country-specific emission factor is also challenging, requiring long-term measurements of representative WWTPs and a methodology to normalize the EF.

Similarly for Tier 3 monitoring, the 2019 Refinement does not provide a methodology to develop a Tier 2 country-level EF from an in-country dataset. The basis for statistical assessment of EF data from multiple WWTPs is not well established in research to date – for example whether data is normally distributed, the most suitable analysis to apply (e.g. linear regression) and basis for analysis and EF (e.g., total nitrogen load influent or total nitrogen removed).

4.3.2.2.1.3 Tier 1 – global EFs

If country-specific EFs are not available, the global EF derived in the 2019 Refinement should be applied (i.e., Tier 1). By implication, under the Paris Agreement, the first global emission inventory report in 2023 from signatory Parties should be either in line with this Tier 1 as international best practice, set out by the IPCC, or by a derived country-specific EF.

The implications of the change in the Tier 1 EF are significant, especially as many water utilities that account for process emissions as part of their carbon footprint apply the IPCC methodology to estimate N_2O emissions. An increasing number of water utilities in countries with centralized WWTPs are working towards Scope 1 process emissions reduction to provide emissions reduction in their sector in alignment with the Paris Agreement, as reflected in country-level carbon reduction targets. It is important to note that applying a global EF will provide little value in supporting these water utilities to quantify and mitigate their contribution to National emissions. A Tier 2 or Tier 3 approach is critical to both measure existing emissions and derive EFs but, most importantly, to allow mitigation interventions to be measured and monitored.

4.3.2.2.2 Nitrous oxide emissions from wastewater effluent

The IPCC also provides the methodology to estimate N_2O emissions from wastewater after disposal of untreated or treated wastewater effluent into aquatic environments by accounting for the removal of nitrogen through treatment processes prior to wastewater effluent disposal. Similarly, this methodology has been in place since the 2006 Guidelines, Volume 5: Waste, Chapter 6: Wastewater Treatment and Discharge, and has been revised in the 2019 Refinement to incorporate wastewater discharge into nutrient-impacted aquatic environments (eutrophic or hypoxic), where emissions can be significantly higher in comparison to relatively clean and/or well-oxygenated environments (Bartram *et al.*, 2019).

Table 4.4 Nitrous oxide emission factors by type of discharge aquatic environment with 95%ile limits shown in brackets (Bartram *et al.*, 2019).

Type of discharge environment	$EF_{Effluent}$ (kg N_2O-N/kg N)
Freshwater, estuarine, and marine discharge (Tier 1)	0.005 (0.0005–0.075)
Nutrient-impacted and/or hypoxic freshwater, estuarine and marine environment (Tier 3, if needed)	0.019 (0.0041–0.091)

The following equation (from Equation (6.7) in the 2019 IPCC Refinement) is used to estimate N_2O emissions from the discharge of treated or untreated wastewater into aquatic environments:

$$N_2O_{Effluent,DOM} = N_{Effluent,DOM} \times EF_{Effluent} \times 44/28 \quad (4.8)$$

where, $N_2O_{Effluent,DOM}$ is the N_2O emission from domestic wastewater effluent (kg N_2O/year); $N_{Effluent,DOM}$ is the nitrogen in the effluent discharged to aquatic environments (kg N/year). The 2019 IPCC Refinement provides a methodology to estimate $N_{Effluent,DOM}$ based on total nitrogen influent (TN_{DOM}), degree of utilization of the treatment system (T_j), and fraction of total nitrogen removed (N_{REM}). $EF_{Effluent}$ is the emission factor for N_2O emissions from wastewater discharged to aquatic environments (kg N_2O-N/kg N), as depicted below in Table 4.4. The factor 44/28 is the stoichiometric conversion from N to molecular N_2O.

The 2019 Refinement notes that the EFs are based on limited field data and on specific assumptions regarding occurrence of nitrification and denitrification in rivers and estuaries. For well-oxygenated environments, a total of 62 data points were reviewed, while 59 studies were evaluated for low-oxygen environments.

4.4 NATIONAL GUIDELINES

The majority of signatory countries of the Kyoto Protocol use the IPCC Guidelines as the basis for their national GHG inventory assessment. Some countries have developed country-based methodologies to provide greater clarity on the use of EFs and activity data – examples of this being country-level derivation of country- or facility-specific EFs.

4.4.1 Methane

There are limited national guidelines for estimation of CH_4 emissions from WWTPs at national level.

4.4.1.1 Australia

A Tier 2 approach for Australia is described in the National Greenhouse and Energy Reporting (Measurement) Determination 2008 made under sub-section 10(3) of the *National Greenhouse and Energy Reporting Act 2007.* This legislation provides four methods for GHG emissions assessment (Department of Industry, Science, Energy & Resources, 2021; OPC, 2017b):

The Determination provides three methods for estimating CH_4 emissions from treatment and emissions from flaring in Part 5.3 Wastewater Handling (Domestic and Commercial). The methods, which align with Tier 2 and 3 approaches, are summarized below:

- Method 1: based on the estimated quantity of CH_4 in biogas produced, considering standard per capita COD contributions. This subtracts biogas which is utilized on site, flared or exported and provides a separate emissions calculation for wastewater and for sludge – in each case with a default MCF and default EF for CH_4. Wastewater and sludge MCFs are based on the 2006 IPCC Guidelines correction factors (e.g. from 2006 IPCC Volume 5, Table 6.3). Separate consideration of sludge types – that is volatile solids in primary and waste activated sludge – is given.

- Method 2: considers an approach aligned with Method 1 but with more specific consideration of a facility. This is based on designation of sub-facility levels based on treatment areas and the use of measured data (e.g. COD or BOD) with considerations for the operator in designating a sub-facility, considerations of representative sampling (which must be on at least a monthly basis), and description of requirements for certification of samples taken by accredited laboratories.
- Method 3: aligns with Method 2 but provides for different sampling laboratory certification.

The methods meet IPCC Tier 2 and 3 approaches in part – in that they allow for facility type and site level calculations through Methods 1, 2 and 3. The Determination provides for Method 4 in GHG emissions assessment – defined as facility-specific measurement of emissions by continuous or period emissions monitoring – but this is not included as a method for CH_4 emissions estimation in the present approach.

4.4.1.2 United Kingdom

Water Companies in the UK are required to report their GHG emissions to the economic regulator for water companies in England and Wales, Ofwat, using country-developed EFs and a peer-reviewed industry-wide tool for operational carbon assessment – the Carbon Accounting Workbook (CAW). Sector-level reporting has been required by Ofwat since 2007 (Ofwat, 2010). Emissions reported in the CAW are in part used for compilation in UK National Inventory Reporting.

The CAW provides for calculation of fugitive CH_4 emissions from sludge storage, thickening and treatment in anaerobic digesters. The EFs included for mass of CH_4 per mass of raw dry solids of sewage sludge consider losses from digesters, venting due to ignition failure and downtime at flare stacks, fugitive emissions and secondary digester emissions. They also consider advanced AD processes including the thermal hydrolysis process (THP) and acid phase digestion (APD). A recent review of the applied EFs highlights their derivation is based on theoretical assessments only – and is not from measured datasets. It concludes with the need to further review and revise these which is currently ongoing (United Kingdom Water Industry Research, U 2020).

Whilst considering EFs that have been derived for national use by the water sector, the UK methodology is not well aligned with the IPCC methodology and has been recommended for review and revision through an industry monitoring program (United Kingdom Water Industry Research, U 2020).

4.4.1.3 Other country considerations

Elsewhere, there is evidence of voluntary program approaches to quantify and reduce CH_4 emissions which include the water sector. For example, the Swedish biogas industry have focused on leak detection and operational controls for CH_4 emissions reduction for biogas systems as a voluntary mechanism since 2007 (Holmgren *et al.*, 2012).

Implementation of future regulations, such as the European Union Best Available Techniques 14 (BAT 14) for the waste sector (Commission Implementing Decision (EU) 2018/1147 of 10 August 2018) (European Union, 2020) will likely require interventions for AD sites to reduce fugitive CH_4 emissions through on-site measurements, such as leakage detection and repair (LDAR), including for AD and associated processes at WWTPs. Industry initiatives and regulatory requirements could potentially result in better estimation of CH_4 emissions from WWTPs for more accurate national guidelines for estimation based on bottom-up measurements.

4.4.2 Nitrous oxide

4.4.2.1 Australia

A Tier 2 approach for Australia is described in the National Greenhouse and Energy Reporting (Measurement) Determination 2008 made under sub-section 10(3) of the *National Greenhouse and*

Energy Reporting Act 2007. It provides four methods for GHG emissions assessment (Department of Industry, Science, Energy and Resources, 2021; OPC 2017a):

- Method 1 (known as the default method): derived from the National Greenhouse Accounts methods and based on national average estimates.
- Method 2: a facility-specific method, generally, using industry practices for sampling and Australian or equivalent standards for analysis.
- Method 3: the same as method 2 but based on Australian or equivalent standards for both sampling and analysis.
- Method 4: provides for facility-specific measurement of emissions by continuous or periodic emissions monitoring.

Three of these – Methods 1, 2 and 3 are described for N_2O determination. These three methods included in the Determination provide for a Tier 2, country-specific assessment based on national data, EFs and facility-specific nitrogen loads. Work is ongoing to develop in-country facility-specific emissions measurement for use by the industry to develop improved country-level EFs.

All methods adopt a mass balance approach to calculate the removal of organic material, or nitrogen, considering country-specific factors for protein, nitrogen in sludge and disposal routes (to landfill or other disposal), and emissions differentiated by three types of discharge environment as per Equation (4.9) below. See the published Determination Section 5.31 for full method detail, a summary of which is provided below.

$$E_j = (N_{\text{in}} - N_{\text{trl}} - N_{\text{tro}} - N_{\text{outdisij}}) \times EF_{\text{secij}} + N_{\text{outdisij}} \times EF_{\text{disij}} \tag{4.9}$$

where E_j is the emissions of N_2O released from human sewage treated by the plant during the year, measured in tonnes of N_2O and expressed in CO_2e tonnes, N_{in} is the quantity of nitrogen entering the plant during the year, measured in tonnes of nitrogen and calculated according to whether the plant has treatment to a primary or secondary standard, population served, a per capita protein intake of 0.036 tonnes per year and a nitrogen protein fraction of 0.016 tonnes of nitrogen per tonne of protein. N_{trl} is the quantity of nitrogen in sludge transferred out of the plant and removed to landfill during the year, measured in tonnes of nitrogen and calculated using a mass flow of dry sludge and assumed fraction of nitrogen of 0.05, N_{tro} is the quantity of nitrogen in sludge transferred out of the plant and removed to a site other than landfill during the year, measured in tonnes of nitrogen and calculated using a mass flow of dry sludge and nitrogen fraction of 0.05, N_{outdisij} is the quantity of nitrogen leaving the plant, differentiated by discharge environment as described by EF_{disij} factors to different discharge environments.

The EF for wastewater, EF_{secij}, is currently 4.9 tonnes of N_2O, measured in CO_2e per tonne of nitrogen 'produced' from the wastewater treatment process (i.e. N_2O removed through the WWTP) or 0.016 kg N_2O/kg TN removed in secondary treatment based on the 2017 Determination (with N_2O GWP of 298). This was based historically on international literature sources and work of the IPCC. This is currently in revision and a proposed revised factor following consultation in July 2020 is 2.082 tonnes of N_2O measured in CO_2e per tonne of nitrogen produced or 0.0079 kg N_2O/kg TN.

For Methods 2 and 3, the same EF applies but laboratory sampling methods are specified (see Determination Sections 5.33, 5.34, 5.35, 5.36). The Determination requires that samples be representative, sufficient in coverage, free from bias, sampled in accordance with quoted international or Australian standards and, in the case of wastewater, sampled on a monthly basis.

Work continues in the Australian water sector to estimate in-country EFs and develop in-country monitoring methodologies which might allow for future development of a Method 4 approach.

4.4.2.2 Austria

A Tier 2 country-specific EF of 43 g N_2O/PE/year is used to estimate N_2O emissions from wastewater treatment for their national GHG inventory assessment for WWTPs serving over 2 000 PE. Water utilities are not required to report GHG emissions at the sector-level to their regulator.

The estimation of a country-specific EF was developed based on a national measuring program conducted in 2012–2014 with 20 field measurements at 8 representative activated sludge WWTPs (BMLFUW, 2015). The monitoring campaigns were carried out with long-term online measurements of several weeks, with both off-gas measurement (flux chamber) and liquid measurement (Unisense micro-sensor). The results concluded that nitrification was the main source of N_2O emissions, and an observed correlation with the TN removal degree confirmed the role of the denitrification as N_2O sink. The country-specific EF was derived through linear regression of N_2O emissions and nitrogen removal for 18 of the 20 campaigns and extrapolated to include nitrogen removal consideration for WWTPs with an organic design capacity larger than 5 000 PE (94%) and less than 2 000 PE (~6%) (BMLFUW, 2015). This draws on work previously discussed by Valkova *et al.* (2020) and Parravicini *et al.* (2016) which draws attention to the link between N_2O emissions and the extent of total nitrogen reduction. Based on the Austrian wastewater emission ordinance a 70% minimum reduction degree on annual average basis is required for municipal WWTPs > 5 000 PE (EVO, 1996).

In addition, to estimate the N_2O emissions from the discharge of wastewater to aquatic environments, the Austrian methodology considers country-specific measured/reported values for $N_{\text{Effluent,DOM}}$ (Equation (4.8)) for both WWTP effluent and for effluent of the population not connected to the WWTP (less than 5%). The total N_2O emissions for the inventory is the following:

$$N_2O_{\text{TOTAL}} = N_2O_{\text{WWTP}} + N_2O_{\text{EFFLUENT}} \tag{4.10}$$

where, N_2O_{WWTP} is N_2O emissions from advanced WWTPs for the population connected to WWTPs with controlled nitrification and denitrification; N_2O_{EFFLUENT} is N_2O emissions from WWTPs effluent and from effluent of the population not connected to WWTPs.

4.4.2.3 Denmark

Denmark has completed a national survey of N_2O emissions from representative WWTPs and is applying mitigation strategies to reduce N_2O emissions across the sector (VTU, 2016). This was achieved through monitoring across 10 facilities and analysis of data (including removal of facility N_2O emission pertaining to sidestream treatment) has resulted in a new country-specific EF of 0.84% N_2O based on influent TN (noting that all WWTPs achieve very high degrees of total nitrogen removal). This is significantly higher than the previous in-country EF of 0.32% N_2O based on influent TN (The Danish EPA, 2020). Regulatory incentives are being discussed to reduce emissions to a target value and provide a mechanism for the water companies to fund this, which is likely to result in a focus on online continuous monitoring and mitigation for large facilities.

4.4.2.4 Japan

Different Tier 2 EFs are used to estimate N_2O emissions in Japan based on research conducted in WWTPs in the country for specific treatment types, as detailed in Table 4.5.

The EFs for high load denitrification and membrane separation were derived based on the median value of on-site measurements at 13 WWTPs (National Institute for Environmental Studies (2006)). For other treatment processes, the EF was obtained by dividing the upper limit value for standard denitrification from Tanaka *et al.* (1995) by treated nitrogen concentration in fiscal year 1994 (GIO, 2019).

Table 4.5 Nitrous oxide emission factors by wastewater treatment plant.

Wastewater treatment process	N_2O EF_{WWTP} (kg N_2O-N/kg N load)
High load denitrification	0.0029
Membrane separation	0.0024
Other (including anaerobic, aerobic and standard denitrification treatment processes)	0.0000045

High load denitrification facilities treat 'night soil' and black water 'sludge' from different wastewater treatment configurations through 'high-load denitrification devices, solid-liquid separation devices and flocculation separation devices'; standard denitrification facilities include grey water which results in a more diluted wastewater and comprises a biochemical denitrification process (Ministry of Environment, 2018). Membrane separation processes also appear to be high-load denitrification facilities where membrane separation devices are adopted for solid-liquid separation instead of traditional sedimentation tanks or mechanical devices (Ministry of Environment, 2019).

4.4.2.5 United States of America

In the USA mandatory GHG emissions reporting from water companies can occur at different levels, depending on the State. For instance, in California, water companies emitting from 10 000 to 25 000 tCO_2e/year need to report at a State level to the California Air Resources Board (CARB) Regulation for the Mandatory Reporting of GHG Emissions (MRR), and also choose to report at a sector-level to The Climate Registry (TCR) voluntary reporting program (McGuckin *et al.*, 2013).

For the national level GHG inventory assessment, in addition to using the 2006 IPCC Guidelines EF of 3.2 g N_2O/person/year (0.00035 kg N_2O-N/kg N load) for WWTPs without intentional denitrification (Czepiel *et al.*, 1995), the United States Environmental Protection Agency (USEPA) have introduced a country-developed EF for WWTPs with intentional nitrification and denitrification due to the degree of biological nutrient removal (BNR) WWTPs in the country, which serves a population of 21.3 million people (Scheehle & Doorn, 2001; USEPA, 2019). The EF of 7.0 g N_2O/person/year (0.00074 kg N_2O-N/kg N load) was adopted based on a study conducted in Germany in 1993, and thus not derived from in-country estimates (Schon *et al.*, 1993). Per capita protein intake figures are considered specific to dietary intake in the US whilst the IPCC 2006 estimate of 16 kg N/kg protein is applied. This results in an incoming TN of 16 g N/PE/day.

4.4.2.6 United Kingdom

In the United Kingdom (UK), emissions from the water sector are reported both at the national level based on the Department for Environment, Food & Rural Affairs (Defra) Guidelines, and at the sector level to the economic regulator for water companies in England and Wales, Ofwat.

For the national reporting, N_2O emissions from wastewater treatment are not reported, only indirect N_2O from discharge of effluent based on the 2006 IPCC Guidelines is reported. The NIR specifies that 'the UK GHG inventory mostly follows the UK water industry GHG emission estimation methodology developed by [the UK Water Industry Research] UKWIR and used by all UK water companies to generate their annual emission estimates from all sources/activities' (Brown *et al.*, 2019).

For sector-level reporting, water companies have been required to report their annual operational GHG emissions to the regulator, Ofwat, since 2007 (Ofwat, 2010). The reporting is done using the industry-wide tool, CAW, which provides a framework for harmonized estimation and reporting of the annual GHG emissions from the UK water sector (UKWIR, 2005). Since its first publication by UKWIR in 2005, the CAW has been reviewed each year to include the latest available information and reviews in 2009 and 2020 considered process EFs. The latest review undertaken in 2020 was aimed at addressing the need for an improved understanding of process emissions, specifically CH_4 and N_2O emissions (UKWIR, 2020a, 2020b).

For estimation of N_2O emissions from wastewater treatment, the latest review updated the country-developed EF to its original value of 0.004 kg N_2O-N/kg N load in secondary treatment, which was originally derived from the simple statistical average of nine studies (lab-, pilot-, and full-scale) conducted globally between 1994 and 2002 (UKWIR, 2008; (United Kingdom Water Industry Research, U 2020). The UK water sector have acknowledged that accurate estimation and mitigation of process emissions is one of the main challenges in their pathway to achieving net zero by 2030. Work is underway to develop an approach for industry wide monitoring of N_2O from representative WWTPs to develop country-specific EFs across fixed-film and suspended growth process types (United Kingdom Water Industry Research, U 2020).

REFERENCES

Aboobakar A., Cartmell E., Stephenson T., Jones M., Vale P. and Dotro G. (2013). Nitrous oxide emissions and dissolved oxygen profiling in a full-scale nitrifying activated sludge treatment plant. *Water Research*, **47**, 524–534, https://doi.org/10.1016/j.watres.2012.10.004

Ahn J.-H., Kim S., Park H., Rahm B., Pagilla K. and Chandran K. (2010). Emissions from activated sludge processes, 2008–2009: results of a National Monitoring Survey in the United States. *Environmental Science and Technology*, **44**, 4505–4511, https://doi.org/10.1021/es903845y

Aizawa, T., Sato A. and Nojiri, (2006). National Greenhouse Gas Inventory Report of Japan (August, 2006). National INstitute for Environmental STudies. Available at https://www.nies.go.jp/gio/en/archive/nir/jqjm1000000kannj-att/I069.pdf

Bao Z., Sun S. and Sun D. (2016). Assessment of greenhouse gas emission from A/O and SBR wastewater treatment plants in Beijing, China. *International Biodeterioration & Biodegradation*, **108**, 108–114, https://doi.org/10.1016/j.ibiod.2015.11.028

Bartram D., Short M. D., Ebie Y., Karkas J., Gueguen C., Peters G. M., … Karthik M. (2019). 2019 Refinement to the 2006 IPCC Guidelines for National Greenhouse Gas Inventories, IPCC.

Bellucci F., Kozak J. A., Heraty L., Carbone J., Sturchio N. C., Connor C., … Lanyon R. (2010). Greenhouse Gas Emissions from Three Chicago Wastewater Treatment Plants. Water Environment Federation (p. 417904199). WEFTEC.

BMLFUW (2015). Bundesministerium für Land- und Forstwirtschaft, Umwelt und Wasserwirtschaft. Retrieved from Bundesministerium für Land- und Forstwirtschaft, Umwelt und Wasserwirtschaft: https://www.bmlfuw.gv.at/service/publikationen/wasser/Lachgasemissionen---Kl-ranlagen.html

Brotto A. C., Huosheng L., Muriel D., Gabarro J., Colprim J., Murthy S. and Chandran K. (2015a). Characterization and mitigation of nitrous oxide (N2O) emissions from partial- and full-nitrification BNR processes based on post-anoxic aeration control. *Biotechnology & Bioengineering*, **9999**, 1–7.

Brotto A. C., Kligerman D. C., Andrade S. A., Ribeiro R. P., Oliveira J. L., Chandran K. and de Mello W. Z. (2015b). Factors controlling nitrous oxide emissions from a full-scale activated sludge system in the tropics. *Environmental Science Pollution Research*, **22**, 11840–11849, https://doi.org/10.1007/s11356-015-4467-x

Brown P., Broomfield M., Cardenas L., Choudrie S., Jones L., Karagianni E., … Wakeling D. (2019). UK Greenhouse Gas Inventory, 1990 to 2017. Department of Business, Energy and Industrial Strategy, London.

C40 (2020, October 19). C40 Cities. Retrieved from https://www.c40.org/

Czepiel P., Crill P. M. and Harris R. C. (1995). Nitrous oxide emissions from municipal wastewater treatment. *Environmental Science and Technology*, **29**, 2352–2356, https://doi.org/10.1021/es00009a030

Daelman M. R., van Voorthuizen E. M., van Dongen U. G., Volcke E. I. and van Loosdrecht M. C. (2012). Methane emission during municipal wastewater treatment. *Water Research*, **46**, 3657–3670, https://doi.org/10.1016/j.watres.2012.04.024

Daelman M. R., van Voorthuizen E. M., van Dongen L., Vockle E. I. and van Loosdrecht M. C. (2013). Methane and nitrous oxide emissions from municipal wastewater treatment – results from a long-term study. *Water Science and Technology*, **67**, 2350–2355, https://doi.org/10.2166/wst.2013.109

Daelman M. R., van Voorthuizen E. M., van Dongen U. G., Volcke E. I. and van Loosdrecht M. C. (2015). Seasonal and diurnal variability of N2O emissions from a full-scale municipal wastewater treatment plant. *Science of the Total Environment*, **536**, 1–11, https://doi.org/10.1016/j.scitotenv.2015.06.122

de Haas D. and Ye L. (2021). Nitrous oxide emissions from wastewater treatment: a case for variable emission factors. *e-journal*, **6**(2). https://doi.org/10.21139/wej.2021.008

Delre A., Monster J. and Scheutz C. (2017). Greenhouse gas emission quantification from wastewater treatment plants, using a tracer gas dispersion method. *Science of The Total Environment*, **605–606**, 258–268, https://doi.org/10.1016/j.scitotenv.2017.06.177

de Mello W. Z., Ribeiro R. P., Brotto A. C., Kligerman D. C., Piccoli A. S. and Oliveira J. L. (2013). Nitrous oxide emissions from an intermittent aeration activated sludge system of an urban wastewater treatment plant. *Quimica Nova*, **36**, 16–20. Retrieved from: https://www.scielo.br/j/qn/a/KH9PdV5YM6Fp8fhRxLS3bQR/

Department of Industry, Science, Energy and Resources (2021) National Greenhouse and Energy Reporting (Measurement) Determination 2008 made under subsection 10(3) of the National Greenhouse and Energy Reporting Act 2007, Compilation No. 13, dated 1 July 2021.

Doorn M. R., Strait R. P., Barnard W. and Eklund B. (1997). Estimate of Global Greenhouse Gas Emissions from Industrial and Domestic Wastewater Treatment, EPA-600/R-97-091, U.S. EPA Office of Research and Development, Air Pollution Prevention and Control Division, Research Triangle Park, NC.

Doorn M. R., Towprayoon S., Vieira S. M., Irving W., Palmer C., Pipatti R. and Wang C. (2006). Chapter 6: wastewater treatment and discharge. In: 2006 IPCC Guidelines for National Greenhouse Gas Inventories. Simon E., Leandro B., Kyoko M., Todd N. and Kiyoto T. Volume 5: Waste. IPCC. Japan by Institute for Global Environmental Strategies (IGES), Hayama, Japan. https://www.ipcc-nggip.iges.or.jp/public/2006gl/pdf/0_Overview/V0_0_Cover.pdf

European Union (2020, September 15). EUR_Lex. Retrieved from EUR_Lex: https://eur-lex.europa.eu/eli/dec_impl/2018/1147/oj

EVO (1996). Allgemeine Emissionsverordnung für kommunales Abwasser /emission ordinance for municipal wastewater, adaptation to the EU guideline for municipal wastewater 271/91 (accessed 5 January 2021).

Federici S. (2019). 2019 Refinement to the 2006 IPCC Guidelines for National GHG Inventories. UN Climate Change Conference. IPCC, Madrid.

Foley J., de Haas D., Yuan Z. and Lant P. (2010). Nitrous oxide generation in full-scale biological nutrient removal wastewater treatment plants. *Water Research*, **44**, 831–844, https://doi.org/10.1016/j.watres.2009.10.033

GIO (2019). National Greenhouse Gas Inventory Report of JAPAN, Ministry of the Environment/Japan Greenhouse Gas Inventory Office of Japan (GIO)/Center for Global Environmental Research (CGER)/ National Institute for Environmental Studies (NIES). National Institute for Environmental Studies.

Greenhouse Gas Protocol (2021). GHG Protocol Global Protocol for Community-Scale Greenhouse Gas Inventories. Published by World Resources Institute. Retrieved from: https://ghgprotocol.org/sites/default/files/standards/GPC_Full_MASTER_RW_v7.pdf

Greenhouse Gas Protocol (2015). Revised Edition, A Corporate Accounting and Reporting Standard. World Resources Institute.

Greenhouse Gas Protocol (2020b). Greenhouse Gas protocol. Retrieved from: https://ghgprotocol.org/

Holmgren M., Hellstrom H. and Petersson A. (2012). The Swedish Voluntary Agreement for Control of Methane Emissions From Biogas Plants. Orbit 2012.

IPCC (2006). 2006 IPCC Guidelines for National Greenhouse Gas Inventories, Prepared by the National Greenhouse Gas Inventories Programme. In: H.S. Eggleston, L. Buendia, K. Miwa, T. Ngara and K. Tanabe (eds.). Published: IGES, Japan. https://www.ipcc-nggip.iges.or.jp/public/2006gl/pdf/0_Overview/V0_0_Cover.pdf.

IPCC (2014a). Climate change 2014: synthesis report. In: Contribution of Working Groups I, II and III to the Fifth Assessment Report of the Intergovernmental Panel on Climate Change, Core Writing Team, R. K. Pachauri and L. A. Meyer (eds.), IPCC, Geneva.

IPCC (2014b). 2013 Supplement to the 2006 IPCC Guidelines for National Greenhouse Gas Inventories: Wetlands, T. Hiraishi, T. Krug, K. Tanabe, N. Srivastava, J. Baasansuren, M. Fukuda and T. G. Troxler (eds.), IPCC, Switzerland.

IPCC (2018a). IPCC Factsheet: What is the IPCC. IPCC.

IPCC (2018b). Summary for Policymakers. In: Global Warming of 1.5°C. An IPCC Special Report on the impacts of global warming of 1.5°C above pre-industrial levels and related global greenhouse gas emission pathways, in the context of strengthening the global response to the threat of climate change, sustainable development, and efforts to eradicate poverty. V. Masson-Delmotte, P. Zhai, H.-O. Pörtner, D. Roberts, J. Skea, P.R. Shukla, A. Pirani, W. Moufouma-Okia, C. Péan, R. Pidcock, S. Connors, J.B.R. Matthews, Y. Chen, X. Zhou, M.I. Gomis, E. Lonnoy, T. Maycock, M. Tignor and T. Waterfield (eds.). World Meteorological Organization, Geneva, Switzerland, 32 pp.

IPCC (2020a). AR6 Synthesis Report: Climate Change 2022. Retrieved from IPCC: https://www.ipcc.ch/report/sixth-assessment-report-cycle/#:~:text=The%20IPCC%20is%20currently%20in,due%20for%20release%20in%202022

IPCC (2020b). The Paris agreement. Retrieved from United Nations Climate Change: https://unfccc.int/process-and-meetings/the-paris-agreement/the-paris-agreement (accessed 5 January 2021).

IPCC (2021). Summary for Policymakers. In: Climate Change 2021: The Physical Science Basis. Contribution of Working Group I to the Sixth Assessment Report of the Intergovernmental Panel on Climate Change. V. Masson - Delmotte, P. Zhai, A. Pirani, S.L. Connors, C. Péan, S. Berger, N. Caud, Y. Chen, L. Goldfarb, M.I. Gomis, M. Huang, K. Leitzell, E. Lonnoy, J.B.R. Matthews, T.K. Maycock, T. Waterfield, O. Yelekçi, R. Yu and B. Zhou (eds). Cambridge University Press. In Press.

Kimochi Y., Inamori Y., Mizuochi M., Kai-Qin X. and Matsumara M. (1998). Nitrogen Removal and N_2O Emission in a Full Scale Domestic Wastewater Treatment Plant with Intermittent Aeration. *Journal of Fermentation and Bioengineering*, **86**, 202–206.

Kozak J. A., O'Connor C., Granato T., Kollias L., Bellucci F. and Sturchio N. (2009). Methanes and nitrous oxide emissions from wastewater treatment plant processes. 5347–5361. Available at https://accesswater.org/publications/-296015/methane-and-nitrous-oxide-emissions-from-wastewater-treatment-plant-processes

Kyung D., Kim M., Chang J. and Lee W. (2015). Estimation of greenhouse gas emissions from a hybrid wastewater treatment plant. *ournal of Cleaner Production*, **95**, 117–123.

Lake A., Brotto A., Porro J., Holmen Andersen M. and Sandino J. (2021). Nitrous oxide – no laughing matter – lessons from Europe. unpublished.

Masuda S., Suzuki S., Sano I., Li Y.-Y. and Nishimura O. (2015). The seasonal variation of emission of greenhouse gases from a full-scale sewage treatment plant. *Chemosphere*, **140**, 167–173, https://doi.org/10.1016/j.chemosphere.2014.09.042

Masuda S., Sano I., Hojo T., Li Y.-Y. and Nishimura O. (2018). The comparison of greenhouse gas emissions in sewage treatment plants with different treatment processes. *Chemosphere*, **193**, 581–590, https://doi.org/10.1016/j.chemosphere.2017.11.018

McGuckin R., Oppenheimer J., Badruzzaman M., Contreras A. and Jacangelo J. G. (2013). Toolbox for Water Utility Energy and Greenhoyse Gas Emission Management, Water Research Foundation, Denver.

Metcalf & Eddy (2003). Wastewater Engineering: Treatment and Reuse. McGraw-Hill, New York.

Miljøstyrelsen (English translation Danish Environmental Protection Agency) (2020). MUDP Lattergaspulje Dataopsamling på måling og reduktion af lattergasemissioner fra renseanlæg, Retrieved from https://www2.mst.dk/Udgiv/publikationer/2020/12/978-87-7038-254-0.pdf

Ministry of the Environment (2018). Night Soil Treatment and Decentralised Wastewater Treatment Systems in Japan. Retrieved from https://www.jeces.or.jp/en/downroad/pdf/nightsoilsystems_en.pdf

Ni B.-J., Pan Y., van den Akker B., Ye L. and Yuan Z. (2015). Full-scale modeling explaining large spatial variations of nitrous oxide fluxes in a step-feed plug-flow wastewater treatment reactor. *Environmental Science and Technology*, **49**, 9176–9184.

Ofwat (2010). Playing Our Part – Reducing Greenhouse Gas Emissions in the Water and Sewarage Sectors. Supporting Information, Ofwat, Birmingham.

OPC: Office of Parliamentary Counsel (2017a). National Greenhouse and Energy Reporting (Measurement) Determination 2008 made under subsection 10(3) of the National Greenhouse and Energy Reporting Act 2007, compilation No. 9 dated 1 July 2017, Office of Parliamentary Counsel, Canberra.

OPC (2017b). National Greenhouse and Energy Reporting (Measurement) Determination 2008, Compilation No. 9, Office of Parliamentary Council, Canberra.

Pan Y., van den Akker B., Ye L., Ni B.-J., Watts S., Reid K. and Yuan Z. (2016). Unravelling the spatial variation of nitrous oxide emissions from a step-feed plug flow full scale wastewater treatment plant. *Scientific Reports*, **6**, 20792. https://doi.org/10.1038/srep20792

Parravicini V., Svardal K. and Krampe J. (2016). Greenhouse gas emissions from wastewater treatment plants. *Energy Procedia*, **97**, 246–253, https://doi.org/10.1016/j.egypro.2016.10.067

Pijuan M., Tora J., Rodriguez-Caballero A., Carrera C. E. and Perez J. (2014). Effect of process parmeters and operational mode on nitrous oxide emissions from anitritation reactor treating reject wastewater. *Water Research*, **49**, 23–33, https://doi.org/10.1016/j.watres.2013.11.009

Rodriguez-Caballero A., Aymerich I., Poch M. and Pijuan M. (2014). Evaluation of process conditions triggering emissions of green-house gases from a biological wastewater treatment system. *Sci Total Environ* **493**, 384–391.

Rodriguez-Caballero A., Aymerich I., Marques R., Poch M. and Pijuan M. (2015). Minimizing N_2O Emissions and Carbon Footprint on a Full-Scale Activated Sludge Sequencing Batch Reactor. *Water Research* **71**, 1–10.

Rogelj J., Shindell D., Jiang K., Fifita S., Forster P., Ginzburg V., Handa C., Kheshgi H., Kobayashi S., Kriegler E., Mundaca L., Séférian R. and Vilariño M. V. (2018). Mitigation Pathways Compatible with 1.5°C in the Context of Sustainable Development. In: Global Warming of 1.5°C. An IPCC Special Report on the impacts of global warming of 1.5°C above pre-industrial levels and related global greenhouse gas emission pathways, in the context of strengthening the global response to the threat of climate change, sustainable development, and efforts to eradicate poverty. V. Masson-Delmotte, P. Zhai, H.-O. Pörtner, D. Roberts, J. Skea, P.R. Shukla, A. Pirani, W. Moufouma-Okia, C. Péan, R. Pidcock, S. Connors, J.B.R. Matthews, Y. Chen, X. Zhou, M.I. Gomis, E. Lonnoy, T. Maycock, M. Tignor and T. Waterfield (eds). In Press, pp. 93–174.

Schaum C., Fundneider T. and Cornel P. (2016). Analysis of methane emissions from digested sludge. *Water Science and Technology*, **73**, 1599–1607, https://doi.org/10.2166/wst.2015.644

Scheehle E. A. and Doorn M. R. (2001). Improvements to the US Wastewater Methane and NItrous Oxide Emissions Estimate.

Schon M., Walz R., Angerer G., Bohm E., Hillenbrand T., Hiessl H. and Reichert J. (1993). Emissionen der Treibhausgase Distickstoffoxid und Methan in Deutschland. Emissionsbilanz, Identifikation von Forschungs- und Entwicklungsbedarf Sowie Erarbeitung von Handlungsempfehlungen. Phase 1. Erich Schmidt Verlag,

Berlin, Germany. Retrieved from: http://publica.fraunhofer.de/documents/PX-46292.html (accessed 5 January 2021)
Tanaka M., Inoue Y., Matsuzawa Y. and Watanabe I. (1995). B2-(1)Research into Volumes Released from Waste Treatment Plants, 1994 Global Environment Research Fund Outcome Report.
UKWIR (2005). Workbook for Quantifying Greenhouse Gas Emissions, 05/CL/01/3, UKWIR, London.
UKWIR (2008). Carbon Accounting in the UK Water Industry: Methodology for Estimating Operating Operational Emissions, 08/CL/01/5, UKWIR, London.
UKWIR (2020a, September 15). How do we remove more carbon than we emit by 2050. Retrieved from UK Water Industry Research: https://ukwir.org/How-do-we-remove-more-carbon-than-we-emit-by-2050 (accessed 5 January 2021)
UKWIR (2020b). Quantifying and Reducing Direct Greenhouse Gas Emissions From Waste and Water Treatment Processes, CL/01/E/201, UKWIR, London.
United Kingdom Water Industry Research (2020). Quantifying and Reducing Direct Greenhouse Gas Emissions From Waste and Water Treatment Processes, UKWIR, London.
United Nations (2020, September 15). What is the kyoto protocol. Retrieved from United Nations Climate Change: https://unfccc.int/kyoto_protocol (accessed 5 January 2021)
USEPA (2019). Inventory of the U.S Greenhouse Gas Emissions and Sinks. 1990–2017, US Environmental Protection Agency.
Valkova T., Parravicini V., Saracevic E., Tauber J., Svardal K. and Krampe J. (2020). A method to estimate the direct nitrous oxide emissions of municipal wastewater treatment plants based on the degree of nitrogen removal. *Journal of Environmental Management*, **279**, 1–10.
Vasilaki V., Massara T. M., Stanchev P., Fatone F. and Katsou E. (2019). A decade of nitrous oxide (N2O) monitoring in full-scale wastewater. *Water Research*, **161**, 392–412, https://doi.org/10.1016/j.watres.2019.04.022
VTU (2016). Intelligent og miljømæssig bæredygtig energieffektivisering af renseanlæg – Emissionsminimering af drivhusgasser.
Wang J., Chen N., Yan W., Wang B. and Yang L. (2015). Effect of dissolved oxygen and nitrogen on emission of N2O from rivers in China. *Atmospheric Environment*, **103**, 347–356, https://doi.org/10.1016/j.atmosenv.2014.12.054
Wang Y., Fang H., Zhou D., Han H. and Chen J. (2016). Characterization of nitrous oxide and nitric oxide emissions from a full-scale biological aerated filter for secondary nitrification. *Chemical Engineering Journal*, **299**, 304–313.
Woess-Gallash S., Bird S., Enzinger P., Jungmeier G., Padinger R., Pena N. and Zanchi G. (2010). Greenhouse gas benefits of a biogas plant in Austria, Joanneum Research Forschungsgesellschaft mbH, Resources – Institute of Water, Energy and Sustainability, Graz, Austria.
World Resources Institute (2021). Global Protocol for Community Scale Greenhouse Gas Inventories. World Resources Institute: Washington DC, USA, https://ghgprotocol.org/sites/default/files/standards/GPC_Full_MASTER_RW_v7.pdf (accessed 20 March 2022)
World Resources Institute and World Business Council for Sustainable Development, (2011). Corporate Value Chain (Scope 3) Accounting and Reporting Standard. Place of publication WRI – Washington DC, USA. ISBN 978-1-56973-772-9 Retrieved from: https://ghgprotocol.org/sites/default/files/standards/Corporate-Value-Chain-Accounting-Reporing-Standard_041613_2.pdf. (Accessed 20 March 2022)

NOMENCLATURE

AD	Anaerobic digestion
BAT	Best available technology
BNR	Biological nutrient removal
BOD	Biological oxygen demand
CARB	California Air Resources Board Regulation for the Mandatory Reporting of GHG Emissions
COD	Chemical oxygen demand

EF	Emission factor
EU	European Union
GHG	Greenhouse gas
GWP	Global warming potential
IPCC	Intergovernmental Panel on Climate Change
LDAR	Leakage detection and repair
MCF	Methane correction factor
NDC	Nationally determined contributions
NIR	National Inventory Report
PE	Population equivalents
TCR	The Climate Registry
TN	Total nitrogen
UK	United Kingdom
UKWIR	United Kingdom Water Industry Research
UNFCC	United Nations Framework Convention on Climate Change
USA	United States of America
USEPA	United States Environmental Protection Agency
WWTP	Wastewater treatment plant

doi: 10.2166/9781789060461_091

Chapter 5

Full-scale quantification of N_2O and CH_4 emissions from urban water systems

Vanessa Parravicini[1], Ahlem Filali[2], Antonio Delre[3], Oriol Gutierrez[4,5] and Haoran Duan[6]

[1]*TU Wien, Institute for Water Quality and Resource Management, Karlsplatz 13/226, 1040 Vienna, Austria. E-mail: vparravi@iwag.tuwien.ac.at*
[2]*Université Paris-Saclay, INRAE, PROSE, 1 rue Pierre Gilles de Gennes, 92160 Antony, France. E-mail: ahlem.filali@inrae.fr*
[3]*Department of Environmental Engineering, Technical University of Denmark, Bygningstorvet, Bygning 115, 2800 Kgs., Lyngby, Denmark. E-mail: antodel@env.dtu.dk*
[4]*Catalan Institute for Water Researchearch (ICRA), Emili Grahit 101-17003, Girona, Spain, E-mail: ogutierrez@icra.cat*
[5]*The University of Girona, 17003 Girona, Spain. E-mail: ogutierrez@icra.cat*
[6]*School of Chemical Engineering, The University of Queensland, St Lucia, Queensland 4072, Australia. E-mail: h.duan@uq.edu.au*

SUMMARY

The quantification of direct greenhouse gas (GHG) emissions from sewers and wastewater treatment plants is of great importance for urban sustainable development. In fact, the identification and assessment of anthropogenic sources of GHG emissions (mainly nitrous oxide and methane) in these engineered systems represent the first step in establishing effective mitigation strategies. This chapter provides an overview of the currently available nitrous oxide and methane quantification methods applied at full-scale in sewers and wastewater treatment plants. Since the first measurement campaigns in the early 90 s were based on spare grab sampling, quantification methodologies and sampling strategies have evolved significantly, in order to describe the spatio-temporal dynamics of the emissions. The selection of a suitable quantification method is mainly dictated by the objective of the measurement survey and by specific local requirements. Plant-wide quantification methods provide information on the overall emissions of wastewater treatment plants, including unknown sources, which can be used for GHG inventory purposes. To develop on-site mitigation strategies, in-depth analysis of GHG generation pathways and emission patterns is required. In this case, process-unit quantifications can be employed to provide data for developing mechanistic models or to statistically link GHG emissions to operational conditions. With regard to sewers, current available methods are not yet capable of capturing the complexity of these systems due to their geographical extension and variability of conditions and only allow the monitoring of specific locations where hotspots for GHG formation and emission have been identified.

Keywords: Greenhouse gas; quantification method; sewers; wastewater treatment

TERMINOLOGY

Term	Definition
Carbon footprint	A carbon footprint is the total greenhouse gas emissions caused by an individual, event, organization, service, or product, expressed as carbon dioxide equivalent.
Greenhouse gas (GHG)	Gas that absorbs and emits radiant energy within the thermal infrared range.
Hydraulic retention time (HRT)	HRT is a measure of the average length of time that a volume of wastewater remains in a given sewer section or a process unit.
K_La	K_La describes the rate of mass transfer process. "K_L" is the mass transfer coefficient while "a" refers to the liquid-gas interface area per volume (A/V). Due to the difficulties in separating the two parameters experimentally, the two are combined in a term K_La, and measured as an overall parameter.
Mass transfer	In this chapter, mass transfer refers to the liquid-to-gas or gas-to-liquid transport process of a gaseous species such as nitrous oxide. The rate of mass transfer is proportional to the difference between the equilibrium concentration and the concentration of concern. The rate of transfer reduces to zero, when the equilibrium is reached.
Negative pressure	In this chapter, negative pressure refers to a pressure under the hood that is lower than the atmospheric pressure.
Off-gas	Refers to any gas that is emitted from a given unit-process.
Sewer	A network of artificial underground conduits that convey and transport wastewater and/or stormwater from its origin to its treatment point.
Sewer rising main pipes	A rising main is a type of drain or sewer through which sewage and/or surface water runoff is pumped from a pumping station to an elevated point. Rising main pipes are fully pressurized and anaerobic conditions prevail in these sections of sewers.
Sewer gravity pipes	Opposite to rising main pipes, gravity sewer pipes are conduits that use a difference in elevation points, from high to low, and gravity to transport wastewater. Gravity pipes have a liquid and a gas phase which implies a certain reaeration of wastewater.

5.1 INTRODUCTION

A key to formulating strategies to control and reduce greenhouse gas (GHG) emissions to the atmosphere is the identification and quantification of all sources. This chapter describes the current existing quantification methodologies that are capable of quantifying fugitive GHG emissions from engineered urban water systems. Special focus is given to full-scale quantification in sewer systems and wastewater treatment plants (WWTPs), which have been revealed so far to be anthropogenic sources of direct N_2O and CH_4 emissions. The impact of these engineered systems is lower than some other natural and anthropogenic sources on a global scale. According to the estimations of the United Nation Framework Convention on Climate Change (UNFCCC), in 2018 N_2O and CH_4 emissions from wastewater treatment and discharge in industrialized (Annex I) countries contributed 2.6% and 3.6% to the total CO_2e emissions, respectively (UNFCCC, 2018). However, mitigation strategies aiming to achieve a sustainable development of urban areas must address these emission pathways. In response to this, measurement methods for source identification and quantification of overall N_2O and CH_4 emissions in these systems have been developed. In addition, tailored methodological approaches that provide a deeper insight in the GHG production and emission mechanisms in sewers and WWTPs can be applied when the focus is on the implementation of mitigation measures at process-unit scale. In this way the operation of single process-units can be optimized to reduce the overall carbon footprint.

To the authors' knowledge, GHG emission measurements at WWTPs were firstly performed by Czepiel P.M., Crill P.M., and Harries R.C. at Durham, New Hampshire (US) and date back to 1993. The measurement campaign aimed to quantify CH_4 emissions from the primary and secondary wastewater

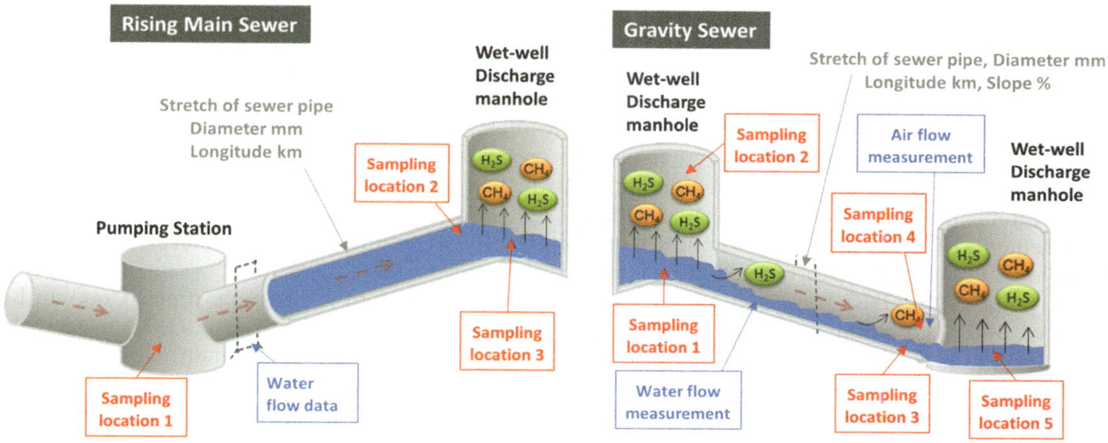

Figure 5.1 Sampling points and data required to quantify CH_4 emissions depending on the typology of sewer (adapted from Liu *et al.*, 2015a).

treatment processes and correlated these emissions with fluctuations in wastewater temperature (Czepiel *et al.*, 1993). The results of a second measurement campaign which targeted N_2O emissions were published two years later (Czepiel *et al.*, 1995). These pioneering works represented, till 2019, the scientific basis supporting the default emission factors suggested by the IPCC guidelines for national GHG Inventories (IPCC, 1996, 2006). Since then, the number of GHG emission measurements conducted at full-scale has increased steadily and hence the applied quantification methodology has been significantly improved. As a matter of fact, first measurement campaigns were performed using a grab sampling approach and therefore did not capture the temporal and seasonal variability of emissions, as clearly indicated by later full-scale surveys and experimental research works. The development of more rigorous quantification protocols was multilateral proceeding from grab sampling to online monitoring, from short-term to long-term measurement campaigns, and from process-unit to plant-wide quantification methodologies. With the "2019 Refinement" of the IPCC Guidelines (IPCC, 2019), default emission factors were revised using state-of-the-art knowledge and also CH_4 emissions from the sewers were considered. Nevertheless, the provided estimation of GHG emissions for inventory protocols still remains questionable, as it relies on fixed and generic emission factors that do not depict the wide variability of emission pattern against time, local process specification and operating conditions. Consequently, in most cases, the quantification and monitoring of GHG emissions at full-scale remains the only possibility to accurately describe emission loads and patterns.

This chapter is intended to give a general overview on the most widely applied methods to quantify GHG emissions from full-scale sewers and WWTPs. Sections 5.2 and 5.3 are dedicated to the quantification of GHG emissions in sewers and in WWTPs, respectively. Most of the methods applied to these engineered systems can quantify both N_2O and CH_4 emissions.

5.2 QUANTIFICATION OF GHG EMISSIONS IN SEWERS

Even though GHG emissions from sewers have been long realised, current quantification methods do not account for the complexity of the sewer systems due to their geographical extension and the high variability of conditions that exist within these systems (changing flows, temperatures, turbulence, loads, etc.). Ideally, longitudinal GHG concentrations in sewer networks (both dissolved and in the headspace) should be assessed to capture the spatio-temporal variability of GHG production under different conditions. However, available methods are not sufficiently advanced and are only applied to monitor specific locations that have been identified as hotspots for GHG formation and emission.

GHG emissions from sewers depend highly on the configuration of the sewer sections. Operationally, sewer systems can be divided into two categories: (i) fully filled pressure sewers (rising main sewers), which are predominantly anaerobic, and (ii) partially filled gravity sewers, where re-aeration processes can take place (Figure 5.1). In sewers, microbial processes that lead to GHG production mainly take place in wetted biofilms and sediments, with little contribution from the suspended biomass in the water phase or in the gas phase (see Chapter 3). Thus, quantification should typically include measurements of both liquid and gas phases, combined with water and air flow measurements to close the balance between GHG produced and emitted.

CH_4 is the main GHG produced in sewer systems. For instance, in rising main pipes, CH_4 can be produced and accumulated even beyond saturation concentrations in the transported sewage and then released to the atmosphere at ventilated locations such as pumping stations, manholes or influent headworks of WWTPs. CH_4 produced in gravity sewers is usually released into the gas phase along the sewer pipe, with more intensive emissions at locations with higher sewage turbulence (Liu *et al.*, 2015a, 2016). These aspects need to be considered when identifying sampling points for a measuring campaign.

N_2O has also been detected in a few field-scale sewer sampling campaigns, which could contribute to the overall GHG inventory (Short *et al.*, 2014). The methods used for N_2O quantification are similar to those traditionally used for CH_4 detection. To date, little information exists regarding the role that sewers play in the production and emission of N_2O due to the low number of studies and limited monitoring of this compound in wastewater collection networks.

To date, the primary method for GHG measurement in sewers is by manual sampling at regular intervals over several hours followed by offline gas chromatography (GC) analysis (Foley *et al.*, 2011a; Guisasola *et al.*, 2008; Short *et al.*, 2014). This approach has several limitations as production and emission of CH_4 in sewers displays a significant temporal and spatial variation which is difficult to capture with this approach. Thus, continuous and extended online monitoring of CH_4 is recommended although the number of options is still limited, especially with regard to the measurements of dissolved CH_4 and N_2O. Although there is no fixed duration for sampling campaigns, their length should be sufficient to include variations in sewer conditions, which typically occur over several days to weeks. In sewers, there is a diurnal flow pattern with mostly quiescent conditions overnight due to lower flows followed by higher turbulence during the daytime due to higher flows. Ideally, samples should be taken across several time points of the day to include daily variations. In addition, it is also recommended to perform measurement campaigns in warm and cold seasons of the year as sewage temperature, among other parameters, can play a significant role in CH_4 production (Liu *et al.*, 2015b).

5.2.1 Quantification methods of CH_4 emissions in sewers

Due to the operational complexity of sewer systems and dynamic nature of CH_4 emissions it is impractical to estimate overall CH_4 emissions from large networks through either online or offline measurements. The large number of GHG forming and emission points makes full large network monitoring practically impossible. The most common monitoring approach consists of first identifying the main GHG hotspots in sewers and then carrying out individual measurements on those points. This approach assumes that the measurements will include the majority of emissions and will reduce the error of overall estimated emissions.

As stated above, the majority of the CH_4 is formed in rising mains and then completely stripped to the atmosphere via ventilation in manholes, gravity sewers or at WWTPs. This is also supported by the fact that biological CH_4 oxidation in gravity sewer conditions is expected to be a slow process (Valentine and Reeburgh, 2000). CH_4 estimations from rising main sewers are simpler and more accurate because CH_4 generated along the pipe will be released only in the upstream discharge point. Therefore, rising main data can be used to calculate the maximal potential overall CH_4 emission rates of a particular rising main section of a sewer. The CH_4 load in a rising main pipe can be calculated from the following equation:

$$M_{CH4} = C_{l,CH4} \cdot Q_d \tag{5.1}$$

where M_{CH4} is the mass of CH_4 potentially emitted to the atmosphere per time unit (g/d), $C_{l,CH4}$ is the dissolved CH_4 contained in the bulk liquid in mass/volume (g/m^3) and Q_d is the flowrate of wastewater in the rising main pipe (m^3/d).

Flow measurements are usually carried out by means of flowmeters or can also be estimated from the functioning regime of the pump stations upstream of rising mains. Dissolved CH_4 is calculated by applying the headspace method because of the lack of practical methods to directly measure the dissolved concentration in wastewater. A sewage sample is placed in a vacuumed partially-filled container where dissolved CH_4 is stripped from the liquid to the gas phase. Once under equilibrium, CH_4 gas can be measured and converted back to the liquid phase concentration using Henry's Law gas-liquid equilibrium and mass balance as described in the following equations:

$$C_{l,CH4} = \frac{(V_c - V_s) \cdot C_{g,eq} + V_S \cdot C_{l,eq}}{V_S} \tag{5.2}$$

$$C_{l,eq} = H \cdot R \cdot T \cdot C_{g,eq} \tag{5.3}$$

where $C_{l,CH4}$ is the dissolved methane concentration in the sewage sample (mol/L), V_S is the volume of the liquid sample (L), V_C is the volume of the sample container (L), $C_{g,eq}$ is the methane concentration in gas under equilibrium (mol/L), $C_{l,eq}$ is the methane concentration in water under equilibrium (mol/L), H is the Henry's Law constant (mol/L·atm), R is the ideal gas constant (0.0821 L atm/mol·K) and T is the temperature (K).

On the other hand, quantification in gravity sewers is complex and still highly impractical. Gravity sewers combine a liquid and a gas phase, which are highly dynamic since CH_4 can be formed and stripped heterogeneously over an extensive distance. A comprehensive analysis would require simultaneous long-term measurements in the gas and liquid phase combined with reliable data of wastewater and airflow in multiple locations of a network (Figure 5.1). Due to this constraint, studies to date have focused on the quantification of CH_4 emissions by direct measurement of CH_4 gas flux from single discharge manholes (Willis *et al.*, 2011). However, this methodology is expected to underestimate emissions as CH_4 could also be emitted at several other locations in the network.

5.2.2 Measurement of CH_4 in the liquid phase

Dissolved CH_4 sampling in fully-filled rising main sewers is mainly carried out through tappings connecting a sampling tap at ground level to the tapping arrangement of the underground pipe. Wastewater samples are collected from the pipe using a hypodermic needle and plastic syringe to prevent exposure of sampled wastewater to the atmosphere and oxygen, as shown in Figure 5.2 below (Foley *et al.*, 2009). Dissolved CH_4 is then measured and calculated by applying the headspace method for GC using Henry's Law and mass balance as described in Section 5.2.1.

For sampling dissolved CH_4 in gravity sewers, manholes, wet wells and pumping stations, wastewater samples are usually collected with a sampling device consisting of an open-head cylindrical container which is lowered and filled below the water level, and then gently retrieved. Within the container, sample aliquots are extracted with a plastic syringe from ca. 5 cm below the water surface to avoid contact with air (Foley *et al.*, 2011b). Alternatively, a submersible pump can be used to collect a sample from below-ground at low speed in order to avoid turbulence. Sub-samples are subsequently extracted into evacuated Exetainer® tubes (Labco, Wycombe, UK) or a pre-treated serum bottle (Daelman *et al.*, 2012). The contents of the tube or bottle are mixed overnight to reach gas–liquid equilibrium. CH_4 concentration in the headspace is again measured by GC, and the dissolved CH_4 concentration of the sample is then calculated using Henry's Law and mass balance. A more accurate method using evacuated Exetainer® tubes for both gas and liquid phase CH_4 sampling and measurement was proposed

Figure 5.2 Collection of dissolved CH_4 sample directly from the rising main into an airtight syringe, adapted from Foley *et al.* (2009). Reprinted from Water Science & Technology, volume 60, issue number 11, pages 2963–2971, with permission from the copyright holders, IWA Publishing.

by Sturm *et al.* (2014) which uses nitrogen gas to thoroughly flush the tubes before evacuating and sampling, to minimize the residual CH_4 present in the Exetainer® tubes.

A limited number of commercial sensors are available for online, dissolved CH_4 measurement (Boulart *et al.*, 2010; Camilli and Hemond, 2004). However, these are mainly designed for measuring CH_4 in clean water, using gas-permeable membranes to extract CH_4 gas from water, and cannot be used in sewage containing a large amount of impurities as well as high sulfide concentrations (Boulart *et al.*, 2010). Liu and co-authors (2015b) developed an online, dissolved CH_4 sensor that uses an online gas phase CH_4 sensor to measure CH_4 under equilibrium conditions after stripping from the sewage. The data is then converted to liquid phase, dissolved CH_4 concentrations according to Henry's Law. The detection limit (ca. 0.24 mg/L) and range (ca. 0–24.2 mg/L) are both suitable for sewer application, and can be adjusted by varying the ratio of liquid-to-gas phase volume settings according to specific applications, i.e., at a ratio of 4, a resolution of 0.09 mg/L can be achieved at the expense of a reduced measurement range of 0 to 9.3 mg/L. The sensor demonstrated good performance over a six-week period when positioned at the end of a rising main sewer network (Figure 5.3).

5.2.3 Measurement of CH_4 in the gas phase

Several online sensors for gas phase CH_4 monitoring are available but most are not applicable in sewer conditions due to interference from hydrogen sulfide which is simultaneously produced and emitted from sewers (Deng *et al.*, 1993; Schierbaum *et al.*, 1992). Infrared (IR) spectroscopy is the most promising method for online CH_4 measurement in sewer conditions (Foley *et al.*, 2011b). Particular sampling arrangements are required for measuring gas phase CH_4 concentrations. Gas may be sampled from a ventilation point (Shah *et al.*, 2011) or from a purpose-built sampling chamber connected to the sewer headspace (Liu *et al.*, 2015b). For grab sampling, gas bags or evacuated Exetainer® tubes can be used. The gas samples can be then analysed using GC equipped with a flame ionization detector (FID).

A key feature of sewer air is the high humidity, typically in the range 80–100% RH (relative humidity) (Joseph *et al.*, 2012), which could potentially interfere with IR CH_4 measurement (You-Wen *et al.*, 2011). Liu *et al.* (2015b) evaluated the suitability of IR spectroscopy-based online sensors

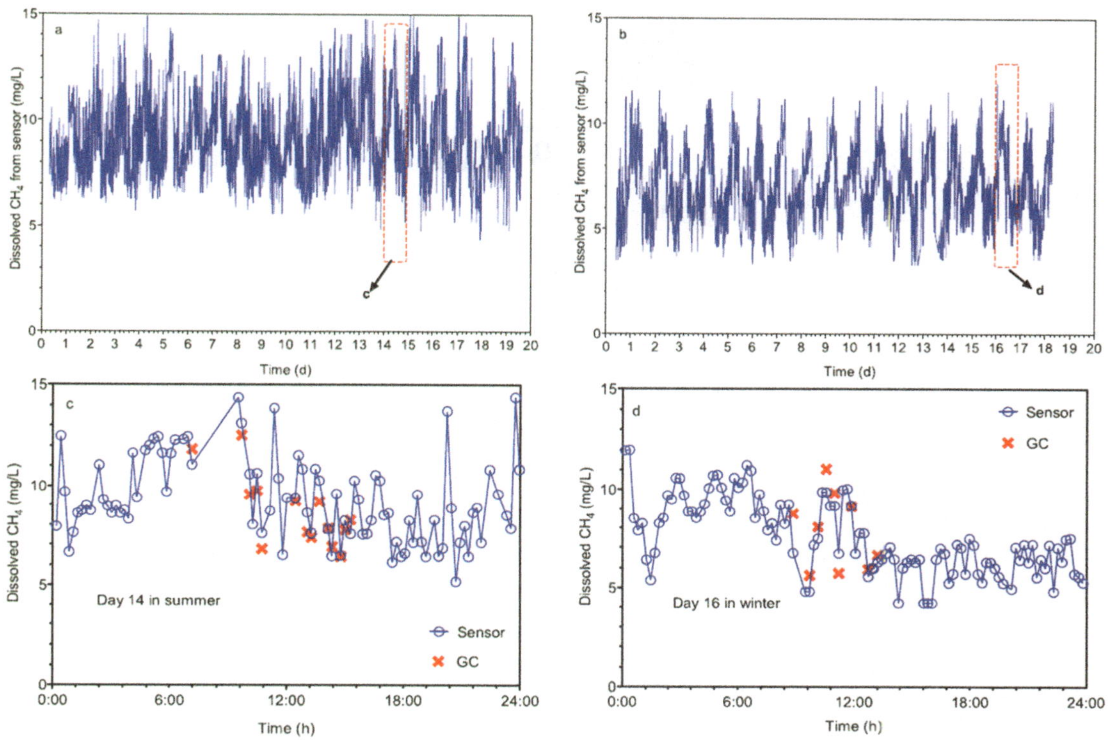

Figure 5.3 Three-week field CH_4 measurement with the online CH_4 sensor at the end of a rising main sewer network at Gold Coast: (a) summer; (b) winter. The agreement between the sensor measured results and those obtained through manual sampling and offline GC measured results is shown in (c) and (d). This figure was published in Water Researchearch, Vol number 68, Y. Liu, Keshab R. Sharma, M. Fluggen, K. O'Halloran, S. Murthy, Z. Yuan, Online dissolved methane and total dissolved sulfide measurement in sewers, Page Nos 109–118, Copyright Elsevier (2015).

for measuring CH_4 gas in humid and condensing sewer air. An IR sensor with external power supply was extremely robust in variable and high humidity. A battery-operated IR sensor was sensitive to changes in humidity, but the problem was resolved by maintaining the humidity on the sensor probe surface at 50–70% RH through increasing surface temperature or refrigeration (Figure 5.4). Both sensors exhibited excellent linearity and can be applied with factory calibration. The detection limit of sensors i.e., ca. 0.023–0.110% vol, corresponds to a dissolved CH_4 range of 0.005 to 0.026 mg/L under equilibrium conditions at 20°C and 1 atm, which was suitable for measuring CH_4 gas in sewers. In-sewer application (with external power supply) for nearly one month confirmed accuracy and longevity of the sensor. In the future, infrared spectroscopy will be a powerful tool for accurate quantification of CH_4 emission from sewers.

Another system proposed by Kim *et al.* (2009) consists of an innovative and fully automated sewer gas monitoring device based on a floating and drifting embedded sensor platform (Sewer Snort). This sensor float can be introduced upstream and drift to the end of the network, collecting location-tagged gas measurements, thus providing a gas concentration profile along the sewer line. However, to date, the experiments have been based on a dry land emulator, and verification in actual sewers is needed before field application.

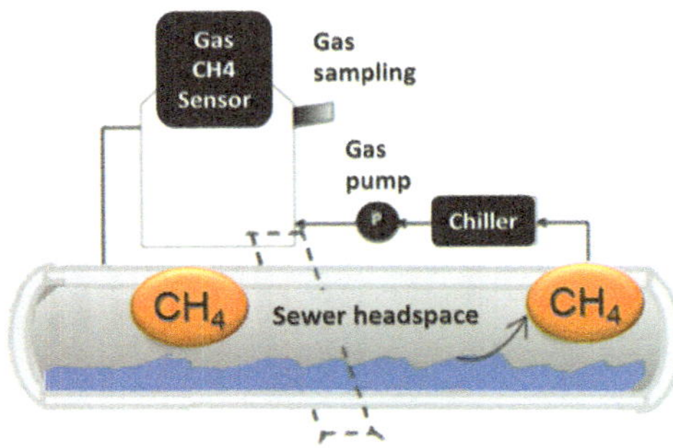

Figure 5.4 A purpose-built device for gas sampling or infrared (IR) gas sensor application in sewer headspace: a gas pump continuously recycles the gas from the sewer headspace to the chamber and then back to the sewer. A chiller is used in the gas line feeding the chamber to maintain the desired level of 50–70% relative humidity (RH) for the IR sensor (adapted from Liu *et al.*, 2015a).

5.2.4 Recommended measurement practice

Direct measurements of GHG in sewers can only be carried out in specific location-sections (such as rising main pipes, gravity pipes, manholes or pumping stations) due to the limited tools available to date. Those individual sections of sewers make up only a small part of a much larger network and hence more GHG production is expected when the wastewater is transported through the remaining parts of the network before reaching the WWTP (Pikaar *et al.*, 2014). Due to the operational complexity of sewer systems and dynamic nature of CH_4 emissions, it is not recommended to estimate overall CH_4 emissions from large networks with online or offline measurements. However, the combination of measurements in selected hotspots (rising mains, for instance) with mathematical modelling of GHG production is a viable solution to obtain estimations of full-network emissions (Willis *et al.*, 2019). See Chapter 8 for further information.

Sampling campaigns in those selected spots should encapsulate diurnal flow variations (Figure 5.3) with samples taken from the whole range of HRTs and also, should be carried out in different seasons of the year to include differences due to temperature.

Also, there is a need to develop GHG monitoring equipment able to work in harsh sewer conditions. These conditions limit the capacity to carry out more comprehensive sampling campaigns, having to rely on assumptions that need to be always carefully taken.

5.3 QUANTIFICATION OF GHG EMISSIONS IN WASTEWATER TREATMENT PLANTS

Over the past two decades there have been intensive efforts to quantify and investigate GHG emissions from WWTPs. The majority of the measurement campaigns were research related and their objectives varied from quantifying and understanding potential emissions under different WWTP conditions (e.g., Ahn *et al.*, 2010; Daelman *et al.*, 2013; Foley *et al.*, 2010) to mechanistic modelling of GHG production and emission from full-scale WWTPs (e.g., for N_2O emissions, Guo and Vanrolleghem, 2014; Ni *et al.*, 2013).

Although, the floating hood method is the most frequently applied measurement method to date, specific local requirements and measurement objectives have led to the development and application of alternative measurement approaches for full-scale GHG quantification. In general, quantifying methods can be classified into plant-wide and process-unit measurement approaches. Plant-wide quantification enables the determination of the overall GHG emissions of the plant including sources that might be difficult to investigate (accessibility) or might be missed by process-unit methods (unknown sources such as e.g., biogas leakages). However, the contribution from each single emission source to the overall emission cannot be differentiated. In contrast, the process-unit approach identifies and quantifies single GHG emission sources, allowing a deeper understanding of the mechanisms of GHG production and emission patterns at the plant. This information is essential not only for research and modelling purposes, but also required for the development of mitigation measures for GHG emissions at WWTPs.

The aim of the chapter is to provide researchers and practitioners a general overview on the methodologies currently available for the quantification of N_2O and CH_4 emissions at full-scale WWTPs, highlighting the field of applicability, instrumental requirements, and strengths and limitations of those methods that have been already successfully applied. Methods in the development stage are not presented. Due to the focus of the chapter being on quantification methodology, analytical methods for the detection of the GHGs N_2O and CH_4 will only be briefly described. References to analytical methods will be provided for more information.

5.3.1 Plant-wide quantification of N_2O and CH_4 emissions

N_2O and CH_4 can be emitted from almost all stages of the wastewater and sewage sludge treatment (please refer to Chapters 2 and 3). GHG emissions occur from several small sources located in a large area, have different shapes (e.g., small leaks from biogas holding units and large liquid surfaces from biological reactors), and take place at different heights. These emission features result in a complex diffusive and fugitive emission pattern (Delre *et al.*, 2017). The emission pattern is diffusive because emissions are scattered throughout the WWTP, and it is fugitive, because gases escape unintendedly from process units (Delre, 2018). The complexity of the GHG emission pattern from WWTPs is increased by operational conditions that produce different emissions over time (Delre *et al.*, 2017; Yoshida *et al.*, 2014).

The literature offers several methods that allow the quantification of emission rates from area sources: mobile tracer gas dispersion method (MTDM), inverse dispersion modelling method (IDMM), solar occultation flux (SOF), differential absorption light detecting and ranging (DIAL), and radial plume mapping (RPM) (Mikel and Merrill, 2011). All these methods calculate the emission rate of the target gas through two main steps: (1) describing the plume generated by the target area, and (2) defining the atmospheric dispersion that the target gas undergoes travelling downwind from the target area. The plume is described by measuring downwind atmospheric gas concentrations from the ground. For this reason, these methods are called ground-based remote sensing methods. In the majority of the cases, the atmospheric dispersion of the target gas is defined by using local atmospheric models (e.g., backward Lagrangian stochastic model used in the IDMM). Only in the case of the MTDM, is the atmospheric dispersion of the target gas obtained by releasing a tracer gas from the target area, without deploying any atmospheric model.

Among the ground-based remote sensing methods, only the MTDM was implemented for quantifying N_2O and CH_4 emissions from WWTPs. The MTDM was applied at eight WWTPs with different plant layouts, using different process units and technologies. Investigated WWTPs were located in Denmark, Sweden and France (Delre *et al.*, 2017; Samuelsson *et al.*, 2018; Yoshida *et al.*, 2014; Yver Kvok *et al.*, 2015). Although the use of the other ground-based remote sensing methods is potentially possible, the literature still lacks in applications of these methods at WWTPs. The IDMM was used for quantification of CH_4 emissions from biogas plants, while SOF, DIAL and RPM where applied for CH_4 emission quantifications from industrial sites and landfills (Mikel and Merrill, 2011).

All ground-based remote sensing methods are highly dependent on the analytical technology used, because it affects the measurable type of target gas and the quality of the atmospheric plume description at a proper distance from the emitting area. Any ground-based remote sensing method can only be successfully deployed if the analytical technology is capable of distinguishing properly, at a suitable distance from the emitting area, the atmospheric plume concentrations from the background values.

In the following section, the application of the MTDM at WWTPs is described. Detailed information about other ground-based remote sensing methods can be found in Mikel and Merrill (2011).

5.3.1.1 Mobile tracer gas dispersion method (MTDM)

The MTDM uses a controlled release of a tracer gas with measurements of atmospheric gas concentrations taken downwind of the target area. Additionally, the MTDM benefits from the features of gases with long atmospheric lifetimes to keep a constant concentration ratio during transportation and mixing in the atmosphere (Lamb *et al.*, 1995; Stiversten, 1983). Thus, when the tracer gas is released at a constant rate from the emitting area, the target gas emission rate can be calculated in real-time by relating the measured plume traverse concentrations of the target and tracer gases, as shown in Equation (5.4).

$$M_{tg} = Q_{tr} \cdot \frac{\int_{\text{plume start}}^{\text{plume end}} \left(C_{tg} - C_{tg\ \text{baseline}}\right) dx}{\int_{\text{plume start}}^{\text{plume end}} \left(C_{tr} - C_{tr\ \text{baseline}}\right) dx} \cdot \frac{MW_{tg}}{MW_{tr}} \tag{5.4}$$

where M_{tg} is the target gas emission in mass per time, Q_{tr} is the known tracer release in mass per time, C_{tg} and C_{tr} are the detected plume concentrations within the plume traverse in parts per billion (ppb), $C_{tg\ \text{baseline}}$ and $C_{tr\ \text{baseline}}$ are baseline concentrations of the target and the tracer gas (ppb), and MW_{tg} and MW_{tr} are the molecular weights of the target gas and tracer gas, respectively (Scheutz *et al.*, 2011). Acetylene (C_2H_2) is usually used as a tracer gas, due to there being very few possible interfering sources and its long atmospheric lifetime (Delre *et al.*, 2017).

The mobile measurement platform used in most of the studies was a vehicle equipped with two gas analysers for measurements of atmospheric gas concentrations and a global navigation satellite system for recording measurement locations. Each of the gas analysers detected target and tracer gases simultaneously. Atmospheric gas was sampled from the roof of the vehicle and analysed, while screens displayed detected concentrations in real time. The tracer gas was constantly released using gas cylinders with calibrated flowmeters.

Figure 5.5 shows the key phases of a measurement campaign, which consists of a screening phase, carried out off-site and on-site, and a quantification phase. The screening phase starts outside the facility, to guarantee the absence of off-site sources that could interfere with the target and the tracer gas. Later, screening inside the facility allows identification of on-site emitting sources. The on-site screening (Figure 5.5 A) allows the right placement of the tracer gas cylinders (Figure 5.5 B1), so that the target gas emission pattern is properly simulated by the tracer release. During the quantification phase, tracer gas is constantly released (Figure 5.5 B2) while the plume is crossed multiple times at a suitable distance away from the emitting source (Figure 5.5 C). The measuring distance should guarantee enough mixing between target and tracer gas and produce a proper signal-to-noise ratio in the concentration of gases recorded along the plume traverses. Proper simulation of the target gas emission is continuously checked through a good correlation between target and tracer gases within a plume traverse.

The success of the MTDM relies on mutual dependence among the following factors: (1) features of the analytical instrument, (2) size of the emitting source, (3) emission rate of the target gas, (4) atmospheric stability and target gas dispersion, (5) measurement distance from the emitting source, and (6) simulation of the target gas emission pattern.

A suitable analytical instrument should have good precision and high detection frequency when measuring concentrations of target and tracer gases. Such features allow a better plume definition and a faster measurement execution, resulting in smaller measurement uncertainties and lower method application costs (Delre *et al.*, 2018). An analytical instrument with a good precision is especially relevant when emissions are to be quantified from large area sources. In this situation, measurements must be carried out at a long distance from the emitting area to obtain proper mixing between target and tracer gases at the measuring location. At long distances, atmospheric gas dispersion produces,

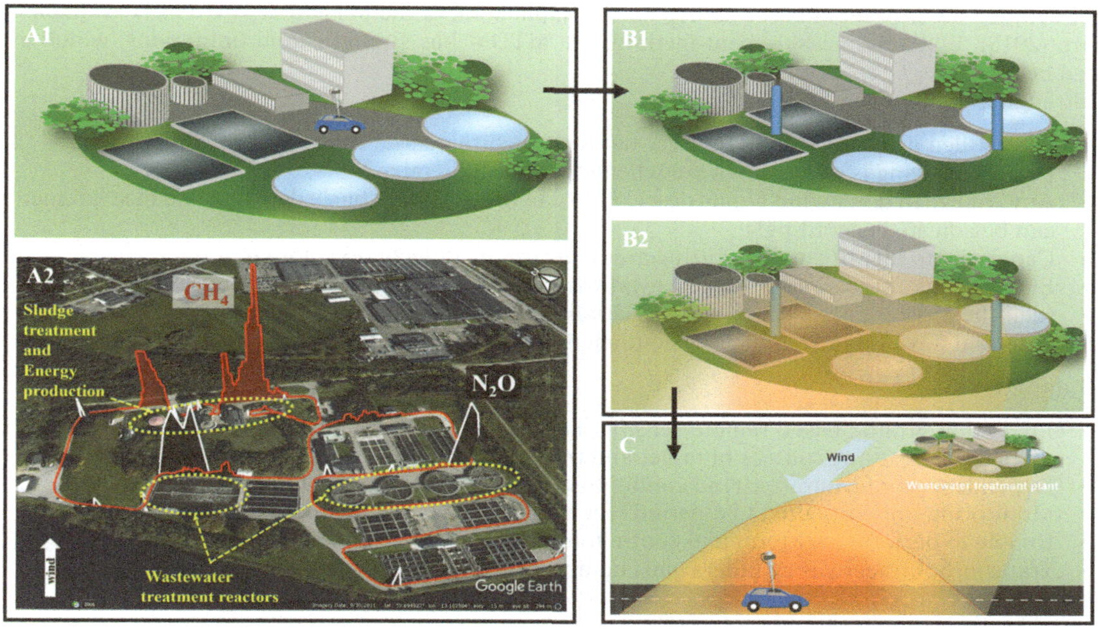

Figure 5.5 Illustration of the tracer gas dispersion method applied at wastewater treatment plants. (a) The initial screening phase with A1 showing on-site measurements of atmospheric concentrations of target and tracer gases and A2 showing an example of on-site screenings performed at Källby (SE) visualized on a Google Earth © image. CH_4 (marked in red) and N_2O (marked in white) concentrations are shown above the background level. The white arrow shows the wind direction. (b) Tracer placement with B1 showing the location of the tracer gas for source simulation and B2 showing the release of the tracer gas into the atmosphere. (c) The quantification phase showing downwind gas concentrations measurement performed along a plume transect. This figure was published in in Science of Total Environment, Vol number 605–606, Delre A., Mønster J., Scheutz C., Greenhouse gas emission quantification from wastewater treatment plants, using a tracer gas dispersion method, Page Nos 258–268, Copyright Elsevier (2017).

within the same plume traverse, small differences between gas concentrations detected inside and outside the plume. A precision of 0.7, 3.8 and 0.5 ppb when measuring N_2O, CH_4 and C_2H_2, respectively, was found to be sufficient for measuring at WWTPs (Delre, 2018). Reported values of instrument precision are defined as three times the standard deviation of six minutes' constant concentration reading (Delre *et al.*, 2018). The magnitude of the target gas emission rate also influences the success of the quantification, because if the emission rate is too low, the plume cannot be distinguished from the background at a proper measuring distance. In this case, a detection limit can be estimated (Delre *et al.*, 2017). Stable weather conditions produce a lower vertical atmospheric dispersion of gases compared to unstable situations. Thus, stable conditions are usually preferred because this allows better plume definition within a plume traverse. This is mainly relevant for the target gas rather than the tracer gas, because the downwind signal-to-noise ratio of the tracer gas can be improved by increasing the flow rate of the release. Correct tracer placement and consequent proper simulation of the target gas emissions is central when applying the MTDM (Delre *et al.*, 2018; Mønster *et al.*, 2014). As long as the employed analytical instrument can detect the tracer gas, any long-lived atmospheric gas can be used as tracer gas in MTDM application (Delre *et al.*, 2018). However, this statement does not consider price and environmental issues, which could be important constraints in the choice of the tracer gas.

A detailed description of best practice for the application of the MTDM at WWTPs is available in Delre (2018). Like any measurement method, the MTDM has strengths and limitations, which are listed below.

Strengths:

- One skilled operator alone can carry out the measurements;
- Data processing is straightforward when gases are fully mixed;
- Any change in downwind plume description can be instantaneously detected and measurements can be adjusted accordingly;
- Capability to identify main emitting areas, especially if occurring close to ground level;
- Emission quantification is possible even without locating specific on-site emitting sources;
- Identification of possible emission variation within the measurement campaign;
- Potential flexibility of moving the equipment around using different means of transportation.

Limitations:

- Like other ground-based remote sensing methods, emission quantifications are not possible if there are interfering sources of target and tracer gases upwind of the target area;
- Dependence on favourable wind conditions combined with road access;
- Monitoring time is limited to a period when wind blows with favourable conditions;
- Unable to perform long term and continuous monitoring;
- Transport of tracer gas cylinders must comply with specific regulations.

5.3.2 Process-unit quantification of N_2O and CH_4 emissions

Process-unit GHG quantification methods are designed to measure emissions from single process units as opposed to a large whole-of-plant footprint. Plant-wide quantification of emissions is however achievable with these methods, provided that all individual emission sources are quantified separately and aggregated. Process-unit quantification is essential when process specific emissions need to be characterized, for example, for calibration of mechanistic models or to link emissions to the operating conditions of the plant. The development of mitigation measures for GHG emission also requires the identification and quantification of each single source at the plant.

The first N_2O monitoring campaigns at WWTPs were based on grab-sampling methods but due to the large temporal fluctuations in emissions occurring in most cases, the continuous online monitoring methodology is favoured. The floating hood method is the most common approach – among others – to sample the off-gas leaving the surface of activated sludge tanks with bubble aeration. The off-gas stream captured by the hood is usually fed to an online gas analyser for quantification of N_2O or CH_4. This method has also been employed in biofilm-based reactors (Bollon *et al.*, 2016a, 2016b; Gruber *et al.*, 2020; Vieira *et al.*, 2019; Wang *et al.*, 2016). In activated sludge tanks equipped with surface aerators the liquid-to-gas mass transfer method needs to be adopted, which allows N_2O or CH_4 emissions to be calculated from liquid phase measurements. In covered tanks, continuous or discontinuous off-gas sampling and analyses can be performed directly from the ventilation system. Furthermore, ground-based remote sensing methods (e.g., tracer gas dispersion method) can also be applied for process-unit GHG quantification.

All these measurement approaches are described in Sections 5.3.2.1 to 5.3.2.4. A summary of these methods is given in Section 5.3.3. Finally, Section 5.3.4 provides general recommendations for minimum data requirements.

5.3.2.1 Floating hood

The WERF protocol (Chandran, 2009) gives guidance on planning and performing an N_2O measuring campaign based on the floating hood method. It was completed by Chandran *et al.* (2016) notably to highlight the major considerations for carrying out representative sampling of off-gas emissions. As measuring campaigns have been carried out in previous years by different research groups around the world, additional know-how has been gained that has led to alternative sampling hood designs with adapted flux calculation, novel monitoring approaches in gas and liquid phase, as well as a better understanding of relevant requirements of sampling procedures (e.g., tubing, hood placement, etc.). This section summarizes (with no claim to completeness) the most common options.

Emissions from open surface process units were, most of the time, monitored using floating hoods (also called floating chambers, isolation flux chambers, gas-collecting chambers or closed chambers). They are floating devices that are maintained at a given process-unit position, usually using ropes, to collect and sample the gases emitted at the water-air interface. In some cases, they were also designed to measure the off-gas flowrate. Such devices are floating versions of hoods that are used for measuring emissions from soils or landfill. The first floating hoods were employed to measure volatile organic compounds from wastewater treatment plants (Tata *et al.*, 2003) using a surface emission isolation flux chamber (SEIFC). The SEIFC hood is one of the few devices approved by the United States Environmental Protection Agency (U.S. EPA) and is used in U.S. EPA method EPA/600/8-86/008 (1986) to measure gaseous emission rates from land surfaces. Chandran (2009, 2011) adapted this method for measuring N_2O emissions from biological nutrient removal (BNR) plants, which has resulted in a comprehensive field measurement protocol certified by the U.S. EPA (Chandran, 2009). Although the protocol was specifically developed for SEIFC hoods, many researchers successfully applied the guideline to alternative methods for measuring N_2O and CH_4 emissions from different process units of WWTPs including those operated with advective gas flow (aerated units) and those having a passive liquid surface (non-aerated units). Application examples are provided in Tables 5.1 and 5.2.

While some research teams used commercial hoods (such as SEIFC and AC'SCENT® Flux Hood), many others used custom-built floating hoods. This resulted in a great variety in hood shape (cuboid, half-spherical, cylindrical, etc.), material (stainless steel, aluminium, wood, plastic) and size (surface area covered ranging from 0.03 m^2 to 2 m^2). In most cases, hoods are submersed (by a few centimetres) to prevent lateral movement and introduction of external air. This can be achieved by placing the floating system (e.g., polystyrene float or inner tube) above the bottom of the hood. Since atmospheric carbon dioxide (CO_2) concentration in the off-gas is much higher than that of the atmosphere, its measure in the off-gas can be used to check that no external air enters the measuring loop (Valkova *et al.*, 2020). Despite the variety in hood design, they can be classified into two categories: closed flux chambers and open flux chambers. Figure 5.6 presents the basic scheme of the most common configurations of open and closed flux chambers.

5.3.2.1.1 Closed flux chambers

The basic principle of closed flux chambers is to isolate a given surface area from the atmosphere, thus allowing for the accumulation of the gas inside the hood over time. The emission rate is then determined by the change in gas concentration over time. Gas mixing is usually achieved by installing a fan inside the hood or by recirculating the gas flow between the hood and the GHG analyser (Figure 5.6a and b). If the chamber is operated without a gas flow (without recirculation) it could be referred to as a "static chamber", otherwise, it is a "dynamic chamber". This technique was originally developed to measure gas emissions from natural soils where surface emissions are controlled by diffusion (Mønster *et al.*, 2019). Likewise, it was successively applied to measure GHG emissions from non-aerated unit processes of WWTPs, such as equalization tanks (Masuda *et al.*, 2015), anoxic and anaerobic tanks (Mello *et al.*, 2013; Ren *et al.*, 2013; Wang *et al.*, 2011; Yan *et al.*, 2014), primary and secondary settlers (Caniani *et al.*, 2019; Czepiel *et al.*, 1993, 1995; Masuda *et al.*, 2015; Ren *et al.*, 2013; Wang *et al.*, 2011; Yan *et al.*, 2014) and sludge storage tanks (Oshita *et al.*, 2014; Ren *et al.*, 2013). Build-up of high concentrations in the hood is not recommended as it may reduce the emission rate during the course of the experiment which would result in underestimating the actual emission rate

Table 5.1 Examples of floating hoods employed in advective flow conditions.

Reference	Process unit	GHG	Method	Floating hood					Gas line	
				Type/shape	Material	Area (m^2)	Sweep/tracer gas	Mixing	$Q_{g,hood}$	Parameters monitored
Benckiser *et al.* (1996); Sümer *et al.* (1995)	AS	N_2O	OFC	Cuboid	PVC	*0.24*		Perforated plates	Measured	
Kimochi *et al.* (1998)	AS	N_2O	OFC	Cuboid					Q reactor	
Ahn *et al.* (2010); Chandran (2011)	AS	N_2O	OFC	1		0.13	He		He mass balance	P, T
Desloover *et al.* (2011)	PN, AS	N_2O, CH_4	OFC	Lindvall hood, Cuboid	Al	0.864	Air	–	Measured	T
Aboobakar *et al.* (2013a); Aboobakar *et al.* (2013b)	AS	N_2O, CH_4	OFC	2		0.34			Q reactor	
Ren *et al.* (2013)	AGC, AS	N_2O, CH_4	CFC	1	SST	0.13		Gas circulation	–	
Mello *et al.* (2013)	AS	N_2O	OFC	Funnel attached to a pipe	Plastic	0.071			Q reactor	
Rodriguez-Caballero *et al.* (2015); Rodriguez-Caballero *et al.* (2014)	AS	N_2O, CH_4	OFC	3	SST	0.13			Q reactor	O_2 conc.
Brotto *et al.* (2015)	AS	N_2O	OFC	Funnel attached to a pipe	PVC	0.071	No		Q reactor	
Masuda *et al.* (2015)	AS	N_2O, CH_4	OFC		PVC	0.196			Q reactor	
Marques *et al.* (2016)	AS	N_2O	OFC	3	SST	0.13			Q reactor	
Bollon *et al.* (2016a); Bollon *et al.* (2016b)	BAF	N_2O	OFC	Cuboid	Wood	1.6	No	No	Measured	
Duan *et al.* (2020); Pan *et al.* (2016a)	AS	N_2O	OFC	Modified hopper tank	Plastic	0.22	No	No	Measured	P, T
Yang *et al.* (2016)	PN/A	N_2O	OFC	Cuboid		0.81		Fan	Q reactor	

Wang *et al.* (2016)	BAF	N_2O	OFC	Half spherical		0.0314	No	Fan	Not clear	T, liquid level
Bellandi *et al.* (2017)	AS	N_2O	OFC	Different chambers		0.35–2			Measured or estimated	
Ribeiro *et al.* (2017)	AS	N_2O	OFC		PVC	0.05			Measured	
Spinelli *et al.* (2018)	AS	N_2O	OFC	Different chambers	HDPE, PP	0.125–0.457			Q reactor	
Caniani *et al.* (2019)	AS	N_2O	OFC	Complex shape	SST	0.7			Measured	
Vieira *et al.* (2019)	BAF	N_2O	OFC	3					Q reactor	
Tauber *et al.* (2019)	AD	CH_4	OFC/ CFC	Lindvall hood/Cuboid	Wood	1	Air/No	No/Gas circulation	Measured	P, T
Gruber *et al.* (2020)	AS, BAF	N_2O	OFC	Triangular top and cuboid body	PE	1			Q reactor	

The italics indicate the elements calculated from information provided in the scientific paper or manufacturer's technical documentation.1: Commercial "SEIFC", half circular top and cylindrical body, 2: Commercial hood provided with the N-Tox nitrification toxicity monitoring system (Water Innovate, UK), 3: Commercial "AC'SCENT® Flux Hood", half circular top and cylindrical body.

AD: anaerobic digester; AGC: aerated grit chamber; Al: aluminium; AS: activated sludge; BAF: biologically active (or aerated) filter; CFC: closed flux chamber; GC: grit chamber; HDPE: high density polyethylene; He: helium; OFC: open flux chamber; PN: partial nitritation; PN/A: partial nitritation/anammox; P: pressure; PP: polypropylene; PVC: polyvinyl chloride; SST: stainless steel; T: temperature.

Table 5.2 Examples of floating hoods employed in passive liquid surface conditions.

Reference	Process unit	GHG	Method	Floating hood					Gas line	
				Type/shape	Material	Area (m^2)	Sweep/ tracer gas	Mixing	$Q_{g, hood}$	Parameters monitored
Czepiel *et al.* (1995); Czepiel *et al.* (1993)	SET	N_2O	CFC	Cuboid	Al			Gas circulation	–	T
Kimochi *et al.* (1998)	AS	N_2O	OFC	Cuboid			Ar		Mass balance	
Ahn *et al.* (2010); Chandran (2011)	AS	N_2O	OFC	1		0.13	He, Air	Gas circulation	Mass balance	P, T
Desloover *et al.* (2011)	Anammox, AS	N_2O, CH_4	OFC	Lindvall hood / Cuboid	Al	0.864	Air	–	Measured	T
Ren *et al.* (2013); Wang *et al.* (2011)	IPS, SET, AS, SCT	CH_4	CFC	1 (probably)	SST	0.13		Gas circulation	–	T
Mello *et al.* (2013)	AS	N_2O	CFC	Custom-made	PVC	*0.045*			–	
Rodriguez-Caballero *et al.* (2015); Rodriguez-Caballero *et al.* (2014)	AS	N_2O, CH_4	OFC	2	SST	0.13	Air		Mass balance on oxygen	
Oshita *et al.* (2014)	SCT	N_2O, CH_4	CFC	Cylindrical	PVC	0.15			–	T
Yan *et al.* (2014)	SET, AS	N_2O, CH_4	CFC	Cylindrical	SST	*0.14*		Gas circulation		T
Mikola *et al.* (2014)	SET, AS	N_2O	OFC	Truncated cone		*0.15*			=flow of the analyser	
Masuda *et al.* (2015)	ET, DT, SET	N_2O, CH_4	CFC		PVC	0.16			–	
Marques *et al.* (2016)	AS	N_2O	OFC	2					=flow of the analyser	
Bellandi *et al.* (2017)	AS	N_2O	OFC	Modified Lindvall hood			Air		=flow of the analyser	
Duan *et al.* (2020)	AS	N_2O	OFC	Modified hopper tank	Plastic	0.22			Measured	P, T
Caniani *et al.* (2019)	SET	N_2O	CFC	Complex shape	SST	0.7	Air	Gas circulation	Measured	

The italics indicate the elements calculated from information provided in the scientific paper or manufacturer's technical documentation.1: Commercial "SEIFC", half circular top and cylindrical body, 2: Commercial "AC'SCENT® Flux Hood", half circular top and cylindrical body.
Al: aluminium; Ar: argon; AS: activated sludge; CFC: closed flux chamber; DT: disinfection tank; ET: equalization tank; He: helium; IPS: influent pump station; OFC: open flux chamber; P: pressure; PVC: polyvinyl chloride; SCT: sludge concentration tank; SET: settler; SST: stainless steel; T: temperature.

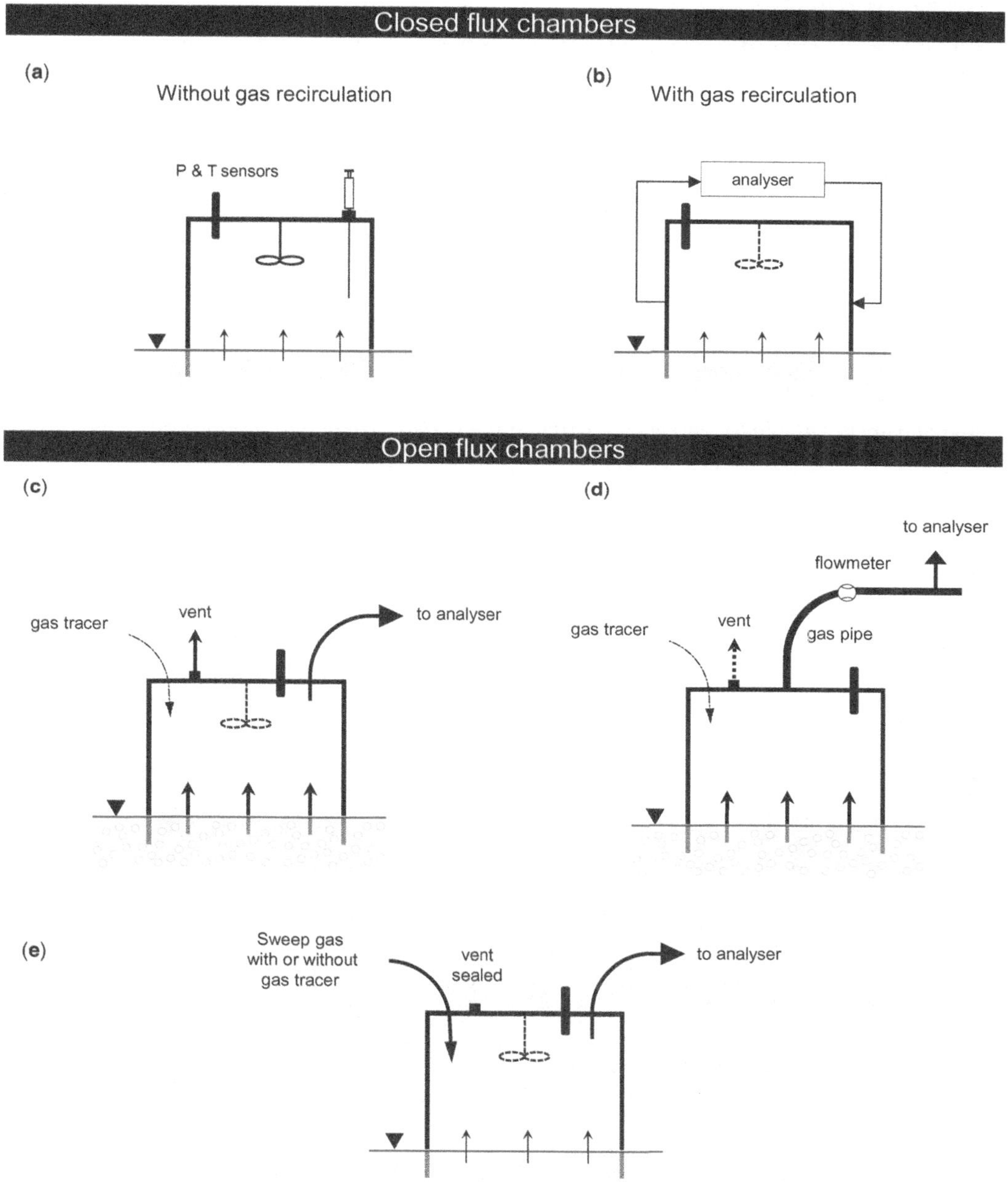

Figure 5.6 Most common hood configurations. Dotted elements are optional.

(Gao and Yates, 1998). To avoid it, short sampling times were applied (generally less than half an hour) and fresh air was introduced in the hood in between two sampling rounds. In most cases, a manual sampling was performed but it is possible to design fully automated devices (Filali *et al.*, 2017; Oshita *et al.*, 2014; Pavelka *et al.*, 2018).

5.3.2.1.2 Open flux chambers

The open flux chambers (also referred as "dynamic chambers") are fitted with tubes and vent ports allowing headspace gas to escape from the hood and/or the introduction of a sweep gas into the hood (Figure 5.6c–e and a photo of an example in Figure 5.7). In aerated zones or tanks, the off-gas sample to be analysed can be directly extracted from the hood, usually at a constant flowrate that is much lower than the gas flowrate entering the hood to avoid negative pressure built-up (Figure 5.6c). The excess gas is exhausted from the hood through vents. The SEIFC and AC'SCENT® flux hoods are designed on this principle. The alternative configuration (Figure 5.6d) would be to let the off-gas escape through a large pipe and direct a small portion of the off-gas to the analyser. The dimensions of the pipe are also very variable ranging from 25 mm to 100 mm (Bollon *et al.*, 2016a, 2016b; Duan *et al.*, 2020; Gruber *et al.*, 2020; Pan *et al.*, 2016a; Spinelli *et al.*, 2018). This configuration is convenient because it allows, with some caution, measurement of the off-gas flow rate.

In non-aerated zones or tanks (Figure 5.6e), a flow of sweep gas can be applied to enhance an effective gas flow through the flux chamber (Ahn *et al.*, 2010; Caniani *et al.*, 2019; Chandran, 2011; Kimochi *et al.*, 1998; Rodriguez-Caballero *et al.*, 2014, 2015). As sweep flow rate was evidenced to influence the estimated emission rate of several compounds in dynamic flux chambers (Gao and Yates, 1998; Prata *et al.*, 2018), some researchers applied a wind-tunnel-type, namely the Lindvall hood (Lindvall *et al.*, 1974), that allows better control of gas velocities inside the hood. In this hood, the sweep gas (usually ambient air) is introduced in a directional way to simulate the action of the wind on the sampled surface (Capelli *et al.*, 2013). It was applied by Desloover *et al.* (2011) to measure N_2O and CH_4 emissions from a full-scale partial nitritation and anammox process. Bellandi *et al.* (2017) applied a modified version to measure N_2O from anoxic zones of two activated sludge plants.

Figure 5.7 Floating hood (configurations C and E in Figure 5.6) and associated equipment for measuring off-gas flowrates and off-gas concentrations. Photos courtesy of Dr. Maite Pijuan (ICRA Catalan Institute for Water Researchearch). These figures were published in Journal of Cleaner Production, Vol number 212, A. Ribera-Guardia, L. Bosch, L. Corominas, M. Pijuan, Nitrous oxide and methane emissions from a plug-flow full-scale bioreactor and assessment of its carbon footprint, Page Nos 162–172, Copyright Elsevier (2019) and in Science of the Total Environment, Vol number 493, A. Rodriguez-Caballero, I. Aymericha, M. Poch, M. Pijuan, Evaluation of process conditions triggering emissions of green-house gases from a biological wastewater treatment system, Page Nos 348–391, Copyright Elsevier (2014).

Prior to concentration measuring, the gases sampled are usually conditioned to remove moisture and/or other gases (such as CO_2) that might interfere with the target gas measurement. Depending on the duration of the monitoring campaign, conditioning methods with different degrees of sophistication can be used. They include the use of condensation moisture traps, silica gel and sodium hydroxide traps or conditioning units (membranes and coolers/condensers).

In the case of grab sampling, gaseous concentrations were mainly measured by gas chromatography equipped with an electron-capture detector (N_2O) and with a flame ionization detector (CH_4). When a continuous sampling was applied, concentrations were mainly measured using optical techniques including non-dispersive infrared spectroscopy (NDIR) and Fourier transform infrared spectroscopy (FTIR). In some cases, photo-acoustic spectroscopy (Desloover *et al.*, 2011) and Clark-type N_2O gas sensors (Marques *et al.*, 2016) were used. More details on the analytical methods for measuring N_2O concentration can be found in Rapson and Dacres (2014).

The spatial variability of the emissions was investigated by sampling different positions of the process units either sequentially (Aboobakar *et al.*, 2013a, 2013b; Ahn *et al.*, 2010; Caniani *et al.*, 2019; Oshita *et al.*, 2014; Rodriguez-Caballero *et al.*, 2014) or (almost) simultaneously using a multi-hood system (Bellandi *et al.*, 2017; Duan *et al.*, 2020; Gruber *et al.*, 2020; Pan *et al.*, 2016b). In the last case, an automated valve system is used to direct the off-gas captured from the individual hoods to the analyser at a short interval time (usually of few minutes). Recently, an automated, wireless and self-moving floating hood "LESSDRONE" was developed within the project LESSWATT (LIFE16 ENV/IT/000486). The automatic positioning of the hood is managed by global positioning system (GPS). In addition to GHG emissions monitoring, the device was designed to allow the real-time monitoring of the oxygen transfer efficiency.

Depending on the flux hood method employed, different approaches were used to estimate the surface emission rate. The surface emission rate $M_{g,hood}$ (kg/(m²*d)) from the closed flux chamber is determined by the change in concentration of the targeted GHG over time ($dC_{g,hood}/dt$, kg/(m³*d)) with reference to the headspace volume (V_{hood}, m³) and surface area of the hood (A_{hood}, m²) using Equation (5.5):

$$M_{g,hood} = \frac{V_{hood}}{A_{hood}} \times \frac{dC_{g,hood}}{dt} \tag{5.5}$$

Unlike closed flux chambers, open flux chambers require determination of the off-gas flowrate ($Q_{g,hood}$, m³/d) to estimate the surface emission rate (Equation (5.6)). The latter was determined following three main methods described below.

$$M_{g,hood} = \frac{Q_{g,hood} \times C_{g,hood}}{A_{hood}} \tag{5.6}$$

(1) The off-gas flowrate during aeration can be directly measured from the gas pipe of the floating hood after closing any other vent port, according to configuration D in Figure 5.6. For a proper measurement of the flowrate, care must be taken to maintain the pressure under the hood close to that of atmospheric pressure. A way to do so would be to regulate the gas extraction rate while monitoring the pressure. When steady state conditions are achieved, the emission flowrate ($Q_{g,hood}$) is estimated to be equal to the extraction flowrate and it can subsequently be measured using a gas flowmeter. This is the typical approach employed for measuring the off-gas flowrate in oxygen transfer testing (ASCE, 1997). As the flowrate can only be measured punctually using this method, a correlation between $Q_{g,hood}$ and the total off-gas flowrate of the tank is established to allow quantification of the surface emission flux continuously (Bollon *et al.*, 2016a).

Pan *et al.* (2016a) suggested another option allowing for a continuous monitoring of the flowrate. The basic difference between this and the previous method is that the hood is submersed by 100mm to 150 mm, to avoid any leakage of the gas from the sides, which generates a slight over pressure (1.0–1.5 kPa) under the hood. Continuous monitoring and recording of pressure in the gas line are performed to correct $Q_{g,\text{hood}}$ accordingly.In the case of the Lindvall hood, the off-gas is exhausted from a vent port on which a gas flowmeter can be installed. The emission rate is estimated considering the dilution of the off-gas with the sweep gas ($Q_{g,\text{sweep}}$) according to Equation (5.7). If ambient air is used as the sweep gas, regular measurements of ambient N_2O and/or CH_4 concentrations ($C_{g,\text{sweep}}$) should be performed.

$$M_{g,\text{hood}} = \frac{Q_{g,\text{hood}} \times C_{g,\text{hood}} - Q_{g,\text{sweep}} \times C_{g,\text{sweep}}}{A_{\text{hood}}} \tag{5.7}$$

(2) The second method employed is known as the "tracer method" which was proposed by Chandran (2009) and applied to the SEIFC hood having a configuration close to that shown in Figure 5.6c (in aerobic zones) and Figure 5.6e (in anoxic zones). The method can be applied only discontinuously during the measurement campaign, the determined off-gas flowrate escaping the hood needs to be linked to the aeration flowrate to capture fluctuations. Briefly, a tracer gas (helium) with a given concentration ($C_{g,\text{He}}$) is introduced into the hood at a known flow rate ($Q_{g,\text{He}}$). Helium concentration in the exhaust gas from the hood ($C_{g,\text{He-hood}}$) is measured using a field gas chromatograph equipped with a thermal conductivity detector. The difference in concentration due to the dilution by the off-gas can be used to calculate the flux. In non-aerated zones, a sweep gas at a known flowrate ($Q_{g,\text{sweep}}$) is introduced to enhance an effective gas flow through the hood as explained above. $Q_{g,\text{hood}}$ can be computed using Equation (5.8) (in aerated zones $Q_{g,\text{sweep}} = 0\ \text{m}^3/\text{d}$).

$$Q_{g,\text{hood}} = \frac{Q_{g,He} \times (C_{g,He} - C_{g,He-\text{hood}})}{C_{g,He-\text{hood}}} - Q_{g,\text{sweep}} \tag{5.8}$$

The surface emission rate at the hood location ($M_{g,\text{hood}}$) can be extrapolated to a given zone i ($M_{g,i}$) of the process unit assuming that the off-gas concentration and emission rate measured with the hood were uniform over that zone (Equation (5.9)).

$$M_{g,i} = M_{g,\text{hood}} \times \frac{A_i}{A_{\text{hood}}} \tag{5.9}$$

where A_i is the surface area of the zone i (m²).

The total emission rate of the process unit ($M_{g,\text{total}}$, kg/d) can be computed considering the contribution of the different sampled zones (Equation (5.10)). Caution must be taken when defining these sampling zones as they can greatly affect the estimated emission rate.

$$M_{g,\text{total}} = \sum_{i=1}^{n} M_{g,i} \times A_i \tag{5.10}$$

(3) An alternative method consists in estimating the total off-gas flowrate of the tank using a gas mass balance for nitrogen and argon gas over the activated sludge tank and considering the intake airflow rate of the blower and the off-gas composition leaving the aerated tank (Valkova *et al.*, 2020). The intake air flowrate can be calculated based on rotation frequency and manufacturer's data. The rotation frequency of the air blower drive motors can be measured and logged with an electricity analyser with a one-minute time lag. This method requires that in the system "blower/tank" the blower (or the group of blowers) provides air to exclusively one single activated sludge tank. If this is not the case, the aeration rate sent to the tank needs to be

accurately measured on site. It can be noted that surface extrapolation is not required with this method because the concentration measured in one or more hoods is referred to the calculated total off-gas airflow rate of the tank. Thereby, uncertainties induced by insufficient spatial sampling of $Q_{g,hood}$, that could be encountered with the two previous methods, are avoided.

The floating hood method was applied to many different process units of the plants ranging from the influent pumping station to the disinfection unit (Tables 5.1 and 5.2). Basically, it can be applied to sample the off-gas of any process unit, in which a part of the water surface, the interphase where the gas-liquid mass transfer takes place, can be covered with the hood. It is difficult to use on process units equipped with surface aerators where aeration is achieved through dispersing water in the air (cf. Section 5.3.2.2). Severe foaming and turbulence can complicate gas collection and hood placement. Hoods are easy to build and relatively simple to deploy onsite. In most cases, the floating hood method does not require specialized skills (depending on the associated measurement system) but requires a good understanding of the operation of the studied process unit to properly design the measurement campaign. Its main strength relies on online and continuous quantification of spot specific emissions, concurrently with monitoring of the plant's operating conditions. Thus, it is highly appropriate when a deep understanding of the triggers of GHG emissions is sought. Additionally, it allows spatial variability in emissions across different zones of the process units to be quantified. A whole-site GHG emission quantification, as with ground-based remote sensing methods, is not achievable.

In the absence of comparative studies, it is difficult to assess which hood method (e.g., open or closed flux chamber, with or without tracer and sweep gas, small or large) can provide higher accuracy for estimating GHG emissions. It is likely that the universal hood does not exist, as its design should be adapted to the experimental conditions, which are by definition site specific. Nevertheless, this section summarizes typical pitfalls when designing and measuring GHG emissions using flux hoods. Additionally, it provides some recommendations for best measurement practice.

5.3.2.1.3 Passive liquid surfaces

In passive liquid surfaces it is believed that the main concern is that the conditions inside the hood do not resemble critical features of the atmospheric flow to which the water surface is exposed in the absence of the hood, such as the boundary layer structure or surface currents and waves (Prata *et al.*, 2018). In closed flux chambers, excessive gas accumulation may alter the diffusional flux, resulting in non-linear gas concentration accumulation curves (Mønster *et al.*, 2019). In that case, it is recommended to select the linear part of the curve (i.e., the starting points from the hood's closure time) to avoid underestimating the emission rate (Pavelka *et al.*, 2018). The duration of the hood closure and the number of samples to collect should be adjusted accordingly on site. Adequate gas mixing must be achieved inside the hood. Spherical shaped hoods are believed to offer the best gas mixing conditions as they lack dead zones. Otherwise, mixing can be enhanced by placing a fan/blower inside the hood, recirculating the headspace gas in a closed loop or applying a flow of sweep gas (open flux chambers). Ideally, the gas mixing achieved inside the hood should be close to wind speed at the water surface level (Caniani *et al.*, 2019). When applying a sweep gas, caution must be taken to ensure that the concentration of the diluted off-gas can still be measured accurately.

5.3.2.1.4 Advective flow conditions

In advective flow conditions, previous work comparing a custom-made, large hood (cuboid) and the SEIFC hood types indicated that hood size and design do not significantly impact N_2O measurements assuming they are properly vented to prevent pressure build-up (Porro *et al.*, 2014). Similar conclusions were reached in the study of Spinelli *et al.* (2018). The authors recommended avoiding the use of fixed hoods (instead of floating hoods) because they showed higher gas compression phenomena in the headspace due to the variation of the water level inside the hood. Thus, it is recommended to fit the hood with adequate vent ports (in number or in size) and to monitor and record the pressure under the hood to correct the off-gas concentration (and the flowrate) accordingly if needed (Chandran *et al.*, 2016). Finally, the volume of the hood should be selected with regard to the gas retention time under the experimental conditions, the dynamics of the process unit investigated and the subsequent use of the collected data. If the gas is conditioned using moisture traps, silica gel columns or condensers, one must consider the additional gas retention time resulting from the introduction of this type of device.

The site-specific measurement plan should address floating hood placement depending on the reactor configuration, operating conditions and the objectives of the measurement campaign. If the reactor presents spatial gradients in concentrations of dissolved oxygen, nitrogen species or biomass along the reactor path (due to its design or to bad mixing), emissions should be sampled at different positions of the reactor to achieve a coherent estimation of the emission rate. In most cases, the positions of the hood (or hoods) are chosen so as to sample zones with contrasting operating conditions (e.g., beginning, middle and end of a plug flow reactor). When using the SEIFC hood, Chandran *et al.* (2016) recommend sampling at least two positions per aerated zone to address any variability in gas fluxes that may result from variations in mixing or flow patterns therein. The study of Caniani *et al.* (2019) is one of the few to fix a quantitative coverage criterion of 2% of the total aerated tank surface in accordance with oxygen transfer measurement practices (ASCE, 1997). In that respect, large hoods offer the advantage of covering a greater surface area and thus provide a better averaging of emissions in a given zone (Porro *et al.*, 2014). On the other hand, small hoods are lighter and thus easier to move around for measuring many different positions and/or lanes.

If not measured in the hood, the off-gas flowrate was estimated considering either a homogenous gas distribution over the surface of the aerated tank (or an aerated zone of it) or a variable gas distribution approximated according to the aerator density in the relative zone. Depending on the tank configuration and design, these assumptions can be a source of great error. Thus, it is recommended to measure both off-gas concentration and flowrate to estimate the local emission rate. Additionally, data from the plant's air flowmeters need to be checked against the calculated intake airflow rate of the air blowers.

5.3.2.2 Liquid-to-gas mass transfer estimation method

N_2O and CH_4 emissions in WWTPs derive from generation processes in the liquid phase. The determination of N_2O and CH_4 transferred from the liquid phase serves as a feasible approach to estimate their emissions to the atmosphere. The liquid-to-gas mass transfer estimation method has been mostly applied to quantify N_2O emissions in WWTPs. The transfer rate of N_2O and CH_4 across the gas-liquid interphase (*dC/dt*) is governed by the N_2O or CH_4 gas-liquid transfer coefficient ($K_L\alpha$) as well as the respective gas and liquid concentrations. The mass transfer of N_2O, CH_4 as well as other soluble gases such as O_2, can be described by Equation (5.11) (Holley, 1973):

$$dC/dt = K_L\alpha * (C_{l,\,eq} - C_{l,\,(t)}) - r \tag{5.11}$$

where *dC/dt* is the dissolved gas concentration in the bulk liquid with time (gN/(m³*d)), $K_L\alpha$ is the volumetric mass transfer coefficient (d^{-1}), r is the uptake rate of the studied substance per unit volume per unit time (gN/(m³*d)), $C_{l,t}$ is the dissolved gas concentration in the bulk liquid at time t (gN/m³), $C_{l,eq}$ is the dissolved gas concentration at the liquid-gas boundary, which is assumed to be in equilibrium with the gas phase as given by Henry's law, calculated by using the unitless Henry's coefficient (H) and the gas concentration (C_g): $C_{l,eq} = C_{g,eq}/\mathrm{H}$, (gN/m³).

When the substance uptake rate is zero, the initial dissolved gas concentration ($C_{l,\,0}$) at time 0 ($t = 0$) and the dissolved gas concentration at time t ($t = t$), can be calculated as Equation (5.12):

$$\frac{C_{l,(t)} - C_{l,eq}}{C_{l,o} - C_{l,eq}} = e^{-K_L a^* t} \tag{5.12}$$

Therefore, the amount of N_2O or CH_4 emissions (M_t) during the period from $t = 0$ to $t = t$ through liquid gas transfer can be estimated by Equation (5.13):

$$M_{(t)} = V * \int_{t0}^{t1} K_L a_{(t)} * (C_{l,(t)} - C_{l,eq})dt \tag{5.13}$$

The estimation of N_2O or CH_4 emissions requires measurements of N_2O and CH_4 liquid concentrations and their volumetric mass transfer coefficient. Online monitoring of liquid N_2O concentrations can be carried out in a WWTP using a modified Clark electrode N_2O probe (Figure 5.8). By comparing the liquid N_2O probe monitoring with simultaneous gas chromatograph analysis of the off-gas, studies have demonstrated the accuracy of liquid N_2O probe monitoring in WWTPs (N_2O gas concentration in the range of 0–1000 ppm) (Baresel *et al.*, 2016; Marques *et al.*, 2016; Myers, 2019). Nevertheless, it should be noted that the liquid N_2O probe is sensitive to disturbances and should be used with care. The electrode N_2O probe requires relatively frequent calibration to ensure accurate measurement, and has an expected life time of 4–6 months (manufacture information). Temperature variation could affect the response of the liquid concentration measurement and thus require corrections to be applied (Marques *et al.*, 2016).

Online probes for dissolved CH_4 are currently not widely employable with wastewater, therefore grab sampling needs to be applied. In this case the concentration of dissolved CH_4 can be measured using the headspace method for gas chromatography as described in Section 5.2.2.

Compared with N_2O or CH_4 liquid measurement, the determination of K_La is more challenging. There are three approaches proposed to determine the volumetric N_2O mass transfer coefficient: theoretical, empirical and oxygen proximity.

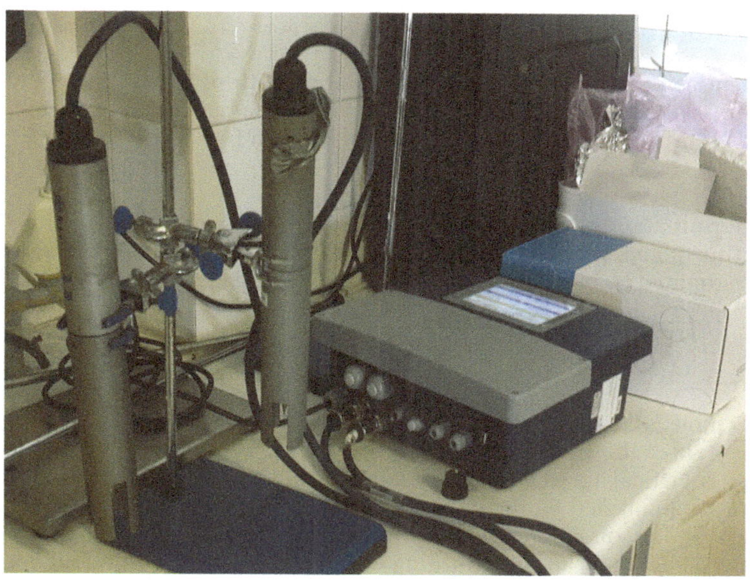

Figure 5.8 Dissolved N_2O probes with protective cover (left), and measurement controller (right). Photo courtesy of Dr. Adrian Oehmen (The University of Queensland).

5.3.2.2.1 Theoretical method

With simultaneous measurement of N_2O concentrations in the gas and liquid phase, the $K_La_{N_2O}$ can be theoretically calculated as Equation (5.14). The equation is derived from the two-film derivation with the assumption that the activated sludge basin is well-mixed with no vertical stratification of dissolved N_2O concentrations. Such an assumption allows a simplified integration with regard to time from the bottom to the surface of the basin (Myers, 2019).

$$C_g = C_{g,\text{in}} * e^{-\frac{K_La*V_L}{H*Q_A}} + H * C_{l,(t)} * \left(1 - e^{-\frac{K_La*V_L}{H*Q_A}}\right) \tag{5.14}$$

where Cg is the gas N_2O concentration (gN/m^3), $Cg_{,\text{in}}$ is the gas N_2O concentration in the aeration bubbles at the bottom of the aeration basin (gN/m^3), V_L is the volume of bulk liquid (m^3), Q_A is the aeration air flowrate (m^3/d), and H is the unitless Henry's coefficient.

At steady state when the changes of dissolved N_2O concentration can be assumed negligible, Equation (5.14) can be simplified and written as Equation (5.15).

$$K_La = \frac{Q_A * C_g}{V_L * (C_{l,eq} - C_{l,(t)})} \tag{5.15}$$

5.3.2.2.2 Empirical method

The empirical determination of K_La was proposed by Foley *et al.* (2010) based on field and laboratory measurements of liquid and off-gas N_2O. Air flow and depth correction were considered in the determination, as shown in Equation (5.16).

$$K_La = \left(\frac{H_R}{H_L}\right)^{-0.49} * 34\,500 * \left(\nu_g\right)^{0.86} \tag{5.16}$$

where H_R is the depth of the field reactor (m), H_L is the depth of the lab stripping column applied in Foley *et al.* (2010), which is 0.815 m, and ν_g is the superficial gas velocity of the field reactor (m^3/(m^2*d)), calculated as air flowrate (Q_A) divided by aerated area (A).

5.3.2.2.3 Oxygen proximity method

The third K_La determination approach is based on Higbie's penetration model (Equation (5.17)) (Higbie, 1935). In this model, when two gases share similar low solubilities and diffusivities, the K_La of one gas can be estimated by measuring the K_La of the other gas under the same conditions.

$$K_La_{N_2O} = K_La_{O_2} * \sqrt{\frac{D_{N_2O}}{D_{O_2}}} \tag{5.17}$$

where D_{N_2O} is N_2O molecular diffusivity in water (1.84×10^{-9} m^2/s at 20°C), and D_{O_2} is O_2 molecular diffusivity in water (1.98×10^{-9} m^2/s at 20°C).

Oxygen transfer is critical to wastewater treatment and therefore often monitored in WWTPs. The K_La for O_2 can be quantified by the in-situ oxygen uptake rate (OUR) method, or by the off-gas method (ASCE, 1997). The off-gas method is based on a gas-phase mass balance, which requires the use of a suitable gas analyser for determining the oxygen concentration and hoods to collect the off-gas. Due to the wide availability of dissolved oxygen (DO) monitoring in WWTPs, and the more straightforward experimental procedure, the in-situ OUR method is more commonly used. The in-situ

OUR method uses the in-situ OUR, and liquid O_2 concentrations ($C_{l,(t)}$) to determine the $K_La_{O_2}$, as described in Equation (5.18) (Moutafchieva *et al.*, 2013). Note that the determination will require online or continuous monitoring of the DO concentrations for a period of time.

$$\frac{dC}{dt} = K_La_{O_2} * (C_{l,eq} - C_{l,(t)}) - OUR \tag{5.18}$$

When the direct online/continuous measurement of DO is not feasible, $K_La_{O_2}$ can still be obtained by oxygen balance analysis, enabling the subsequent estimation of N_2O emissions. In particular, for aeration basins with mechanical aeration systems, such as surface aerators, wastewater is disrupted at the surface to allow the mass transfer of oxygen. In close proximity to the surface aerator there is a high liquid-gas transfer rate while the continuous/online measurement of liquid O_2/N_2O concentration is practically challenging. The turbulent mixing of mechanical aerators creating fast flowing low buoyancy waters presents an unacceptable health and safety risk for measuring the liquid oxygen concentrations. The direct quantification of $K_La_{O_2}$ or $K_La_{N_2O}$ in the intensive aeration area is difficult.

Ye *et al.* (2014) proposed an oxygen balance analysis approach to obtain the $K_La_{O_2}$ in the intensive aeration area. An illustration of the oxygen balance, chemical oxygen demand (COD) and total Kjeldahl nitrogen (TKN) oxidation is shown in Figure 5.9. The oxygen consumption for the entire plant ($M_{O2\ tot}$) is due to the oxidation of COD ($M_{O2\ COD\ tot}$) and TKN ($M_{O2\ TKN\ tot}$) (Equation (5.19)). The oxygen consumption for COD and TKN oxidation can be solved by Equations (5.20) and (5.21), respectively (Ye *et al.*, 2014). Note that CH_4 emission from total COD loss is ignored in the COD balance (Figure 5.9), due to the fact that methanogens can hardly grow with the frequent exposure to oxygen in the aerobic reactor. With the oxygen consumption obtained, the $K_La_{O_2}$ can be solved using

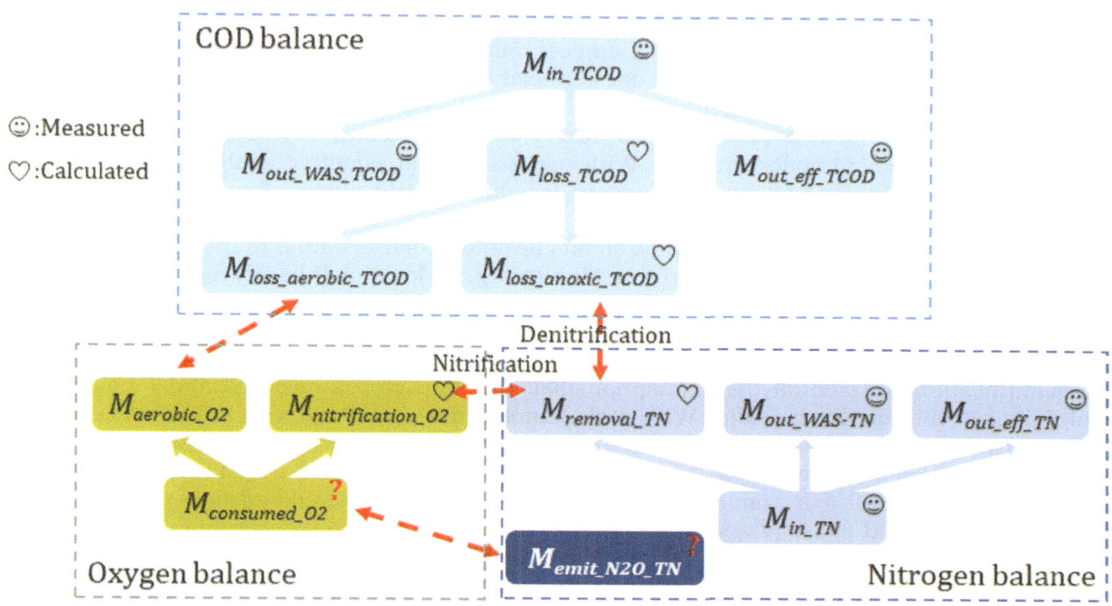

Figure 5.9 Simplified illustration of the methodology to determine the N_2O volumetric transfer coefficient via oxygen mass balance analysis.

Equation (5.13). The $K_La_{N_2O}$ for the surface aerator area can be calculated using Equation (5.17), and the N_2O emission from the surface aerator area is subsequently obtained using Equation (5.13).

$$M_{O2tot} = M_{O2COD_{tot}} + M_{O2TKN_{tot}} \tag{5.19}$$

$$M_{O_2-COD-tot} = \int_{t_0}^{t_1} \begin{bmatrix} Y_A * [Q_{inf}(t) * S_{inf-TKN}(t) - Q_{eff}(t) * S_{eff-TKN}(t) - Q_{WAS}(t) * X_{WAS-TN}(t)] + \\ Q_{inf}(t) * S_{inf-COD}(t) - Q_{eff}(t) * S_{eff-COD}(t) - Q_{WAS}(t) * X_{WAS-COD}(t) \\ -2.86 * [Q_{inf}(t) * S_{inf-TKN}(t) + Q_{inf}(t) * S_{inf-NO_3}(t) - \\ Q_{eff}(t) * S_{eff-TKN}(t) - Q_{eff}(t) * S_{eff-NO_3}(t) - Q_{WAS}(t) * X_{WAS-TN}(t)] \end{bmatrix} dt \tag{5.20}$$

$$M_{O_2-TKN-tot} = \int_{t_0}^{t_1} [Q_{inf}(t) * S_{inf-TKN}(t) - Q_{eff}(t) * S_{eff-TKN}(t) - Q_{WAS}(t) * X_{WAS-TN}(t)] * 4.33 dt \tag{5.21}$$

where $Q_{inf}(t)$ is the daily influent flow rate into the reactor (m^3/d), $Q_{eff}(t)$ is the daily effluent flow rate from the bioreactor (m^3/d), $Q_{WAS}(t)$ is the daily waste activated sludge from plant (wet tonne solids/d), $S_{inf-TKN}(t)$ is the average TKN concentration in the influent (g N/m^3), $S_{eff-TKN}(t)$ is the average TKN concentration in the effluent (g N/m^3), $S_{inf-COD}(t)$ is the average COD concentration in the influent (g COD/m^3), $S_{eff-COD}(t)$ is the average COD concentration in the effluent (g COD/m^3), $S_{inf-NO_3}(t)$ is the average nitrate concentration in the influent (g N/m^3), $S_{eff-NO_3}(t)$ is the average nitrate concentration in the effluent (g N/m^3), $X_{WAS-COD}(t)$ is the COD concentration in the waste sludge (g COD/g wet solids), $X_{WAS-TN}(t)$ is the total N concentration in the waste sludge (g N/g wet solids), and Y_A is the autotrophic yield (g COD/g N).

The mass transfer approach has been applied in WWTPs with different configurations, as summarized in Table 5.3. The mass transfer approach has wide applicability to varying process-unit configurations. It can be applied to any process unit with liquid gas transfer. The mass transfer approach is practically straightforward. It doesn't need the surface area of the investigated process unit to be covered. Therefore, the monitoring configuration is relatively simple. Minimal maintenance is required when moving around different locations. It doesn't necessarily need continuous online monitoring of liquid N_2O concentrations. With options of grab sampling analysis, this method is operator-friendly and incurs relatively low costs. It is particularly suitable for continuous aeration systems at steady state.

However, some limitations of this method should be noted. Firstly, similar to the chamber method, the mass transfer method has a small footprint. Monitoring one location is hardly representative of the overall N_2O emissions of a process unit. Multiple representative monitoring locations must be chosen. Secondly, while grab sampling is feasible to estimate N_2O emissions, the accuracy of the results is questionable. The N_2O liquid concentration in the grab samples analysis needs to be representative. The quantification is based on the assumption that the bioreactor is operated under steady state conditions, which may not be valid. With grab samples, the potentially significant N_2O emission dynamics cannot be captured, so the measured concentration may not be representative of the overall dynamic concentrations. Thirdly, wastewater characteristic changes could also affect the accuracy of mass transfer estimation. For example, increased wastewater salinity could encourage the stripping of N_2O from the liquid phase (Kosse *et al.*, 2017). Finally, considering the dynamic nature of mass transfer with aeration and environmental conditions, using a single representative K_La is inherently problematic. A dynamic K_La should be obtained with simultaneous online measurements. The oxygen proximity method provides opportunities for simultaneous determination of the dynamic K_La for estimating N_2O emissions. This will require further investigation.

Table 5.3 Examples of mass transfer methods employed in estimating GHG emissions in WWTPs, ranked by publication date.

Reference	BNR configuration	GHG	Aeration system	Monitoring frequency	K_La estimation Method	Remarks
Foley *et al.* (2010)	AAO, MLE, Oxidation ditch, Johannesburg and SBR	N_2O	Diffused aeration	Grab samples	Empirical method and oxygen proximity method	N_2O emissions from 7 full-scale WWTPs were quantified based on the mass transfer approach with grab samples.Bioreactors were assumed to be working at near steady-state conditions
Ye *et al.* (2014)	Oxidation ditch	N_2O	Mechanical aeration: surface aerators	Grab samples for liquid N_2O analyses	Oxygen proximity method with oxygen balance analysis	The N_2O emissions from a surface aerator area were quantified with the mass transfer method. K_La was estimated by plant-wide oxygen balance analysis.
Mampaey *et al.* (2015)	SHARON® SBR	N_2O	Gas stripping device	Online gas phase analyser	Theoretical method	A continuous flow of reactor liquid goes through a stripping vessel and the gas concentration in the stripped gas is continuously monitored. This allows liquid phase N_2O or CH_4 concentration determination by gas phase measurements.
Pan *et al.* (2016b)	Sludge drying lagoon	CH_4	No aeration	Grab samples	Oxygen proximity method with COD balance analysis	The mass transfer approach can also be applied to quantify CH_4 emissions from sludge drying lagoon.
Marques *et al.* (2016)	SBR	N_2O	Diffused aeration	Online liquid and gas measurement	Theoretical method	K_La for anoxic zones was calculated based on the dynamic emissions measured by the N_2O gas and liquid sensors.
Baresel *et al.* (2016)	AAO	N_2O	Diffused aeration	Continuous monitoring for two months	Empirical method	Calculated emissions by mass transfer method had good agreement with measured emissions in the off-gas ventilation system from the covered process.
Blomberg *et al.* (2018)	Anoxic/oxic	N_2O	Diffused aeration	Continuous monitoring for 14 days	Empirical method	The mass transfer of N_2O was built into a mathematical model to predict N_2O emissions.
Myers (2019)	Anoxic/oxic	N_2O	Diffused aeration	Continuous monitoring for two weeks	Theoretical, empirical and oxygen proximity methods	Compared the accuracy of the three $K_La_{N_2O}$ determination approaches in a full-scale WWTP. Simultaneous N_2O off-gas concentrations were monitored to validate the liquid-gas transfer model. The findings suggested that the trends of N_2O emissions can be reliably modelled. Of the three $K_La_{N_2O}$ estimation approaches, the oxygen proximity method can best describe the real N_2O gas concentrations.

AAO: anaerobic/anoxic/oxic; MLE: modified Ludzack-Ettinger; SBR: sequencing batch reactor; SHARON®: Single reactor system for High activity Ammonium Removal Over Nitrite.

5.3.2.3 Ground-based remote sensing methods

The layout of the investigated WWTP is an important constraint when applying a ground-based remote sensing method for GHG emission quantifications from a specific process unit. All ground-based remote sensing methods must be applied at a suitable distance from the target emitting area, which should generate an atmospheric plume distinguishable from any other GHG source inside the WWTP.

The mobile tracer gas dispersion method (MTDM) is the only ground-based remote sensing method that has been used for quantifying CH_4 and N_2O emissions from specific process units. Thanks to a specific plant layout, the MTDM was applied on-site at WWTPs (Delre *et al.*, 2017; Samuelsson *et al.*, 2018). Although the inverse dispersion modelling method (IDMM) has never been applied at WWTPs, it could be potentially used for quantifying CH_4 emissions from the sewage sludge treatment area. The literature reports several studies quantifying CH_4 emissions from biogas plants, using the IDMM (Flesch *et al.*, 2011; Groth *et al.*, 2015; Hrad *et al.*, 2014, 2015; Reinelt *et al.*, 2017). Biogas plants have structures and technologies which are very similar to those used for sludge treatment in WWTPs. However, the WWTP layout could be a constraint.

5.3.2.4 Measurements in covered process-units

Process-unit emission measurements of N_2O and CH_4 can be readily employed in covered tanks where the off-gas is extracted and treated prior to its release into the environment. A part of this off-gas stream can be withdrawn and fed to an online gas analyser, as performed in the work of Daelman *et al.* (2012), Carlsson and Lindblom (2015) and Kosonen *et al.* (2016). According to the goal of the measurements, also grab sampling and analysis with GC would be a suitable approach. At fully-covered WWTPs where the off-gas of most components of the plant is constantly withdrawn, a plant-wide quantification can be achieved with this approach. Thanks to the ease of collecting the off-gas samples from the venting pipes, covered activated sludge tanks are suitable candidates to perform long-term measurements. On the other hand, one limitation of the method is its inability to measure spatial variability within a tank, which is essential to identify "hotspots" and hence develop targeted mitigation measures.

By this method, besides the analytical determination of the GHG concentration in the off-gas, the accurate measurement of the off-gas flow rate in the venting pipes is essential. If no online flowmeters are installed, different portable measurement devices such as a hot wire anemometer (Daelman *et al.*, 2012) or a Pitot tube static anemometer (Valkova *et al.*, 2020) can be applied to measure the airflow velocity. Proper calibration and probe positioning inside the off-gas pipe during the measurement need to be ensured. It is also recommended to continuously monitor the operation of the off-gas ventilators on the basis of power consumption data.

The quantification method assumes that the headspace in the covered tank is fully mixed and no airstream short-circuits occur. In order to exclude this, parallel comparative short-term measurements using a floating hood, which can be introduced under the tank covers, are recommended.

Additionally, a static version of the tracer gas dispersion method can be applied where process units are enclosed and indoor air is collected in a ventilation system (Samuelsson *et al.*, 2018). Samuelsson *et al.* (2018) reported a successful application of the static tracer gas dispersion method (STDM) for quantification of N_2O and CH_4 emission rates from the ventilated duct in the building where digestate dewatering and sludge thickening occurred. The STDM deploys a static analytical instrument, as previously reported for CH_4 emission quantifications from leachate wells at landfills (Fredenslund *et al.*, 2010). In the reported measurements, the tracer gas was released in the enclosed ventilated duct upstream of a fan, which facilitated proper mixing of tracer and target gases at the end of the duct, where the air intake of the gas analyser was located. Emission rates of the target gases were calculated by slightly modifying Equation (5.4), whereby, instead of plume integration, the ratio of the two gases was used. Although the hereby reported application of the STDM is limited to process units enclosed in buildings with ventilation system, it can be performed disregarding weather conditions and for a long period of time.

5.3.2.5 Point monitoring of CH_4 leakages

Sources of CH_4 emissions from WWTPs can also be derived from leakages in the biogas valorization system (fugitive emissions). Point monitoring approaches such as a portable flame ionization detector (FID) and an infrared gas imaging camera with an absorption filter within a specific wavelength range (e.g., GasFindIR-camera FLIR GF320) have been used to identify gas leakages at biogas plants (Liebetrau *et al.*, 2013; Reinelt *et al.*, 2017) and can be applied also to anaerobic digesters of WWTPs (Tauber *et al.*, 2019). These methods have a limited spatial and temporal resolution for gas emissions and the application can be difficult for areas with restricted access. After localization, the accessible leakage spots can be encapsulated in a flexible enclosure made, for example, of a gas tight foil, equipped with an input and output pipe and a blower to produce a constant air flow through the enclosed volume as described in Liebetrau *et al.* (2013). The quantification of the CH_4 leakages can be then performed similarly as for Lindvall hoods (see Section 5.3.2.1). Depending on the accessibility of the leakage point, the static chamber method may also be used (Tauber *et al.*, 2019). A method to quantify operational CH_4 emissions from pressure relief valves of biogas plant digestors is described in Reinelt and Liebetrau (2020).

5.3.3 Recommendations for selecting the measurement method

Table 5.4 summarizes (with no claim to completeness) the currently available and most commonly applied methodologies for the quantification of direct N_2O and CH_4 emissions at full-scale WWTPs. The methodologies, their strengths and limitations, and instrumental requirements, as well as some general remarks are presented. Among the ground-based remote sensing methods, the MTDM is considered in this overview, since this is the sole method for plant-wide quantification that has been successfully applied at WWTPs to date.

This overview emphasizes that a universally recommended quantification method does not exist, whereas the choice of the suitable method is mainly dictated by the specific goals of the survey as well as by individual local requirements. Moreover, some objectives can require bottom-up or top-down approaches where the application of more than one method is required. The intention of this section is to provide general recommendations for researchers and practitioners interested in measuring N_2O and CH_4 emissions at full-scale WWTPs. In this context, the most common goals of GHG quantification are:

- Quantification of the overall GHG emissions for a given WWTP to comply with GHG emission protocols/inventories and/or to provide consolidated data for carbon footprint analyses within life cycle assessment (LCA) studies.
- Development of mitigation strategies for direct GHG emissions at WWTPs:
    - Approach 1: Estimation of the N_2O and/or CH_4 emissions for a given WWTP and changing sets of conditions (e.g., operation, load, temperature) with the aim of linking emissions to operational parameters and plant performance in, for example, regression analysis.
    - Approach 2: Calibration and validation of N_2O and/or CH_4 mechanistic models to understand potential generation and emission pathways and be able to accurately describe observed emissions.

When quantification of a WWTP overall GHG emission is the main objective of the survey, plant-wide quantification can be, in many cases, the most suitable approach. When specific local conditions hamper the application of these methods, a bottom-up approach can be followed, by identifying and quantifying the largest emitters at process-unit level and then summing up the single sources. In this case, however, unknown emission sources will remain undetected.

Table 5.4 Comparison of different GHG emissions quantification methodologies for WWTPs.

Methodological approach	Measurement method	Strengths	Limitations	Equipment[1]	Remarks
Plant-wide quantification[2]	Mobile tracer gas dispersion method	On-site screening can identify the major source of GHG emissions. Quantification of the overall GHG emissions, thus reducing the potential overlook of individual on-site emission sources.	Emission quantification is not possible if the facility has an upwind source of the target GHG, which overlaps the target plume. Unable to perform long term and continuous monitoring.	A mobile analytical platform equipped with fast-responding and sensitive gas analyser, a GNSS[5] device, and flow control system from the tracer gas cylinders. The literature reports successful applications using gas analysers with the following features: CH_4 frequency 0.06–2 Hz CH_4 precision[3]: 2.6–3.8 ppb N_2O precision[3]: 0.7–21.1 ppb C_2H_2 precision: 0.3–4.2 ppb CH_4 detection: interval 0–20 ppm N_2O detection: interval 0–400 ppm C_2H_2 detection: interval 0–500 ppb	Minimal detectable emission is related to analytical instrument, size of the target source, weather conditions and measurement distance. Therefore, minimum detectable emission rate is site, time, and equipment specific.
Process-unit quantification	Floating hood	Can investigate the dynamics and spatial variability of GHG emissions. Can establish the links between GHG emission and the nitrogen removal process. Can identify GHG production sources. Relatively straightforward implementation	Relatively small foot-print, challenging to extrapolate to large areas. Unsuitable for surface aeration systems (e.g., surface aerators).	Floating hood and gas analyser, gas flowmeter in some applications. The literature reports successful applications using gas analysers with the following features: For low N_2O conc. range: Detection interval: N_2O 0–50 ppm Detection limit: 0.05 ppm Precision: 1% of reading–For high N_2O conc. range: Detection interval: N_2O 0–1000 ppm Detection limit: 1 ppm Precision: 1% of full-scale[4] –For CH_4: Detection interval: CH_4 0–300 ppm Detection limit: 0.3 ppm Precision: 1% of reading	The most common approach for N_2O quantification, spatial variations however could cause uncertainties for large scale quantification. Other portable sensors can also be mounted to simultaneously measure the wastewater treatment operational parameters, providing opportunities to link operational conditions to the GHG emissions.
	Liquid-to-gas mass transfer method	Suitable for treatment plants with surface aeration systems. Grab samples or online measurement can be carried out.	The investigated system needs to be well defined and in steady state to allow accurate mass balance analysis. Relatively small foot-print, challenging to extrapolate to large areas.	Liquid N_2O sensor: Detection range: 0–14 mg N/L; Detection limit: 28 μg N/L.	The N_2O liquid measurement is sensitive to temperature variations and has a relatively long response time (<20 or <45 s depending on the sensor type). This method can link oxygen consumption, N_2O and CH_4 emissions together.
	Mobile tracer gas dispersion method	Quantification of the GHG emissions from the entire target process unit.	The same as plant-wide quantification.	The same as plant-wide quantification.	The same as plant-wide quantification

[1]Mentioned detection interval and detection limit are manufacturers' specifications.
[2]Only methods currently applied to full-scale WWTPs were considered.
[3]Precision in detecting specific gas is given as three times the standard deviation of six minutes' constant concentration reading (Samuelsson *et al.*, 2018). Acetylene (C_2H_2) is the most commonly used tracer gas.
[4]Precision definition: 1% of full-scale accuracy means that the maximum absolute error of the analyser by measuring a constant concentration is no more than 0.01 times the full-scale value of the analyser (in this case 1000 ppm).
[5]GNSS: global navigation satellite system.

When the focus of the quantification survey is to develop mitigation strategies to reduce GHG emissions at the plant, much deeper insights into the generation and emissions pathways are essential. For this purpose, the process-unit approaches can provide the required information for developing mechanistic model or regression analyses linking operational and emission data. The selection of the suitable method for process-unit quantification will be dictated in many cases by the typology of the targeted process unit itself (in some cases also by the targeted GHG), taking into account the strengths and limitations of each method.

Some examples referring to the most common applications of GHG quantification at WTTPs are given as follows:

Example 1: Decrease the Carbon Footprint of the WWTP.

Possible approach (top-down):

(1) Perform plant-wide quantification and carbon footprint evaluation of the plant. If direct N_2O and/or CH_4 emissions are shown to be contributing significantly to the plant carbon footprint, further investigation will be carried out to quantify emissions from process units.

(2) On-site quantification to identify the largest emission sources applying either ground-based remote sensing or hood methods.

(3) Undertake a long-term study to investigate spatial-temporal dynamics using hoods and liquid sensors and establish links with process parameters and/or implement a model. After having optimized the investigated process towards lower GHG emissions, the generated mitigation potential can be eventually verified by performing new plant-wide measurements.

Example 2: Estimation of N_2O and/or CH_4 emissions for GHG inventories.

Possible approach:

(1) Perform plant-wide quantification of N_2O and CH_4 emissions for different sets of local conditions (e.g., at differing loading conditions and/or water temperature) to generate average values on a yearly basis.

(2) Bottom-up: if largest emitters are already known, process-unit quantification can be used as well and the single sources can be integrated. In this case, unknown emission sources remain undetected. A long-term approach is recommended.

Ultimately, specific local factors such as WWTP design and operation, technical staff resource availability, monetary resources, analytical capacity, and equipment/instrument availability can play a significant role in selecting the quantification method as well as in developing the sampling plan. Therefore, these additional constraints need to be carefully evaluated in the decision-making process to ensure they do not hamper the fulfilment of the measurement goals.

5.3.4 Recommended data requirements

The aim of this section is to provide additional practical guidance for the implementation of measurement campaigns to quantify direct N_2O and CH_4 emissions at full-scale WWTPs. The main objective focuses on the minimum data requirements (in terms of quality and quantity) and duration of a measurement campaign. The minimum data requirements strongly depend on the measurement goals (e.g., GHG inventory versus GHG emission modelling) and will be influenced by several factors such as the site layout, accessibility of sampling points and resources availability (human and monetary) for the campaign.

Multiple factors can impact the accuracy of the estimation of GHG emissions and GHG emission factors of WWTPs. The most relevant of these are summarized in Figure 5.10. Beside issues related to the method implementation and analytical uncertainty, the chosen sampling strategy as well as the plant data availability and quality are of significant concern and need to be thoroughly considered. Plant data are not only essential to compare GHG emissions to influent loads, they also provide

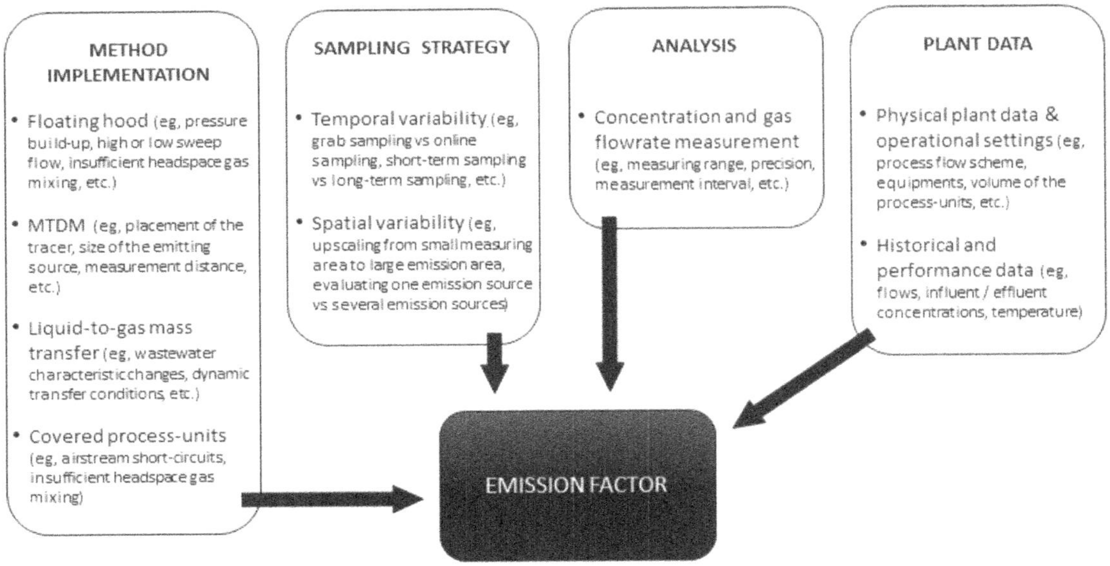

Figure 5.10 Potential sources of errors impacting the estimation of GHG emission factors at WWTPs.

the basis for developing a sound field sampling plan and define the number and duration of the measurement campaigns. Although some protocols (Chandran, 2009; Chandran *et al.*, 2016) provide an overview of the data requirements for the preliminary WWTP assessment and complementary data requirements accompanying off-gas measurements, the design of a GHG quantification campaign remains a challenging task. In general, it can be said that good knowledge of the plant and process unit operation is essential and GHG assessments should be performed in cooperation with the plant operators.

For a preliminary assessment of the GHG emissions, a straightforward grab sampling approach can be carried out to provide an order-of-magnitude estimate of GHG emissions from the sampled location. Discrete samples can be taken for offline analysis of liquid GHG concentrations. Together with the estimated K_La, the emission rates of GHGs can be calculated by the liquid-to-gas mass transfer estimation method (as presented in Section 5.3.2.2). Grab sampling is particularly suitable for continuous aeration systems where the dynamics of K_La are less significant. Due to the spatio-temporal variability of GHG emissions (especially N_2O), this can only serve as a preliminary assessment of the sampled location. Since production and emission of N_2O and CH_4 can occur temporally and/or spatially independently from each other, selection of the sampling locations should reflect this pattern.

In the majority of cases, the minimum number of operating parameters that need to be monitored during the quantification campaign is dictated by the measurement goals. While daily influent load (daily average) and plant removal performance are usually sufficient for GHG emission inventories, more data are needed to establish correlations between the operating conditions and emissions or for model calibration. With regard to the key parameters to be monitored, it is possible to distinguish between those that are essential to estimate N_2O and CH_4 emissions (e.g., plant flows, aeration air flow, nitrogen loading, etc.) and those that influence production/emission pathways (e.g., concentrations of dissolved oxygen, nitrogen species, biomass concentration, sludge age, etc.). The reader interested in data analysis and reconciliation methods is advised to read the procedure proposed by Rieger *et al.* (2013). Additionally, when modelling is targeted, the frequency of the plant data acquisition needs to be increased from composite samples (24-hour average concentrations) that would be suitable for inventories/quantification, to higher resolution sampling (in most cases, online sensors are used). To better correlate N_2O emissions with the influent pollutant load, the 24-hour window for the calculation of the daily N_2O flux should be the same as that for the daily composite influent samples of the WWTP (Valkova *et al.*, 2020).

Owing to the plurality of quantification methodologies, (1) the floating hood method for N_2O emission quantification in activated sludge tanks and (2) plant-integrated GHG emission quantification were chosen as exemplary applications in this section for further discussion. Although the focus is upon these specific applications, general recommendations can also be extrapolated for other quantification methods.

5.3.4.1 Floating hood method for N_2O emission quantification

Spatial variability of GHG emissions was shown to be significant in tanks having spatial gradients of dissolved oxygen/nitrogen species concentrations or different aeration strategies along a lane. Regardless of the objective (inventories or modelling), zones with contrasted conditions should be sampled more intensively, preferably applying multiple floating hoods, as suggested by several authors (Bellandi *et al.*, 2017; Duan *et al.*, 2020; Gruber *et al.*, 2020; Pan *et al.*, 2016a). In the case of modelling, it is also recommended to sample and analyse the bulk liquid at the proximity of the hoods during the course of the monitoring period. Tanks exhibiting complete mixing conditions during aeration do not usually require multiple zone sampling. The variability of emissions between different parallel lanes also needs to be addressed. Differences in key performance indicators (e.g., effluent concentrations, sludge production, energy consumption) can be important indicators of uneven influent loading, leading to differing N_2O emissions (Gruber *et al.*, 2020).

The minimal duration of sampling should cover the diurnal variability of the load (24 h), which also corresponds to the maximum hydraulic retention time of most BNR technologies. However, in practice, a week of sampling including the weekend is advised to capture the temporal variability of N_2O and link it with the operating conditions of the plant.

If the plant treats a proportion of industrial waste, measurement should comprise periods where this load is added to the urban wastewater. If the plant is located in a tourist area, with significant variations in the load over a year, the sampling plan should comprise high and low loading periods. To account for the seasonal variability of GHG emissions, several short-term monitoring campaigns, for example, three to four per year, can be performed. The sampling protocol should cover periods with typical plant loadings and performance (base line) as well as periods with contrasting nitrification and denitrification capacities. Historical plant performance data can help with identifying these periodic patterns.

5.3.4.2 Plant-wide GHG emission quantification

The number of measurement campaigns should properly describe the emissions over one year. To date, no study has investigated the sufficient number of measurement campaigns and the suitable timing of measurements along one year. One quantification per season could be a good compromise in most cases. However, different plants require tailored sampling strategies according to the features of the process units. This is the case for WWTPs where biosolids are stored on site and seasonally applied on land, exhibiting higher GHG emissions when the biosolid storage is full.

The minimum number of plume traverses performed in a single quantification campaign could be set to 10. However, longer measurement campaigns would give the chance to gather potential GHG emission dynamics. In any case, the measurement campaign should last for a period that includes the entire process cycle of specific technologies used at the plant. For example, in WWTPs performing biological nitrogen removal, the N_2O emission quantification should last for a complete cycle of the nitrification/denitrification phases.

Detailed guidelines on how to best perform a measurement campaign, from design to application and data processing, are reported by Delre (2018).

5.4 CONCLUSIONS AND PERSPECTIVES

In the past two decades, the full-scale quantification of N_2O and CH_4 emissions from sewers and WWTPs has been significantly improved. Advances in analytical detection techniques have supported the development and application of plant-wide quantification approaches, capturing the entire emission spectrum of WWTPs, along with sources that were usually overlooked. Moreover, the upgrade from grab sampling towards online monitoring by process-unit applications has contributed to a better understanding of the mechanisms governing the production and emission pathways of both N_2O and CH_4. To estimate N_2O or CH_4 emissions from activated sludge tanks equipped with surface aerators, a tailored methodological approach was implemented. All these methodological improvements are key factors to developing effective mitigation strategies for urban water systems.

Despite the improvements, quantifying GHG in sewers and WWTPs still remains a challenging task. Current quantification methods can only partially depict the high spatio-temporal variability of GHG emissions, which are strongly influenced by environmental and process conditions in sewers and WWTPs, respectively. Extensive sampling of plants and long-term monitoring are necessary to capture the complexity of the targeted systems, thus requiring a significant input of resources on site. However, the improved data quality and quantity achieved through sound sampling and measurement protocols have helped to identify process parameters that trigger GHG emissions and refine models that are able to describe GHG emission profiles from sewers and WWTPs. To further support these achievements, future measurement campaigns at full-scale WWTPs should employ tailored measurement approaches aiming to link emissions to process parameters or performance indicators that can be monitored with less effort. With regard to inventory protocols, such established links would allow estimations of GHG emission intensity based on process data, replacing the current applied fixed and generic emission factors. In addition to this, application of full-scale quantification of GHG emissions will continue to be essential to identify emission sources and to verify the effectiveness of mitigation strategies. Analytical and methodological developments in this field should provide more accurate and resource friendly quantification approaches.

With regard to sewers, current available methods are not yet capable of capturing the complexity of these systems due to their geographical extension and highly varied conditions. However, the combination of measurements in selected hotspots with mathematical modelling of GHG production is a viable solution to obtain estimations of full-network emissions.

ACKNOWLEDGEMENTS

The authors acknowledge Charlotte Scheutz (DTU, Denmark) and Pierre Mauricrace (INRAE, France) for scientific and technical assistance in Sections 5.3.1 and 5.3.2.1, respectively. The authors thank Dr. Ben van den Akker and Dr. Romain Lemaire for reviewing and giving comments that greatly improved the chapter.
All authors contributed equally to this chapter.

REFERENCES

Aboobakar A., Cartmell E., Stephenson T., Jones M., Vale P. and Dotro G. (2013a). Nitrous oxide emissions and dissolved oxygen profiling in a full-scale nitrifying activated sludge treatment plant. *Water Research*, **47**(2), 524–534, doi: 10.1016/j.watres.2012.10.004

Aboobakar A., Jones M., Vale P., Cartmell E. and Dotro G. (2013b). Methane emissions from aerated zones in a full-scale nitrifying activated sludge treatment plant. *Water Air & Soil Pollution*, **225**(1), 1814, doi: 10.1007/s11270-013-1814-8

Ahn J. H., Kim S., Park H., Rahm B., Pagilla K. and Chandran K. (2010). N_2O emissions from activated sludge processes, 2008–2009: results of a national monitoring survey in the United States. *Environmental Science & Technology*, **44**(12), 4505–4511, doi: 10.1021/es903845y

ASCE. (1997). Standard guidelines for in-process oxygen transfer testing ASCE 18-96, 3 45 East 47th Street, New York, NY.

Baresel C., Andersson S., Yang J. and Andersen M. H. (2016). Comparison of nitrous oxide (N_2O) emissions calculations at a Swedish wastewater treatment plant based on water concentrations versus off-gas concentrations. *Advances in Climate Change Research*, **7**(3), 185–191, doi: 10.1016/j.accre.2016.09.001

Bellandi G., Porro J., Senesi E., Caretti C., Caffaz S., Weijers S., Nopens I. and Gori R. (2017). Multi-point monitoring of nitrous oxide emissions in three full-scale conventional activated sludge tanks in Europe. *Water Science and Technology*, **77**(4), 880–890, doi: 10.2166/wst.2017.560

Benckiser G., Eilts R., Linn A., Lorch H. J., Sümer E., Weiske A. and Wenzhöfer F. (1996). N_2O emissions from different cropping systems and from aerated, nitrifying and denitrifying tanks of a municipal waste water treatment plant. *Biology and Fertility of Soils*, **23**(3), 257–265, doi: 10.1007/BF00335953

Blomberg K., Kosse P., Mikola A., Kuokkanen A., Fred T., Heinonen M., Mulas M., Lübken M., Wichern M. and Vahala R. (2018). Development of an extended ASM3 model for predicting the nitrous oxide emissions in a full-scale wastewater treatment plant. *Environmental Science & Technology*, **52**(10), 5803–5811, doi: 10.1021/acs.est.8b00386

Bollon, J., Filali, A., Fayolle, Y., Guerin, S., Rocher, V. and Gillot, S. (2016a). N_2O emissions from full-scale nitrifying biofilters. *Water Research*, **102**, 41–51, doi: 10.1016/j.watres.2016.05.091

Bollon, J., Filali, A., Fayolle, Y., Guérin, S., Rocher, V. and Gillot, S. (2016b). Full-scale post denitrifying biofilters: sinks of dissolved N_2O? *Science of the Total Environment*, **563–564**, 320–328, doi: 10.1016/j.scitotenv.2016.03.237

Boulart, C., Connelly, D. P., Mowlem, M. C. (2010). Sensors and technologies for in situ dissolved methane measurements and their evaluation using Technology Readiness Levels. *TrAC Trends in Analytical Chemistry*, **29**, 186–195. doi: 10.1016/j.trac.2009.12.001

Brotto, A. C., Kligerman, D. C., Andrade, S. A., Ribeiro, R. P., Oliveira, J. L. M., Chandran, K. and de Mello, W. Z. (2015). Factors controlling nitrous oxide emissions from a full-scale activated sludge system in the tropics. *Environmental Science and Pollution Research*, **22**(15), 11840–11849, doi: 10.1007/s11356-015-4467-x

Camilli, R. and Hemond, H. (2004). NEREUS/Kemonaut, a mobile autonomous underwater mass spectrometer. *TrAC Trends in Analytical Chemistry*, **23**, 307–313. doi: 10.1016/S0165-9936(04)00408-X

Caniani, D., Caivano, M., Pascale, R., Bianco, G., Mancini, I. M., Masi, S., Mazzone, G., Firouzian, M. and Rosso, D. (2019). CO_2 and N_2O from Water Researchource recovery facilities: Evaluation of emissions from biological treatment, settling, disinfection, and receiving water body. *Science of the Total Environment*, **648**, 1130–1140, doi: 10.1016/j.scitotenv.2018.08.150

Capelli, L., Sironi, S. and Del Rosso, R. (2013). Odor sampling: techniques and strategies for the estimation of odor emission rates from different source types. *Sensors (Basel, Switzerland)*, **13**(1), 938–955, doi: 10.3390/s130100938

Carlsson A. and Lindblom E. (2015). Monitoring of methane and nitrous oxide emissions from two large wastewater treatment plants. Presented at the 12t Specialized Conference on Design, Operation and Economics of Large Wastewater Treatment Plant, 6–9 September 2015, Prague, Czech Republic.

Chandran K. (2009). Characterization of nitrogen greenhouse gas emissions from wastewater treatment BNR operations – Field protocol with quality assurance plan (U4R07), WERF.

Chandran K. (2011). Methods in enzymology. Research on Nitrification and Related Processes. In: Chapter 16: Protocol for the Measurement of Nitrous Oxide Fluxes from Biological Wastewater Treatment Plants, A. Part and M.G. Klotz, (eds), Academic Press Hardcover. ISBN: 9780123812940, pp. 369–385.

Chandran K., Volcke E. I. P. and van Loosdrecht M. C. M. (2016). Experimental methods in wastewater treatment. In: Chapter 4: Off-gas Emissions Tests, M. C. M. Van Loosdrecht, P. H. Nielsen, C. M. Lopez-Vazquez and D. Brdjanovic (eds), IWA Publishing, London, pp. 177–200.

Czepiel, P. M., Crill, P. M. and Harriss, R. C. (1993). Methane emissions from municipal wastewater treatment processes. *Environmental Science & Technology*, **27**(12), 2472–2477, doi: 10.1021/es00048a025

Czepiel, P., Crill, P. and Harriss, R. (1995). Nitrous oxide emissions from municipal wastewater treatment. *Environmental Science & Technology*, **29**(9), 2352–2356, doi: 10.1021/es00009a030

Daelman, M. R. J., van Voorthuizen, E. M., van Dongen, U. G. J. M., Volcke, E. I. P. and van Loosdrecht, M. C. M. (2012). Methane emission during municipal wastewater treatment. *Water Researchearch*, **46**, 3657–3670. doi: 10.1016/j.watres.2012.04.024

Daelman, M. R. J., van Voorthuizen, E. M., van Dongen, L. G. J. M., Volcke, E. I. P. and van Loosdrecht, M. C. M. (2013). Methane and nitrous oxide emissions from municipal wastewater treatment: results from a long-term study. *Water Science and Technology*, **67**(10), 2350–2355, doi: 10.2166/wst.2013.109

Delre A. (2018). Greenhouse gas emissions from wastewater treatment plants: measurements and carbon footprint assessment. PhD thesis, Department of Environmental Engineering, Technical University of Denmark

(DTU). Available at https://orbit.dtu.dk/en/publications/greenhouse-gas-emissions-from-wastewater-treatment-plants-measure (last visited on 11-09-2020).

Delre A., Mønster J. and Scheutz C. (2017). Greenhouse gas emission quantification from wastewater treatment plants, using a tracer gas dispersion method. *Science of the Total Environment*, **605–606**, 258–268. doi: 10.1016/j.scitotenv.2017.06.177

Delre A., Mønster J., Samuelsson J., Fredenslund A. M. and Scheutz C. (2018). Emission quantification using the tracer gas dispersion method: the influence of instrument, tracer gas species and source simulation. *Science of the Total Environment*, **634**, 59–66. doi: 10.1016/j.scitotenv.2018.03.289

Deng, Y., Nevell, T. G., Ewen, R. J. and Honeybourne, C. L. (1993). Sulfur poisoning, recovery and related phenomena over supported palladium, rhodium and iridium catalysts for methane oxidation. *Applied Catalysis A: General*, **101**, 51–62. doi: 10.1016/0926-860X(93)80137-F

Desloover, J., De Clippeleir, H., Boeckx, P., Du Laing, G., Colsen, J., Verstraete, W. and Vlaeminck, S. E. (2011). Floc-based sequential partial nitritation and anammox at full scale with contrasting N_2O emissions. *Water Research*, **45**(9), 2811–2821, doi: 10.1016/j.watres.2011.02.028

Duan, H., van den Akker, B., Thwaites, B. J., Peng, L., Herman, C., Pan, Y., Ni, B.-J., Watt, S., Yuan, Z. and Ye, L. (2020). Mitigating nitrous oxide emissions at a full-scale wastewater treatment plant. *Water Research*, **185**, 116196, doi: 10.1016/j.watres.2020.116196

Filali, A., Bollon, J., Molle, P., Mander, Ü. and Gillot, S. (2017). High-frequency measurement of N_2O emissions from a full-scale vertical subsurface flow constructed wetland. *Ecological Engineering*, **108**, 240–248, doi: 10.1016/j.ecoleng.2017.08.037

Flesch, T. K., Desjardins, R. L. and Worth, D. (2011). Fugitive methane emissions from an agricultural biodigester. *Biomass and Bioenergy*, **35**, 3927–3935. doi: 10.1016/j.biombioe.2011.06.009

Foley, J., Yuan, Z. and Lant, P. (2009). Dissolved methane in rising main sewer systems: field measurements and simple model development for estimating greenhouse gas emissions. *Water Science and Technology*, **60**, 2963–2971. doi: 10.2166/wst.2009.718

Foley, J., de Haas, D., Yuan, Z. and Lant, P. (2010). Nitrous oxide generation in full-scale biological nutrient removal wastewater treatment plants. *Water Research*, **44**(3), 831–844, doi: 10.1016/j.watres.2009.10.033

Foley, J., Yuan, Z., Keller, J., Senante, E., Chandran, K., Willis, J., Shah, A., van Loosdrecht, M. and van Voorthuizen, E. (2011a). N_2O and CH_4 emission from wastewater collection and treatment systems: state of the science report.

Foley, J., Yuan, Z., Senante, E., Chandran, K., Willis, J., van Loosdrecht, M. and van Voorthuizen, E. (2011b). Global Water Researchearch Coalition: N_2O and CH_4 Emission from Wastewater Collection and Treatment Systems, Technical Report. 2011–30. Global Water Researchearch Coalitionc/o International Water Association, London, UK.

Fredenslund, A. M., Scheutz, C. and Kjeldsen, P. (2010). Tracer method to measure landfill gas emissions from leachate collection systems. *Waste Management*, **30**, 2146e2152. doi: 10.1016/j.wasman.2010.03.013

Gao, F. and Yates, S. R. (1998). Laboratory study of closed and dynamic flux chambers: experimental results and implications for field application. *Journal of Geophysical Research-Atmospheres*, **103**(D20), 26115–26125, doi: 10.1029/98JD01346

Groth, A., Maurer, C., Reiser, M. and Kranert, M. (2015). Determination of methane emission rates on a biogas plant using data from laser absorption spectrometry. *Bioresource Technology*, **178**, 359–361. doi: 10.1016/j.biortech.2014.09.112

Gruber, W., Villez, K., Kipf, M., Wunderlin, P., Siegrist, H., Vogt, L. and Joss, A. (2020). N_2O emission in full-scale wastewater treatment: Proposing a refined monitoring strategy. *Science of the Total Environment*, **699**, 134157, doi: 10.1016/j.scitotenv.2019.134157

Guisasola, A., de Haas, D., Keller, J. and Yuan, Z. (2008). Methane formation in sewer systems. *Water Research*, **42**, 1421–1430, doi: 10.1016/j.watres.2007.10.014

Guo, L. and Vanrolleghem, P. A. (2014). Calibration and validation of an Activated Sludge Model for Greenhouse gases No. 1 (ASMG1) – Prediction of temperature dependent N_2O emission dynamics. *Bioprocess Biosystem Engineering*, **37**, 151–163, doi: 10.1007/s00449-013-0978-3

Joseph, A. P., Keller, J., Bustamante, H. and Bond, P. L. (2012). Surface neutralization and H_2S oxidation at early stages of sewer corrosion: influence of temperature, relative humidity and H_2S concentration. *Water Research*, **46**, 4235–45. doi: 10.1016/j.watres.2012.05.011

Higbie, R. (1935). The rate of absorption of a pure gas into a still liquid during short periods of exposure. *Transactions of the American Institute of Chemical Engineers*, **31**, 365–389.

Holley, E. R. (1973). Diffusion and boundary layer concepts in aeration through liquid surfaces. *Water Research*, **7**(4), 559–573, doi: 10.1016/0043-1354(73)90055-9

Hrad, M., Piringer, M., Kamarad, L., Baumann-Stanzer, K. and Huber-Humer, M. (2014). Multisource emission retrieval within a biogas plant based on inversedispersion calculations-a real-life example. *Environmental Monitoring and Assessment*, **186** (10), 6251–6262, doi: 10.1007/s10661–014–3852-0

Hrad M., Piringer M. and Huber-Humer M. (2015). Determining methane emissions from biogas plants – Operational and meteorological aspects. *Bioresource Technology*, **191** (2015) 234–243. doi: 10.1016/j.biortech.2015.05.016.

IPCC. (1996). 1997 Revised 1996 IPCC Guidelines for National Greenhouse Gas Inventories, Vol. 1: Reporting Instructions, Vol. 2: Workbook, Vol. 3: Reference Manual. Intergovernmental Panel on Climate Change. In: J. T. Houghton, L. G. Meira Filho, B. Lim, K. Tréanton, I. Mamaty, Y. Bonduki, D. J. Griggs and B. A. Callander (eds), Chapter 6, IPPC, Genf.

IPCC. (2006). Doorn, M.R.J., Towprayoon S., Manso Vieira, S. M., Irving W., Palmer C., Pipatti R., Wang C. 2006. Wastewater treatment and discharge. In: IPCC – Intergovernmental Panel on Climate Change, Guidelines for National Greenhouse Gas Inventories, National Greenhouse Gas Inventories Programme, H. S. Eggleston, L. Buendia, K. Miwa, T. Ngara and K. Anabe (eds), Chapter 6, Vol. **5**, IGES, Japan.

IPCC. (2019). Bartram, D., Short, M.D., Ebie, Y., Farkaš, J., Gueguen, C., Peters, G.M., Zanzottera, N.M., Karthik, M., Masuda, S. 2019. Wastewater treatment and discharge. In: 2019 Refinement to the 2006 IPCC Guidelines for National Greenhouse Gas Inventories, E. Calvo Buendia, K. Tanabe, A. Kranjc, J. Baasansuren, M. Fukuda, S. Ngarize, A. Osako, Y. Pyrozhenko, P. Shermanau and S. Federici (eds), Chapter 6, Vol. **5**, IPCC, Switzerland.

Kim, J., Lim, J. S., Friedman, J., Lee, U., Vieira, L., Rosso, D., Gerla, M. and Srivastava, M. B. (2009). SewerSnort: A Drifting Sensor for In-situ Sewer Gas Monitoring. 2009 6th Annual IEEE Communications Society Conference on Sensor, Mesh and Ad Hoc Communications and Networks, pp. 1–9. doi: 10.1109/SAHCN.2009.5168971

Kimochi, Y., Inamori, Y., Mizuochi, M., Xu, K.-Q. and Matsumura, M. (1998). Nitrogen removal and N_2O emission in a full-scale domestic wastewater treatment plant with intermittent aeration. *Journal of Fermentation and Bioengineering*, **86**(2), 202–206, doi: 10.1016/S0922-338X(98)80114-1

Kosonen H., Heinonen M., Mikola A., Haimi H., Mulas M., Corona F. and Vahala R. (2016). Nitrous oxide production at a fully covered wastewater treatment plant: results of a long-term online monitoringcampaign. *Environmental Science & Technology*, **7**, 5547–5554. doi: 10.1021/acs.est.5b04466

Kosse, P., Lübken, M., Schmidt, T. C. and Wichern, M. (2017). Quantification of nitrous oxide in wastewater based on salt-induced stripping. *Science of the Total Environment*, **601–602**, 83–88, doi: 10.1016/j.scitotenv.2017.05.053

Lamb, B. K., Mcmanus, J. B., Shorter, J. H., Kolb, C. E., Mosher, B., Allwine, E., Blaha, D., Westberg, H. A. L., Zimmerman and P.A.T., 1995. Development of atmospheric tracer methods to measure methane emissions from natural gas facilities and urban areas. *Environmental Science & Technology*, **29**, 1468–1479, doi: 10.1021/es00006a007

Liebetrau J., Reinelt T., Clemens J., Hafermann C., Friehe J. and Weiland P. (2013). Analysis of greenhouse gas emissions from 10 biogas plants within the agricultural sector. *Water Science Technology*, **67**(6), 1370–1379, doi: 10.2166/wst.2013.005

Lindvall, T., Noren, O. and Thyselius, L. (1974). Odor reduction for liquid manure systems. *Transactions of the ASAE*, **17**(3), 508–512, doi: 10.13031/2013.36894

Liu, Y., Ni, B. J., Sharma, K. R. and Yuan, Z. (2015a). Methane emission from sewers. *Science of the Total Environment*, **524–525**, 40–51. doi: 10.1016/j.scitotenv.2015.04.029

Liu, Y., Sharma, K. R., Fluggen, M., O'Halloran, K., Murthy, S. and Yuan, Z. (2015b). Online dissolved methane and total dissolved sulfide measurement in sewers. *Water Research*, **68**, 109–118. doi: 10.1016/j.watres.2014.09.047

Liu, Y., Tugtas, A. E., Sharma, K. R., Ni, B. J. and Yuan, Z. (2016). Sulfide and methane production in sewer sediments: field survey and model evaluation. *Water Research*, **89**, 142–150. doi: 10.1016/j.watres.2015.11.050

Mampaey, K. E., van Dongen, U. G. J. M., van Loosdrecht, M. C. M. and Volcke, E. I. P. (2015). Novel method for online monitoring of dissolved N_2O concentrations through a gas stripping device. *Environmental Technology*, **36**(13), 1680–1690, doi: 10.1080/09593330.2015.1005029

Marques, R., Rodriguez-Caballero, A., Oehmen, A. and Pijuan, M. (2016). Assessment of online monitoring strategies for measuring N_2O emissions from full-scale wastewater treatment systems. *Water Research*, **99**, 171–179, doi: 10.1016/j.watres.2016.04.052

Masuda, S., Suzuki, S., Sano, I., Li, Y.-Y. and Nishimura, O. (2015). The seasonal variation of emission of greenhouse gases from a full-scale sewage treatment plant. *Chemosphere*, **140**, 167–173, doi: 10.1016/j.chemosphere.2014.09.042

Mello, W. Z. D., Ribeiro, R. P., Brotto, A. C., Kligerman, D. C., Piccoli, A. D. S. and Oliveira, J. L. M. (2013). Nitrous oxide emissions from an intermittent aeration activated sludge system of an urban wastewater treatment plant. *Química Nova*, **36**, 16–20, doi: 10.1590/S0100-40422013000100004

Mikel, D. K. and Merrill, R. (2011). EPA Handbook: Optical Remote Sensing for Measurement and Monitoring of Emissions Flux. Research Triangle, North Carolina, 27711. Available on-line at https://www3.epa.gov/ttn/emc/guidlnd/gd-052.pdf (Last visited on August 28th, 2020).

Mikola, A., Heinonen, M., Kosonen, H., Leppänen, M., Rantanen, P. and Vahala, R. (2014). N_2O emissions from secondary clarifiers and their contribution to the total emissions of the WWTP. *Water Science and Technology: A Journal of the International Association on Water Pollution Research*, **70**(4), 720–728. doi: 10.2166/wst.2014.281

Mønster, J., Samuelsson, J., Kjeldsen, P., Rella, C. W. and Scheutz, C. (2014). Quantifying methane emission from fugitive sources by combining tracer release and downwind measurements - a sensitivity analysis based on multiple field surveys. *Waste Management*, **34**, 1416–1428. doi: 10.1016/j.wasman.2014.03.025

Mønster, J., Kjeldsen, P. and Scheutz, C. (2019). Methodologies for measuring fugitive methane emissions from landfills – a review. *Waste Management*, **87**, 835–859, doi: 10.1016/j.wasman.2018.12.047

Moutafchieva, D., Popova, D., Dimitrova, M. and Tchaoushev, S. (2013). Experimental determination of the volumetric mass transfer coefficient. *Journal of Chemical Technology and Metallurgy*, **48**(4), 351–356.

Myers, S. (2019). Nitrous Oxide and Gas Transfer in Full-Scale Activated Sludge Basins. Master's thesis, Aalto University, Espoo, Finland.

Ni, B. J., Ye, L., Law, Y., Byers, C. and Yuan, Z. (2013). Mathematical modeling of nitrous oxide (N_2O) emissions from full-scale wastewater treatment plants. *Environmental Science & Technology*, **47**, 7795–7803, doi: 10.1021/es4005398

Oshita, K., Okumura, T., Takaoka, M., Fujimori, T., Appels, L. and Dewil, R. (2014). Methane and nitrous oxide emissions following anaerobic digestion of sludge in Japanese sewage treatment facilities. *Bioresource Technology*, **171**, 175–181, doi: 10.1016/j.biortech.2014.08.081

Pan, Y., van den Akker, B., Ye, L., Ni, B.-J., Watts, S., Reid, K. and Yuan, Z. (2016a). Unravelling the spatial variation of nitrous oxide emissions from a step-feed plug-flow full scale wastewater treatment plant. *Scientific Reports*, **6**(1), 20792, doi: 10.1038/srep20792

Pan, Y., Ye, L., van den Akker, B., Ganigué Pagès, R., Musenze, R. S. and Yuan, Z. (2016b). Sludge-drying lagoons: a potential significant methane source in wastewater treatment plants. *Environmental Science & Technology*, **50**(3), 1368–1375, doi: 10.1021/acs.est.5b04844

Pavelka, M., Acosta, M., Kiese, R., Altimir, N., Bruemmer, C., Crill, P., Darenova, E., Fuß, R., Gielen, B., Graf, A., Klemedtsson, L., Lohila, A., Bernard, L., Lindroth, A., Nilsson, M., Marañón-Jiménez, S., Merbold, L., Montagnani, L., Peichl, M. and Kutsch, W. L. (2018). Standardisation of chamber technique for CO_2, N_2O and CH_4 fluxes measurements from terrestrial ecosystems. *International Agrophysics*, **32**, 569–587, doi: 10.1515/intag-2017-0045

Prata, A. A., Lucernoni, F., Santos, J. M., Capelli, L., Sironi, S., Le-Minh, N. and Stuetz, R. M. (2018). Mass transfer inside a flux hood for the sampling of gaseous emissions from liquid surfaces – experimental assessment and emission rate rescaling. *Atmospheric Environment*, **179**, 227–238, doi: 10.1016/j.atmosenv.2018.02.029

Pikaar, I., Sharma, K. R., Hu, S., Gernjak, W., Keller, J. and Yuan, Z. (2014). Reducing sewer corrosion through integrated urban water management. *Science*, **80**(345), 812–814. doi: 10.1126/science.1251418

Porro, J., Kampschreur, M.J., Pijuan, M., Volcke, E., Daelman, M., Guo, L., Nopens, I., Vanrolleghem, P.A., Yuan, Z., Chandran, K. and Murthy, S. (2014). Measuring Nitrous Oxide Emissions from Biological Wastewater Treatment, Art or Science? In: Global Challenges: Sustainable Wastewater Treatment and Resource Recovery, Kathmandu, Nepal, IWA Specialist Conference, 2014-10-26-30, Papers, International Water Association (IWA).

Rapson T. D. and Dacres H. (2014). Analytical techniques for measuring nitrous oxide. *Trends in Analytical Chemistry*, **54**, 65–74, doi: 10.1016/j.trac.2013.11.004

Ren, Y. G., Wang, J. H., Li, H. F., Zhang, J., Qi, P. Y. and Hu, Z. (2013). Nitrous oxide and methane emissions from different treatment processes in full-scale municipal wastewater treatment plants. *Environmental Technology*, **34**(21), 2917–2927, doi: 10.1080/09593330.2012.696717

Reinelt, T. and Liebetrau, J. (2020). Monitoring and mitigation of methane emissions from pressure relief valves of a biogas plant. *Chemical Engineering & Technology*, **43**(1), 7–18, doi: 10.1002/ceat.201900180

Reinelt, T., Delre, A., Westerkamp, T., Holmgren, M. A., Liebetrau, J. and Scheutz, C. (2017). Comparative use of different emission measurement approaches to determinemethane emissions from a biogas plant. *Waste Management*, **68**, 173–185. doi: 10.1016/j.wasman.2017.05.053

Ribeiro, R., Freitas Bueno, R., Piveli, R., Kligerman, D., Mello, W. and Oliveira, J. (2017). The response of nitrous oxide emissions to different operating conditions in activated sludge wastewater treatment plants in Southeastern Brazil. *Water Science and Technology*, **76**, 2337–2349, doi: 10.2166/wst.2017.399

Rieger, L., Gillot, S., Langergraber, G., Ohtsuki, T., Shaw, A., Takacs, I. & Winkler, S. (2013). Guidelines for Using Activated Sludge Models. IWA Publishing, London, UK.

Rodriguez-Caballero, A., Aymerich, I., Poch, M. and Pijuan, M. (2014). Evaluation of process conditions triggering emissions of green-house gases from a biological wastewater treatment system. *Science of the Total Environment*, **493**, 384–391, doi: 10.1016/j.scitotenv.2014.06.015

Rodriguez-Caballero, A., Aymerich, I., Marques, R., Poch, M. and Pijuan, M. (2015). Minimizing N_2O emissions and carbon footprint on a full-scale activated sludge sequencing batch reactor. *Water Research*, **71**, 1–10, doi: 10.1016/j.watres.2014.12.032

Samuelsson J., Delre A., Tumlin S., Hadi S., Offerle B. and Scheutz C. (2018). Optical technologies applied alongside on-site and remote approaches for climate gas emission quantification at a wastewater treatment plant. *Water Research*, **131**, 299–309. doi: 10.1016/j.watres.2017.12.018

Scheutz, C., Samuelsson, J., Fredenslund, A. M. and Kjeldsen, P. (2011). Quantification of multiple methane emission sources at landfills using a double tracer technique. *Waste Management*, **31**, 1009–1017. doi: 10.1016/j.wasman.2011.01.015

Schierbaum, K. D., Weimar, U. and Göpel, W. 1992. Comparison of ceramic, thick-film and thin-film chemical sensors based upon SnO_2. *Sensors and Actuators B: Chemical*, **7**, 709–716. doi: 10.1016/0925-4005(92)80390-J

Shah, A., Willis, J. and Fillmore, L. (2011). Quantifying methane evolution from sewers: results from WERF/Dekalb Phase 2 continuous monitoring at honey creek pumping station and force main. *Proceedings of the Water Environment Federation*, **475–485**, doi: 10.2175/193864711802836841

Short, M. D., Daikeler, A., Peters, G. M., Mann, K., Ashbolt, N. J., Stuetz, R. M. and Peirson, W. L. (2014). Municipal gravity sewers: an unrecognised source of nitrous oxide. *Science of the Total Environment*, **468–469**, 211–218, doi: 10.1016/j.scitotenv.2013.08.051

Spinelli, M., Eusebi, A.L., Vasilaki, V., Katsou, E., Frison, N., Cingolani, D. and Fatone, F. (2018). Critical analyses of nitrous oxide emissions in a full scale activated sludge system treating low carbon-to-nitrogen ratio wastewater. *Journal of Cleaner Production*, **190**, 517–524, doi: 10.1016/j.jclepro.2018.04.178

Stiversten, B. (1983). Estimation of diffuse hydrocarbon leakages from petrochemical factories. *Air Pollution Control Association*, **33**, 323–327, doi: 10.1080/00022470.1983.10465581

Sturm, K., Yuan, Z., Gibbes, B., Werner, U. and Grinham, A. (2014). Methane and nitrous oxide sources and emissions in a subtropical fresh water researchervoir, South East Queensland, Australia. *Biogeosciences*, **11**, 5245–5258, doi: 10.5194/bg-11-5245-2014

Sümer, E., Weiske, A., Benckiser, G. and Ottow, J. C. G. (1995). Influence of environmental conditions on the amount of N_2O released from activated sludge in a domestic waste water treatment plant. *Experientia*, **51**(4), 419–422, doi: 10.1007/BF01928908

Tata P., Witherspoon J. and Lue-Hing C. (eds). (2003). VOC Emissions from Wastewater Treatment Plants. Lewis Publishers, Boca Raton, FL.

Tauber, J., Parravicini, V., Svardal, K. and Krampe, J. (2019). Quantifying methane emissions from anaerobic digesters. *Water Science and Technology*, **80**(9), 1654–1661, doi: 10.2166/wst.2019.415

United Nation Framework Convention on Climate Change (UNFCCC). (2018). GHG Emission Database, Annex I Countries, including LULUCF, inventory year 2018. https://di.unfccc.int/comparison_by_gas (accessed 26 February 2021).

Valentine, D. L. and Reeburgh, W. S. (2000). New perspectives on anaerobic methane oxidation. *Environmental Microbiology*, **2**, 477–484. doi: 10.1046/j.1462-2920.2000.00135.x

Valkova, T., Parravicini, V., Saracevic, E., Tauber, J., Svardal, K. and Krampe, J. (2020). A method to estimate the direct nitrous oxide emissions of municipal wastewater treatment plants based on the degree of nitrogen removal. *Journal of Environmental Management*, **279**, 111563. doi: 10.1016/j.jenvman.2020.111563

Vieira, A., Marques, R., Galinha, C., Povoa, P., Carvalho, G. and Oehmen, A. (2019). Nitrous oxide emissions from a full-scale biological aerated filter (BAF) subject to seawater infiltration. *Environmental Science and Pollution Research*, **26**(20), 20939–20948, doi: 10.1007/s11356-019-05470-x

Wang, J., Zhang, J., Xie, H., Qi, P., Ren, Y. and Hu, Z. (2011). Methane emissions from a full-scale A/A/O wastewater treatment plant. *Bioresource Technology*, **102**(9), 5479–5485, doi: 10.1016/j.biortech.2010.10.090

Wang, Y., Fang, H., Zhou, D., Han, H. and Chen, J. (2016). Characterization of nitrous oxide and nitric oxide emissions from a full-scale biological aerated filter for secondary nitrification. *Chemical Engineering Journal*, **299**, 304–313, doi: 10.1016/j.cej.2016.04.050

Willis, J., Fillmore, L., Shah, A., Yuan, Z. and Sharma, K. (2011). Quantifying Methane Evolution from Sewers: Results from WERF/DeKalb Phase 2- Continuous Monitoring. 84th Annual Water Environment Federation Technical Exhibition and Conference (WEFTEC 2011), Los Angeles, California, USA.

Willis, J., Brower, B., Graf, W., Murthy, S., Peot, C., Regmi, P., Sharma, K. and Yuan, Z. (2019). New GHG methodology to quantify sewer methane. 91st Annu. Water Environ. Fed. Tech. Exhib. Conf. WEFTEC 2018, pp. 4745–4752. doi: 10.2175/193864718825139465

Yan, X., Li, L. and Liu, J. (2014). Characteristics of greenhouse gas emission in three full-scale wastewater treatment processes. *Journal of Environmental Sciences*, **26**(2), 256–263, doi: 10.1016/S1001-0742(13)60429-5

Yang, J., Trela, J. and Plaza, E. (2016). Nitrous oxide emissions from one-step partial nitritation/anammox processes. *Water Science and Technology*, **74**(12), 2870–2878, doi: 10.2166/wst.2016.454

Ye, L., Ni, B.-J., Law, Y., Byers, C. and Yuan, Z. (2014). A novel methodology to quantify nitrous oxide emissions from full-scale wastewater treatment systems with surface aerators. *Water Research*, **48**, 257–268, doi: 10.1016/j.watres.2013.09.037

Yoshida, H., Mønster, J. and Scheutz, C. (2014). Plant-integrated measurement of greenhouse gas emissions from a municipal wastewater treatment plant. *Water Research*, **61**, 108–118. doi: 10.1016/j.watres.2014.05.014

You-Wen, S., Wen-Qing, L., Yi, Z., Shi-Mei, W., Shu-Hua, H., Pin-Hua, X. and Xiao-Man, Y. (2011). Water vapor interference correction in a non dispersive infrared multi-gas analyzer. *Chinese Physics Letters*, **28**, 73302, doi: 10.1088/0256-307X/28/7/073302

Yver Kwok, C. E., Müller, D., Caldow, C., Lebègue, B., Mønster, J. G., Rella, C. W., Scheutz, C., Schmidt, M., Ramonet, M., Warneke, T., Broquet, G. and Ciais, P. (2015). Methane emission estimates using chamber and tracer release experiments for a municipal waste water treatment plant. *Atmospheric Measurement Techniques*, **8**, 2853–2867. doi: 10.5194/amt-8-2853-2015

NOMENCLATURE

∅	Diameter
AD	Anaerobic digester
Al	Aluminium
Ar	Argon
AS	Activated sludge
BAF	Biologically active (or aerated) filer
BNR	Biological nutrient removal
CFC	Closed flux chamber
CH_4	Methane
DT	Disinfection tank
ET	Equalization tank
FTIR	Fourier transform infrared spectroscopy
GC	Grit chambers
HDPE	High density polyethylene
He	Helium
IPS	Influent pump station
NDIR	Non-dispersive infrared spectroscopy
N_2O	Nitrous oxide

OFC	Open flux chamber
P	Pressure
PE	Polyethylene
PN/A	Partial nitritation/anammox
PN	Partial nitritation
PP	Polypropylene
PVC	Polyvinyl chloride
SCT	Sludge concentration tank
SEIFC	Surface emission isolation flux chamber
SET	Settler
SST	Stainless steel
T	Temperature
WERF	Water Environment Research Foundation
UNFCCC	United Nations Framework Convention on Climate Change
U.S.EPA	United States Environmental Protection Agency
WWTP	Wastewater treatment plant

doi: 10.2166/9781789060461_133

Chapter 6

Full-scale emission results (N_2O and CH_4)

Vasileia Vasilaki[1], Maite Pijuan[2,3], Haoran Duan[4,5] and Evina Katsou[1]

[1]Department of Civil & Environmental Engineering, Brunel University London, Uxbridge UB8 3PH, UK. E-mail: Vasileia.Vasilaki@brunel.ac.uk; Evina.Katsou@brunel.ac.uk
[2]Catalan Institute for Water Research (ICRA), Emili Grahit 101, Girona 17003, Spain
[3]The University of Girona, Girona 17003, Spain. E-mail: mpijuan@icra.cat
[4]School of Chemical Engineering, The University of Queensland, St Lucia, QLD 4072, Australia
[5]Advanced Water Management Centre, The University of Queensland, St Lucia, QLD 4072, Australia. E-mail: h.duan@uq.edu.au

SUMMARY

This chapter reviews the studies from N_2O and CH_4 monitoring campaigns in full-scale wastewater treatment plants (WWTPs) and sewer networks. The focus is on greenhouse gas (GHG) emissions from WWTPs as more literature is available. The analysis classifies quantified N_2O and CH_4 emission factors (EFs), triggering operational conditions and formation pathways for different configurations. Control strategies to minimize N_2O emissions are proposed for different process groups. The main reasons for EF discrepancies are discussed. Overall, N_2O emission factors for processes treating low-strength wastewater streams range between 0.003 and 5.6% of the N-load (average equal to 0.9% of the N-load). Emissions higher than mainstream process average emissions have been reported in sequencing batch reactors (average equal to 3.6% of the influent N-load) and step-fed plug flow reactors. In full-scale sidestream processes, less than 15 monitoring campaigns have reported EFs (average equal to 2.5% of the N-load). Differences in the EFs among the process groups are partially attributed to disparities in the control strategies (i.e. aeration control), configuration, and operational and environmental conditions that favour the preferred enzymatic pathways. Overall, triggering operational conditions for elevated N_2O emissions in full-scale wastewater treatment processes include (i) increased NH_4^+ concentrations leading to a high ammonia oxidation rate (AOR) and increased production of intermediates (e.g. NH_2OH, NO^-, etc.), (ii) improper aeration control (i.e. inadequate aeration and non-aeration duration, over-aeration, under-aeration), (iii) NO_2^- accumulation triggering the nitrifier denitrification pathway, and (iv) sudden shifts in incomplete heterotrophic denitrification (i.e. due to excess dissolved oxygen (DO), chemical oxygen demand (COD) limitation etc.). The N_2O monitoring strategies can also influence the reliability of the quantified EFs. Due to temporal variation of N_2O emissions, short-term studies are not sufficient to quantify annual EFs. The analysis showed that the average EF for processes treating low-strength streams monitored for less than a week is 0.66% of the influent N-load. On the other hand, processes monitored over 6 months have an average EF equal to 1.74%. Compared with N_2O, CH_4 quantification from full-scale WWTPs is less investigated, while it also contributes significantly to the overall plant carbon footprint. The results of full-scale CH_4 quantification studies are summarized in this chapter. Emissions of CH_4 in WWTPs mainly originate from the influent, anaerobic wastewater treatment and anaerobic sludge handling processes. The amount of CH_4 emissions varies greatly with different configurations of WWTPs. For WWTPs without anaerobic sludge handling processes, the CH_4 emissions can mainly be traced back to the CH_4 dissolved in the influent. When anaerobic treatment is applied in WWTPs for wastewater COD removal, its CH_4 emissions might substantially increase the overall plant carbon footprint. GHG monitoring campaigns carried out in WWTPs should include the monitoring of fugitive CH_4 emissions. Finally, CH_4 and N_2O emissions reported from sewer networks are also summarized in this chapter.

The last part of the chapter summarizes some mitigation strategies applied at full-scale to control fugitive CHG emissions from WWTPs and sewers.

Keywords: Full-scale greenhouse gas emissions, methane, nitrous oxide, sewer networks, wastewater treatment plants

TERMINOLOGY

Term	Definition
Activated sludge	Flocs of sludge particles containing microbes, which are formed in the presence of oxygen in aeration tanks.
Activated sludge process	The wastewater treatment process which applies activated sludge to speed up the decomposition of contaminants in wastewater. Oxygen is provided in the aeration tank in favour of the metabolization of activated sludge, to convert contaminants into harmless products. After the aeration tank, the mixed activated sludge goes to a clarifier to separate the sludge and treated water. The treated water will undergo further treatment.
Aeration	The introduction of air into the aeration tank for the oxidation of organic, nitrogenous and phosphorous compounds by microbes, and also for keeping the activated sludge suspended and well mixed.
Aerobic	Conditions with free oxygen in the wastewater.
Ammonia monooxygenase	An enzyme catalysing NH_4^+ oxidation to NH_2OH.
Anaerobic	Conditions without atmospheric or dissolved molecular oxygen in the wastewater.
Anoxic	Conditions of oxygen deficiency and presence of oxidized nitrogen species.
Biomass	A clump of organic material consisting of living organisms, which lives on the substrates in wastewater, or the dead organism debris.
Chemical oxygen demand	An indication of the amount of organic materials in wastewater. It refers to the amount of oxygen equivalent consumed in the chemical oxidation of organic matter by a strong oxidant such as potassium dichromate.
Dissolved oxygen	Molecular oxygen dissolved in wastewater.
Greenhouse gas	Gas that absorbs and emits radiant energy within the thermal infrared range and contributes to the global warming effect.
Heterotrophic denitrification	A series of reduction reactions from nitrate to nitrogen gas by heterotrophic denitrifiers under anoxic conditions, with organic carbon as the electron donor for the reactions.
Nutrient	Substances such as nitrogenous compounds and phosphorous or organic matter that can be assimilated by microbes to promote the metabolism and growth of microbes in the reactor.
Organic matter	The organic waste of plant or animal origin from homes or industry, mainly volatile fraction of solids.

Oxidation	Oxidation is the addition of oxygen, removal of hydrogen, or the removal of electrons from an element or compound. In wastewater treatment, organic matter is oxidized to more stable substances.
pH	An indication of the acidity or alkalinity of solutions.
Reactor	Containers of different size or design which can hold the activated sludge to conduct wastewater treatment processes.
Wastewater	The used water and solids from a community that flow into a treatment plant. Storm water, surface water, groundwater infiltration and a fraction of industrial wastewater also may be included.
EF	Emission factor

6.1 INTRODUCTION

Nitrous oxide (N_2O), is a potent greenhouse gas (GHG), 298 times stronger than CO_2 in terms of global warming potential (IPCC, 2013). N_2O can be generated in large amounts and stripped in the atmosphere during biological nutrient removal (BNR) at wastewater treatment plants (WWTPs). In the past few years, concern regarding the quantification and investigation of N_2O, from full-scale wastewater treatment processes has increased. There are three main biological pathways for N_2O production in BNR systems. N_2O can be formed during the autotrophic oxidation of ammonia to nitrite/nitrate through the activity of ammonia oxidizing bacteria (AOB) under aerobic conditions (nitrification/nitritation). The N_2O production by AOB can be due to the autotrophic denitrification of nitrite (nitrifier denitrification pathway) and due to incomplete oxidation of hydroxylamine (NH_2OH) (NH_2OH oxidation pathway). N_2O is also an intermediate during the reduction of nitrate/nitrite to nitrogen gas through the activity of heterotrophic denitrifying bacteria under anoxic conditions (heterotrophic denitrification pathway). There is a wide variety of different BNR processes applied at wastewater facilities to treat the incoming wastewater (i.e. with different numbers of compartments/zones for nitrification and denitrification, recirculation flows, flow-patterns and feeding strategies). Studies have shown that the direct N_2O emissions of BNR processes in WWTPs can contribute up to ~78% of the operational carbon footprint (Daelman *et al.*, 2013). There are recent studies reporting even higher percentages; for example, N_2O contributes up to 86% of the carbon footprint in the study of Kosonen *et al.* (2016), compared to direct methane emissions (CH_4).

Significant N_2O emissions have been reported from the biological treatment of high-strength wastewater streams. The anaerobic supernatant is a by-product from the treatment of the primary and secondary sludge via anaerobic digestion when the digestate is dewatered. This stream is small in volume (1–2% compared to the mainstream line), but very concentrated in nutrients and is conventionally recycled back to the primary treatment increasing the loads (and thus, the energy requirements and costs) of the mainstream biological treatment (i.e. contains 10–20% of the WWTP nitrogen load). For this purpose, BNR technologies (such as partial-nitritation-anaerobic ammonium oxidation (PN-anammox), nitritation-denitritation, etc.) have been developed to treat high-strength streams in a cost and energy efficient way (Lackner *et al.*, 2014; Zhou *et al.*, 2018). In the sidestream biological processes, favourable conditions for N_2O generation can prevail (i.e., NO_2^- accumulation, elevated NH_4^+ concentrations, etc.). Studies have shown that biological processes treating high-strength streams can contribute over 90% of the total direct N_2O emissions compared to the mainstream BNR processes (Schaubroeck *et al.*, 2015).

The recent mitigation roadmap to carbon neutrality in urban water published by the Water and Wastewater Companies for Climate Mitigation (WaCCliM) project and the International Water Association (IWA) (Ballard *et al.*, 2018), states that direct N_2O emissions in water utilities, should be considered for carbon footprint assessment, reporting and mitigation. However, in practice, the quantification of direct N_2O emissions at WWTPs via monitoring campaigns is not a regulatory requirement. Therefore, wastewater utilities usually estimate N_2O emissions via theoretical methods, that is based on the population equivalent of the WWTP (IPCC, 2006); the latter can significantly underestimate the actual emissions (Cadwallader & VanBriesen, 2017). The 2019 IPCC Refinement of the 2006 IPCC Guidelines has significantly increased the suggested default EF; they propose a value equal to 1.6% of the influent N-load.

Full-scale monitoring campaigns have been implemented in full-scale BNR processes to provide insights into the dynamics and triggering mechanisms for N_2O generation. However, results were variable and there is still not a consensus to explain the exact causes. The application of different WWTP configurations and different biological treatments is a main reason that explains the variation in results. The sampling strategy and duration also play an important role. Most of the studies were performed over a short-term (days–weeks) showing only diurnal emission patterns. The sampling strategy (grabbing samples or online monitoring) is also a factor that can lead to an over or underestimation of the N_2O emissions. Additionally, N_2O fluxes were characterized by significant spatial and temporal variability due to the different interacting biological processes that consume or produce N_2O and the variation in operational conditions (Daelman *et al.*, 2015; Gruber *et al.*, 2020). Mechanistic process-based models have been developed over recent years aiming to integrate N_2O emissions generation of different processes in the design, operation and optimization of biological processes (Domingo-Félez *et al.*, 2017; Mannina *et al.*, 2016; Massara *et al.*, 2017). However, their online integration for the reliable quantitative estimation of N_2O emissions and offline integration for long-term quantitative purposes remain challenging (Haimi *et al.*, 2013; Mampaey *et al.*, 2019).

WWTPs also emit CH_4 (Daelman *et al.*, 2013; Ribera-Guardia *et al.*, 2019). Emissions of CH_4 in WWTPs mainly originate from the influent, anaerobic wastewater treatment and anaerobic sludge handling processes and can present large variations from plant to plant. For WWTPs without anaerobic sludge handling processes, the majority of the CH_4 emitted originates from the dissolved CH_4 in the influent formed in sewer networks. For WWTPs with anaerobic sludge handling processes, anaerobic sludge treatment and handling facilities contribute the most to the CH_4 emissions in plants. CH_4 emissions can substantially contribute to the carbon footprint of a WWTP, especially in those facilities with low N_2O emissions. Despite of its importance in the overall emitted GHG, there are only a few studies in the literature reporting CH_4 emissions from full-scale systems.

Finally, sewer systems also present fugitive greenhouse gas emission, with CH_4 being the main greenhouse gas produced although N_2O has also been reported. The reporting of emissions from sewers is much more scarce as compared to WWTPs but its important contribution to the overall CH_4 emissions of wastewater systems cannot be neglected.

6.2 N_2O EMISSIONS FROM FULL-SCALE WWTP MONITORING RESULTS

This chapter reports emission factors (EFs) for the main BNR processes for wastewater treatment and proposes mitigation measures (Table 6.1). Monitoring campaigns to quantify and mitigate N_2O emissions have been performed over recent years in different WWTP configurations. Our observations to date confirm that due to differences in monitoring strategies (i.e. length of monitoring period) and design and operational conditions, universally acceptable configuration-based or performance based EF estimation modes are not yet available. The challenge of evaluating and mitigating N_2O emissions from BNR processes is further complicated by practical and technological hurdles that are related with the little field data regarding N_2O emissions for several BNR processes and other operational constraints.

Mainstream process groups include biological nutrient removal systems targeting N-removal (N-BNR) (aerobic/anoxic compartments), biological nutrient removal systems targeting both N and P removal (NP-BNR) (anaerobic/anoxic/aerobic compartments) and conventional activated sludge (CAS) systems (only aerobic reactors). Oxidation ditch (OD) reactor types and sequencing batch reactor (SBR) types have been considered as distinct process groups. Sidestream processes including partial-nitritation reactors, 1-step and 2-step PN-anammox and nitritation-denitritation configurations are also categorized as a distinct process group. Other processes with fewer than two case studies

Table 6.1 Methods and main findings of studies resulting in mitigation measures (The abbreviations are explained below*).

Process	EF range (% N-load)	Main findings	Mitigation measures	Source
Partial Nitritation – Anammox (1 reactor)	0.17–3.9	• Smoother aeration transitions during normal reactor operation connected with lower N_2O emissions; comparison with experiments • Prolonged anoxic periods leading to increased N_2O emissions • Over-aeration significantly impacting on N_2O emissions • Nitritation-denitritation SBR: the accumulation of N_2O at the end of the SBR anoxic phase is stripped in the subsequent aerobic phase and can have a significant impact on the amount of N_2O emitted • DO and conductivity have been linked with emissions in nitritation-denitritation SBR systems	• Optimize the aeration regime by introducing aeration control and ensuring smooth shift patterns in the aeration • Preferably operate under shorter cycles and short aeration intervals to avoid accumulation of NO_2^- • Step feeding and use of conductivity as a surrogate to estimate the effluent NH_4-N concentration of the reactor and optimize the anaerobic supernatant feeding load (avoid either FA accumulation or high AOR that trigger N_2O) • Frequent alternation of aerobic/anoxic phases to avoid nitrite accumulation	Castro-Barros *et al.* (2015); Kampschreur *et al.* (2009a, b); Weissenbacher *et al.* (2010); Joss *et al.* (2009); Christensson *et al.* (2013)
Partial Nitritation/–Anammox (2 reactors) and nitritation-denitritation systems	2.3–7.6	• Nitritation: N_2O formation higher during non-aerated periods • Splitting the anoxic period: average anoxic N_2O formation rate decreased • Shorter cycles can reduce the N_2O EF at the expense of higher NO_3^- concentrations • Anammox reactor: NO_2^- accumulation potentially increasing N_2O emissions	• Apply continuous aeration in nitritation reactor • Operate under lower DO setpoint and control the aeration rate. It is preferred that DO > 1.3 mg/L. Lower DO levels have been linked with elevated N_2O generation in nitritation-denitritation SBR systems. • Avoid anoxic phases in nitritation reactors • Operate a one-reactor nitritation-anammox process; potentially emitting less N_2O due to limited NO_2^- accumulation	Mampaey *et al.* (2016); Kampschreur *et al.* (2008); Vasilaki *et al.* (2020); Gustavsson and la Cour Jansen (2011); Ahn *et al.* (2010)

(Continued)

Table 6.1 Methods and main findings of studies resulting in mitigation measures (The abbreviations are explained below*) (*Continued*).

Process	EF range (% N-load)	Main findings	Mitigation measures	Source
N-BNR	0.018–4	• N_2O emissions have been correlated with increased abundances of AOB & lower counts of N_2O-reducers; AOB abundance favoured by higher NO_3^- & NO_2^- concentrations • DO exhibiting a significant influence on the N_2O production • N_2O production mainly in the aerated zones, minor N_2O consumption & minor stripping effect in the anoxic zones • N_2O emitted directly from the aeration basin: low COD:N ratio limiting denitrification & leading to 5-times higher N_2O emissions • N_2O dynamics not significantly influenced by DO variations (within the range of 1.5–2 mg/L) • Daily N_2O peaks occurring under higher aeration flow rates (more intense N_2O stripping) and under elevated bulk NO_2^- concentrations in the bioreactor and under poor plant aeration performance and insufficient DO • Low EF: diluted influent (groundwater infiltration) as the most probable reason	• Avoid NO_2^- accumulation, low temperatures & excess DO in the anoxic bioreactors to enable complete heterotrophic denitrification & hinder nitrifier denitrification • Apply proper control of DO in aerated compartments • Apply the BP-ANN model as a convenient & effective method for the description of N_2O emissions in an A/O • Studies in full-scale MLE reactors have not suggested process/study specific mitigation measures	Castellano-Hinojosa *et al.* (2018); Sun *et al.* (2017); Kosonen *et al.* (2016); Aboobakar *et al.* (2013); Samuelsson *et al.* (2018); Masuda *et al.* (2018); Rodriguez-Caballero *et al.* (2014); Pan *et al.* (2016); Ahn *et al.* (2010); Bellandi *et al.* (2018); Spinelli *et al.* (2018); Townsend-Small *et al.* (2011); Foley *et al.* (2010); Foley *et al.* (2015)

(*Continued*)

Table 6.1 Methods and main findings of studies resulting in mitigation measures (The abbreviations are explained below*) (*Continued*).

Process	EF range (% N-load)	Main findings	Mitigation measures	Source
NP-BNR	0.068–3.4	• N_2O emitted mainly from the oxic zone, with the emission levels increasing greatly from the beginning of the oxic zone towards the anoxic zone • NO_2^- accumulation directly triggering N_2O production • Both diurnal & seasonal N_2O emission levels fluctuating strongly; N_2O generated & emitted more in summer than in winter • Other factors influencing the N_2O emission: low DO/temperature • Microbial population & aeration strategy as key factors of N_2O generation & emission • Risk of elevated emissions in processes with plug-flow pattern with step feeding • Significant spatial variability of N_2O generation within the reactor	• Increase DO availability for both AOB & NOB and improve the growth conditions of AOB • Apply a step-stage aeration mode with varying aeration intensities (location-specific emission patterns for a plug-flow process) • Ensure better mixing via a higher horizontal plug-flow rate combined with an appropriate vertical airflow flux; the large cross-section widths reduced using partition walls to elevate flow velocities under a constant A^2/O tank working volume • Avoid incomplete/intermittent nitrification & over-aeration during the aerobic processes to achieve lower N_2O emissions • Apply uniform spatial DO profiles to promote SND that probably leads to less N_2O emissions • Perform flow equalization to control the peaking factor of the influent N-loading to the AS • Ensure a sufficiently long SRT to prevent NO_2^- accumulation during nitrification • Avoid the COD limitation of the denitrification process by minimizing the pre-sedimentation of organic carbon in the influent & dosing additional organic carbon • Avoidance of over-aeration (DO 1–2 mg/L) and control of mixed liquor recirculation rates to exceed 500% has been shown to reduce N_2O emissions	Wang *et al.* (2016b); Li *et al.* (2016); Ren *et al.* (2013); Foley *et al.* (2010); Foley *et al.* (2015); Yan *et al.* (2014); Zaborowska *et al.* (2019)

(*Continued*)

Table 6.1 Methods and main findings of studies resulting in mitigation measures (The abbreviations are explained below*) (*Continued*).

Process	EF range (% N-load)	Main findings	Mitigation measures	Source
SBR	0.58–5.6	• NH_4^+ accumulation leading to a high AOR during the aerobic SBR phases and, finally, to the increased production of intermediates (e.g. NH_2OH) • Low DO during nitrification majorly influencing N_2O production • Cycles with long aerated phases showing the largest N_2O emissions, with a consequent increase in the carbon footprint • Transient NH_4^+ & NO_2^- concentrations & transition from anoxic to aerobic possibly involved in the increased N_2O production • Spatial variability of N_2O generation within the reactor has been reported due to elevated NH_4^+ concentration in the feeding point	• Apply intermittent aeration and reduce the aerated periods to decrease the NO_2^- accumulation • Increase the aeration rate during the feeding period & decrease it to a proper level for nitrification in the aerobic stage • Alternatively, change the operational SBR mode (from feeding under synchronous aeration to feeding with anoxic stirring) to ensure enough COD provision/better utilization of influent COD for denitrification • Allow the system to consume N_2O through denitrification. Extend denitrification length or if required supply external carbon source during denitrification • Continuous aeration at DO equal to ~0.5 mg/L favouring simultaneous nitrification-denitrification can also reduce the N_2O	Ni *et al.* (2013); Rodriguez-Caballero *et al.* (2015); Sun *et al.* (2013); Duan *et al.* (2020); Gruber *et al.* (2020); Foley *et al.* (2010); Foley *et al.* (2015)
OD	0.03–2.8	• Both nitrifying & denitrifying zones are potential hotspots of N_2O production • Relatively low emissions due to strong dilution effect (relatively long HRT), AOB≈NOB (minor NO_2^- accumulation, less likely N_2O production via nitrifier denitrification), more uniform DO profile in the OD process (SND promoted) • Aerated zones: N_2O fluxes correlated with location-specific pH, AS mixed liquor temperature, DO, NH_4^+ & NO_2^- concentrations & interactive combinations • Anoxic zones: N_2O fluxes correlated with location-specific sCOD, pH, AS mixed-liquor temperature, DO, NO_2^- & NO_3^- concentrations & interactive combinations	• Essential to control DO at proper level during nitrification/denitrification & enhance the utilization rate of influent organic carbon for denitrification • Multivariate analysis can be applied (i.e. clustering, classification) to investigate the combined effect of operational variables on N_2O emissions	Sun *et al.* (2015); Ren *et al.* (2013); Daelman *et al.* (2015); Vasilaki *et al.* (2018); Ahn *et al.* (2010); Yan *et al.* (2014); Masuda *et al.* (2018); Chen *et al.* (2019); Foley *et al.* (2010); Foley *et al.* (2015); Ekström *et al.* (2017)

(*Continued*)

Table 6.1 Methods and main findings of studies resulting in mitigation measures (The abbreviations are explained below*) (*Continued*).

Process	EF range (% N-load)	Main findings	Mitigation measures	Source
CAS	0.36–1.8	• Nitrification is the main driving force behind N_2O emission peaks • Compared to other parameters (e.g. sludge concentration/retention time), air flow-rate variations possibly influencing the N_2O emissions; high N_2O emissions under conditions of over-aeration or incomplete nitrification along with NO_2^- accumulation • The treatment of the anaerobic supernatant in mainstream CAS systems can trigger significant N_2O emissions	• Reduce the aeration rate to decrease the N_2O emission rate and energy consumption required for aeration • Add an anoxic zone & recirculation to a non-BNR system for nitrification; otherwise, high N_2O emissions expected in case of increased DO • Control the DO; dynamic changes in DO concentrations reported as being responsible for N_2O emission peaks • Avoid the concurrence of decreased DO & NO_2^- accumulation	Chen *et al.* (2016); Ribeiro *et al.* (2017); Ahn *et al.* (2010); Gruber *et al.* (2020); Brotto *et al.* (2015)
Other/ generic	0.004–2.8	• Investigate possible links between WWTP operating conditions & N_2O emission fluxes • Aerobic zones: N_2O fluxes correlated with location-specific pH, AS mixed liquor temperature, DO, NH_4^+ & NO_2^- concentrations & interactive combinations • Anoxic zones: N_2O fluxes correlated with location-specific sCOD, pH, AS mixed-liquor temperature, DO, NO_2^- & NO_3^- concentrations & interactive combinations • In cases of low external C-source availability, internally stored compounds (e.g. polyhydroxyalkanoates (PHAs)) can be alternatively utilized. The latter is likely to increase the N_2O production during denitrification	• BNR processes: Avoid high NH_4^+ & NO_2^- & DO concentrations & transients • Aerobic processes: avoid incomplete/ intermittent nitrification & over-aeration • Rely on more uniform spatial DO profiles to promote SND • Minimize peak N-flow (flow equalization) • pH maintained $6 \leq pH \leq 7$ • Provision of sufficient C-source to increase the possibility of N_2O consumption through denitrification. • DO must be controlled at approximately 2 mg/L while aeration minimized to avoid stripping • Perform advanced N-removal (e.g. nitritation-denitritation or partial nitritation-anammox) only after optimizing the process parameters. • BP-ANN and data-driven models suitable for the description of N_2O emissions in other WWTPs with different configurations (e.g. A^2/O, SBR & nitrification-anammox), if influent/environmental parameters & N_2O emission data can be investigated through full-, pilot- or lab-scale experiments	Wang *et al.* (2016a); Ahn *et al.* (2010); Samuelsson *et al.* (2018); Kosonen *et al.* (2016); Mello *et al.* (2013); Filali *et al.* (2013); Foley *et al.* (2010); Foley *et al.* (2015); Baresel *et al.* (2016); Gruber *et al.* (2020)

*SBR: sequencing batch reactor; DO: dissolved oxygen; FA: free ammonia; AOR: ammonia oxidation rate; AOB: ammonia oxidizing bacteria; NOB: nitrite oxidizing bacteria; COD: chemical oxygen demand; BP-ANN: back propagation-artificial neural network; A/O: anoxic/oxic reactor; MLE: modified Ludzack-Ettinger; N-BRN: nitrifying – biological nitrogen removal; NP-BNR: biological nitrogen and phosphorus removal; A2/O: anaerobic/anoxic/oxic reactor; SND: simultaneous nitrification denitrification; AS: activated sludge; SRT: sludge retention time; HRT: hydraulic retention time; OD: oxidation ditch; sCOD: soluble chemical oxygen demand; CAS: conventional activated sludge.

which do not belong to the aforementioned process groups are categorized separately. These include intermittently aerated or simultaneous nitrification-denitrification reactors (i.e., Filali *et al.*, 2013; Gruber *et al.*, 2020; Mello *et al.*, 2013), systems with external carbon dosage (Ahn *et al.*, 2010) and biofilm reactors for C (i.e. Townsend-Small *et al.*, 2011) or N removal (i.e., Bollon *et al.*, 2016).

In total ~67% of the analysed mainstream reactors, have reported the quantified EFs in terms of the %N-load. Approximately 12% of the studies have reported the EFs in terms of N-removed.

There is a significant variation in the N_2O emissions of full-scale wastewater treatment processes. The N_2O emissions range reported in literature is between 0.003% of the influent N-load for a mainstream BNR system treating municipal low-strength wastewater, diluted by groundwater and marine intrusions and 7.6% of the NH_4-N load for a sidestream short cut enhanced nutrient abatement (SCENA) SBR treating anaerobic digestion supernatant. Generally, BNR processes treating high strength streams have been associated with high risk of elevated N_2O emissions. This is mainly due to the high ammonia oxidation rate (AOR) and NO_2^- accumulation typically observed in such systems (Desloover *et al.*, 2011; Gustavsson & la Cour Jansen, 2011; Kampschreur *et al.*, 2008). Discrepancies in the EFs observed in the different process groups can, to some extent, be attributed to variations in operational characteristics and control parameters. In addition to reactor configuration, emission rates depend on the operational/environmental conditions and preferred enzymatic pathways (Wan *et al.*, 2019).

Figure 6.1 shows boxplots of the observed EFs (with respect to the influent N-load) of mainstream processes in different countries. The width of the violin plot outlines surrounding the boxplots represents the data kernel density distribution of the EFs. Overall, ~60% of the monitoring campaigns in processes treating low-strength streams have been performed in China (18%), the United States (18%), Australia (14%) and Sweden (10%). Overall, the highest EFs have been reported in Australia. The median EF in Australia is 1.35% of the N-load (average equal to 1.6%). The lowest EFs have been

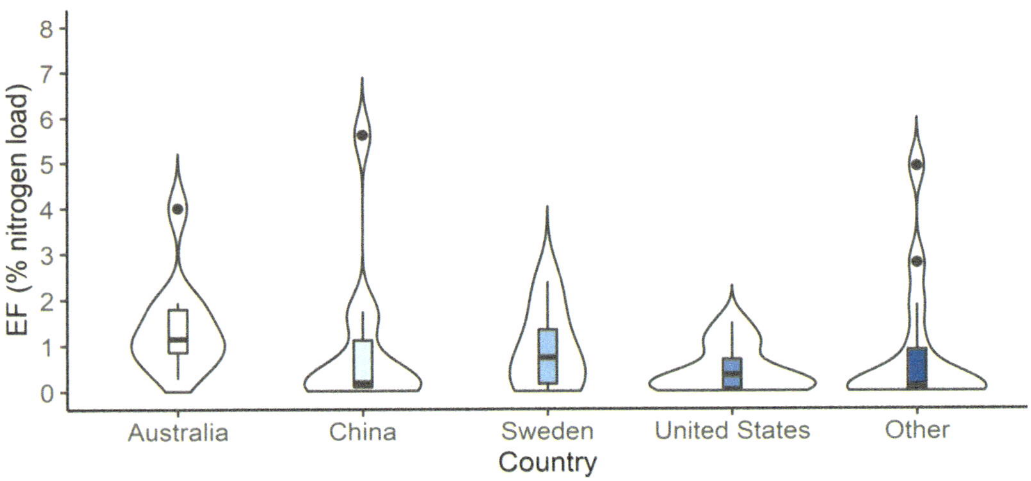

Figure 6.1 Boxplots of the reported EFs (% N load) with respect to the WWTP in different countries using violin plot outlines. The rectangles represent the interquartile range. The median is denoted by the black horizontal line dividing the box into two parts. The dots represent the values exceeding 1.5 times the interquartile range. The upper and lower whiskers stand for values higher or lower than the interquartile range, respectively (within 1.5 times the interquartile range above and below the 75th and 25th percentile, respectively). The violin plot outlines show the kernel probability density of the EF in mainstream and sidestream processes; the width of the violin plot outlines represents the proportion of the data located there.

reported in China; the median EF is equal to 0.2% of the N-load (average 0.8% of the N-load). In the United States the median EF is 0.3%, while in Sweden the median EF is 0.74% of the N-load (averages equal to 0.4% and 0.9% of the N-load, respectively.)

The majority of the processes monitored in Australia are step-feed reactors. Higher than average N_2O emissions have been reported for step-feed reactors. Moreover, the majority of the WWTPs studied in China do not have anaerobic digestion on-site. The anaerobic supernatant is a by-product from the treatment of the primary and secondary sludge via anaerobic digestion when the digestate is dewatered. This stream is small in volume (1–2% compared to the mainstream line), but very concentrated in nutrients and is conventionally recycled back to the primary treatment increasing the loads (and thus, the energy requirements, costs and potentially the N_2O emissions) of the mainstream biological treatment. The majority of the studied processes in Sweden and Australia belong to WWTPs with anaerobic digestion on-site. It is possible that WWTPs that recycle anaerobic supernatant that contains 10–20% of the WWTP nitrogen load, have a higher risk of increased N_2O emissions. The sampling protocols and duration of monitoring campaigns also vary significantly between the different countries. For instance, long-term monitoring of N_2O emissions (>6 months) has been performed mainly in China via grab-samples collected bi-monthly. Vasilaki *et al.* (2019) showed that low-frequency (i.e. bimonthly) grab-sampling might underestimate emissions due to limitations in sampling duration (i.e. it does not occur during night-time) or due to short-term process perturbations triggering elevated emissions not coinciding with the sampling days.

Figure 6.2 shows the EF range for the different groups of mainstream processes and sidestream processes. As a general remark, the majority of the EFs in processes treating low-strength wastewater

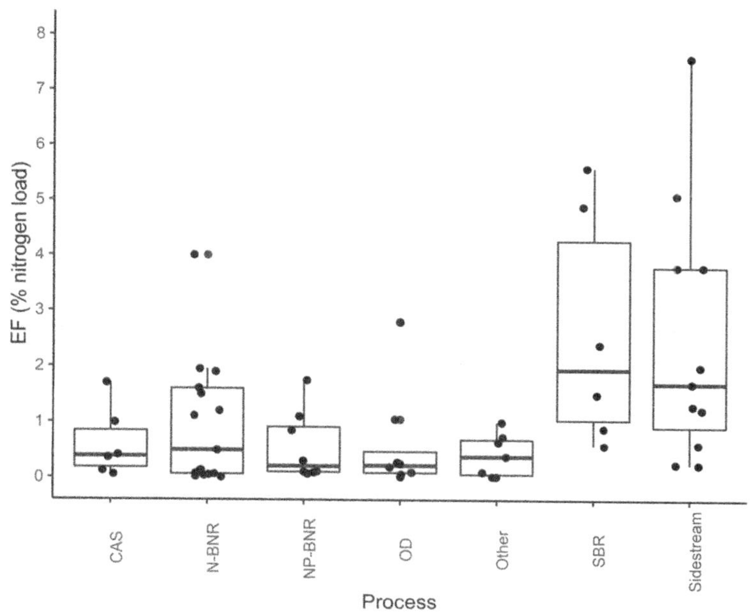

Figure 6.2 Boxplots visualizing the EF range for the different groups of mainstream processes and sidestream processes (adapted from Vasilaki *et al.* 2019). The rectangles represent the interquartile range. The median is denoted by the black horizontal line dividing the box into two parts. The dots represent values exceeding 1.5 times the interquartile range. The upper and lower whiskers represent values higher or lower the interquartile range, respectively (within 1.5 times the interquartile range above and below the 75th and 25th percentile, respectively).

range from 0.1% to 2% of the influent N-load. Higher than average emissions have been reported in SBRs and step-fed plug-flow reactors. The potential for N_2O emissions from reactors treating high-strength wastewater streams is considered higher compared to the mainstream BNR processes. This is mainly because the nitritation/partial-nitritation occurring during sidestream treatment is linked with higher ammonia oxidation rate (AOR) and NO_2^- accumulation (Desloover *et al.*, 2011; Gustavsson & la Cour Jansen, 2011; Kampschreur *et al.*, 2008).

The benchmarking of the EF for groups of processes remains challenging, mainly due to differences in the strategies applied during monitoring, the operational and environmental conditions and the duration of monitoring campaigns. Additionally, limited information exists on the N_2O emissions for several other processes (e.g. biofilm-based processes or partial-nitritation-anammox systems, etc.) (Sabba *et al.*, 2018; Vasilaki *et al.*, 2019).

Process characteristics, EFs, N_2O triggering mechanisms, operational conditions and mitigation measures for processes treating low-strength and high-strength wastewater streams are analysed in Sections 6.2.1 and 6.2.2, respectively.

High sensitivity of the quantified EF between different monitoring strategies and monitoring campaign durations has been reported (Vasilaki *et al.*, 2019). When considering the duration of the monitoring campaign, studies lasting over a year result in a median EF equal to 1.7% of the N-load. On the other hand, most of the monitoring campaigns lasting less than one month have reported EFs less than 0.3% of the N-load. Therefore, short-term monitoring periods may fail to capture underlying seasonal variations in the N_2O formation (or be affected by short-term process perturbations) and, consequently, result in unreliable EFs. Similarly, the studies monitoring N_2O emissions in mainstream wastewater processes continuously (i.e. online via gas analysers), have quantified higher N_2O EFs than studies monitoring N_2O emissions discontinuously (i.e. offline via grab samples). The average EFs of mainstream wastewater processes monitored continuously and discontinuously are 1.2% and 0.44% of the N-load, respectively. Low-frequency sampling campaigns have a high risk of not sufficiently capturing short-term changes in pollutant concentrations, operational conditions and system disturbances impacting N_2O generation.

The reliability of the monitoring campaigns also depends on the amount and location of the sampling points (Gruber *et al.*, 2020). Significant spatial variations of the N_2O emissions have been reported in complete mixing reactors (Duan *et al.*, 2020). The variability was attributed to gradients in the nutrients within the reactor and elevated NH_4^+ concentrations close to the feeding area causing increased AOR and triggering N_2O emissions. The use of one gas chamber for N_2O emissions collection in complete mixing reactors might result in unreliable quantification of N_2O EFs. The latter can have significant implications, since one gas chamber is conventionally used for sampling in complete mixing reactors, whereas several sampling points are suggested for reactors operating in plug-flow mode (Duan *et al.*, 2020). On the other hand, Gruber *et al.* (2020) observed negligible spatial variability of N_2O emissions in a complete mixing reactor monitored with three gas chambers in different locations within the reactor. Therefore, additional studies are required to determine the optimum N_2O sampling points and understand under which conditions nutrient gradients are observed.

Differences in the N_2O emissions have been also reported in parallel reactors. Chen *et al.* (2019), studied parallel OD reactors and observed deviations in the N_2O emissions behaviour under similar NH_4^+, NO_3^- and dissolved oxygen (DO). They suggested that the reliable quantification of WWTP N_2O EFs requires monitoring of all plant reactors. The opposite has been reported by Daelman *et al.* (2015) who observed similar N_2O emission patterns in two parallel OD reactors.

Generally, the quantification of reliable annual EFs requires sampling campaigns lasting at least 1 year. Additionally, a decision tree for the selection of the monitoring strategy has been developed by Gruber *et al.* (2020). They define specific criteria for the selection of sampling points and location. Influent compositions, feeding locations and homogeneity, and the key performance indicators (i.e. removal efficiencies) should be considered to decide whether similar N_2O emissions are expected in parallel reactors. Similarly, plug-flow type reactors featuring spatial variability of concentrations and aeration intensity require multiple sampling points.

The variability of EF reported in full-scale wastewater treatment processes can be attributed to complex relationships between emitted N_2O and operational conditions and different configurations (i.e., SBR, continuous systems), loads (i.e., NH_4^+ concentrations), feeding strategies and operational control (i.e., DO set-points).

The conditions leading to elevated N_2O emissions or N_2O generation are usually associated with N-forms build-up in the reactor (i.e., NH_4^+, NH_2OH, NO-, NO_2^-). Depending on the BNR process and the acclimatized biomass in the reactor, the accumulation of N intermediates does not necessarily have to be very high to trigger N_2O pathways. The accumulation mainly depends on the influent dynamics or on improper process operation and/or design.

During nitrification, NH_3/NH_4^+ concentration can significantly affect the N_2O emissions (Law *et al.*, 2012; Wunderlin *et al.*, 2012). High NO_2^- accumulation, that is the toxic product of aerobic NH_3 oxidation in AOB, has also been linked with elevated N_2O emissions, especially under low DO concentrations (Desloover *et al.*, 2011; Kampschreur *et al.*, 2008; Law *et al.*, 2012; Massara *et al.*, 2017; Peng *et al.*, 2015; Tallec *et al.*, 2006). Different N_2O production dynamics can be potentially triggered under the same NO_2^- concentration depending on the type of AOB. It has been also reported that AOB can adapt to different environments with different NO_2^- concentrations. Overall, N_2O generation has been associated with higher NO_2^- concentrations in wastewater treatment processes (Foley *et al.*, 2010). DO is also considered an important parameter affecting N_2O emissions (Kampschreur *et al.*, 2009b), with sub-optimum DO concentrations generally increasing N_2O emissions. AOB can use nitrite instead of oxygen as an electron acceptor (Kampschreur *et al.*, 2009a, b), in oxygen limiting conditions, generating N_2O emissions. At present, establishing a generic optimum DO concentration threshold to minimize N_2O emissions for nitrifying systems is not possible since other compounds (i.e. N compounds discussed above) have a simultaneous effect on N_2O generation. An optimal DO level for minimal N_2O emissions can be established for each system taking into consideration the concentration of other compounds that affect these emissions. Overall, in aerated reactors/zones, higher emissions are expected under high NH_4^+ concentrations, high AOR, sub-optimum DO (under or over-aeration) or NO_2^- build-up (Desloover *et al.*, 2011; Kampschreur *et al.*, 2008). Sub-optimum pH and short solids retention times (SRTs) have been reported to influence N_2O production in AOB.

Additionally, feeding mainstream reactors with high-strength streams (i.e., anaerobic supernatant) can create peak nutrient loadings increasing the risk of elevated N_2O emissions. In the studied processes, WWTPs that have anaerobic digestion on-site have median EFs equal to 1.5% of the N-loading (average equal to 1.47%). On the other hand, processes that are not fed with anaerobic supernatant (i.e., WWTPs applying sludge dewatering and drying) have a median EF equal to 0.11% of the N-load (average 0.47% of the N-load).

Sub-optimum DO, chemical oxygen demand (COD), pH and SRT can also result in nitrite and N_2O accumulation during denitrification (Schulthess *et al.*, 1994; Yang *et al.*, 2012). Low values of COD/N can result in incomplete denitritation and, therefore, N_2O accumulation via the heterotrophic denitrification pathway (Wunderlin *et al.*, 2012).

Seasonal environmental variations, can influence the bacterial community structure in WWTPs (Flowers *et al.*, 2013) and the N_2O emissions. Temperature can significantly affect the AOB specific growth rate during nitrification (Van Hulle *et al.*, 2010). The higher temperature also decreases the N_2O solubility, thus intensifying the N_2O stripping to the atmosphere (Reino *et al.*, 2017). Adouani *et al.* (2015) reported an increased sensitivity of the N_2O reductase activities at lower temperatures compared to other denitrification enzymes and, therefore, to incomplete denitrification. Other seasonal variations (e.g., influent loading, wet and dry season) can affect the enzymatic reactions and the emissions. Vasilaki *et al.* (2018) observed peaks of N_2O emissions coinciding with precipitation events, at low temperatures. Further investigation is required to understand the impact of seasonal effects on the N_2O emissions (Gruber *et al.*, 2020; Vasilaki *et al.*, 2019).

Disturbances in the process can affect short-term (i.e., 1 day) or even longer period (i.e., >1 week) N_2O generation (Vasilaki *et al.*, 2018). Gruber *et al.* (2021) observed in an SBR reactor, that N_2O

emission peaks, nitrification failure, poor activated sludge settleability and high turbidity of treated effluent, were all linked to a less diverse microbial community and changes in community mixture. Specifically, a decrease in abundance of filamentous and nitrite oxidizing bacteria was reported.

6.2.1 Processes treating low strength streams

6.2.1.1 N-BNR and NP-BNR processes

This section discusses findings regarding N_2O generation in BNR processes. The Modified Ludzack-Ettinger (MLE) process is the most studied N-removal configuration. In total, 41% of the N-BNR systems are MLE processes. The MLE process consists of anoxic and aerobic tanks and a secondary settler. The influent wastewater is first fed to the anoxic tank for denitrification and next to the aerobic zone for nitrification. The process uses an internal recycle flow from the aerobic tank to the head of the anoxic tank providing nitrate for denitrification. After anoxic and aerobic processes, the wastewater is fed to the secondary settler. A part of the sludge, the return activated-sludge, returns to the head of the anoxic zone to increase the mixed liquor volatile suspended solids (MLVSS) concentration in the reactor. In total, the N-BNR configurations consist of a broad category of processes with anoxic and oxic compartments. Step-feed plug-flow reactors with alternating anoxic/oxic zones and reactors with small anoxic compartments (for predenitrification) and aerobic compartments with and without recirculation of nitrates belong to this category.

Similarly, the anaerobic/anoxic/aerobic (A^2/O) process is the most studied N and P-removal configuration. In total 64% of the NP-BNR systems are A^2/O processes. The A^2/O process is a modification of the MLE process. The process consists of an anaerobic zone followed by the same configuration of MLE. The return activated sludge goes to the head of the anaerobic tank. The anoxic tank is used to decrease the amount of nitrate, in the anaerobic tank, that returns from the activated sludge. Overall, the NP-BNR process group includes configurations with anaerobic, anoxic and aerobic compartments, such as reversed-A^2/O configurations (A^2/O systems where the anaerobic and anoxic compartments are reversed) or A^2/O systems with a predenitrification zone.

The median EF of N-BNR processes is 0.5% of the influent N-Load, while the median EF of NP-BNR processes is 0.2% of the influent N-Load. In N-BNR configurations, the N_2O emissions range between 0.003% and 4% of the influent N-load (Foley *et al.*, 2010; Spinelli *et al.*, 2018). In NP-BNR configurations, the N_2O emissions range between 0.07% and 1.75% of the influent N-load (Wang *et al.*, 2016b; Yan *et al.*, 2014). MLE and A^2/O are the most studied configurations; around 54% of the monitoring campaigns have been performed in these two systems.

Overall, in N-BNR and NP-BNR systems, N_2O emission peaks have been reported during the transition from non-aerated to aerated zones/compartments (i.e. Rodriguez-Caballero *et al.* 2014; Sun *et al.* 2017). This can be partially due to incomplete denitrification and accumulation of dissolved N_2O under anoxic conditions. Elevated emissions have been also linked with excess DO in anoxic compartments, inhibiting complete denitrification (Castellano-Hinojosa *et al.*, 2018). Therefore, process control in the anoxic compartments should target the minimization of NO_2^- accumulation and excess DO and the avoidance of COD limitation. This will facilitate complete heterotrophic denitrification and N_2O consumption.

In aerobic compartments, peak N_2O fluxes have coincided with peak nutrient loads and low DO concentrations (Wang *et al.*, 2011, 2016b); the integration of flow equalization can control the influent N-loading peaks to the systems. Moreover, close to the inlet of aerobic compartments with a plug-flow pattern, AOB abundances and high NO_2^- concentrations can result in an increase in the N_2O emissions. Risk of elevated emissions has also been reported in processes with plug-flow pattern and step feeding. Pan *et al.* (2016) showed an EF equal to 0.7% of the influent N-load in the first step of a plug-flow reactor and 3.5% in the second step. The increased N_2O emissions in the second step were attributed to the recirculated stream being directed only at the first step causing dilution; the MLVSS concentration in the second step was 40% lower than that in the first step (70% less biomass compared to the first step). The higher specific AOR in this stage triggered the N_2O generation. It is important to note that in reactors with plug-flow pattern, the effect of the N-load, DO concentration and temperature on N_2O emissions varies along the reactor (Aboobakar *et al.*, 2013). Thus, the dominant N_2O triggering conditions can also vary.

Low EFs have been reported in reactors treating diluted low-strength wastewater (i.e. due to groundwater infiltration) (Bellandi *et al.*, 2018; Spinelli *et al.*, 2018). Low EFs have also been reported in the majority of the A^2/O and reversed A^2/O processes, with the median N_2O EF ~0.11% of the influent N-load. However, it must be noted that the seasonal variability of the N_2O emissions in A^2/O rectors has not been studied adequately. The majority of the monitoring campaigns lasted less than 3 months. Wang *et al.* (2016b) showed that the EF of an A^2/O process has strong temporal patterns and varied between 0.1 and 3.4% of the influent N-load between different months within 1 year. The effect of environmental conditions on N_2O generation is discussed in Section 6.2.

Both the nitrifier denitrification pathway and the NH_2OH oxidation pathway have been suggested as major contributors to the N_2O emissions in aerated compartments/zones. The nitrifier denitrification pathway is considered the main triggering mechanism in aerobic compartments (i) when NO_2^-, NH_4^+ and O_2-limiting conditions co-exist (Wang *et al.*, 2016b), (ii) when NO_2^- is correlated with N_2O emissions, (iii) when increasing N_2O emissions are observed under DO limitation where sufficient O_2 is provided to the AOB for the oxygenation of NH_3 to NH_2OH but not for aerobic respiration; NO_2^- is potentially used as alternative electron acceptor to complete nitrification (Aboobakar *et al.*, 2013; Castellano-Hinojosa *et al.*, 2018; Sun *et al.*, 2017; Wang *et al.*, 2011), and (iv) under shock loads of toxic compounds, where the AOB likely activate their denitrification pathway (Rodriguez-Caballero *et al.*, 2014). In anoxic compartments, the nitrifier denitrification pathway has been suggested as the main contributor to N_2O generation, when excess DO is observed (Castellano-Hinojosa *et al.*, 2018). The NH_2OH oxidation pathway is significantly promoted at higher DO concentrations (Blomberg *et al.*, 2018; Zaborowska *et al.*, 2019) and when N_2O emissions increase together with the AOR increase (Ni *et al.*, 2015; Pan *et al.*, 2016). Finally, heterotrophic denitrification is mainly triggered under carbon-limiting conditions (low COD/N ratio) and excess DO in anoxic compartments (Andalib *et al.*, 2017; Wunderlin *et al.*, 2012).

6.2.1.2 Sequencing batch reactors (SBR)

The SBR process uses a fill-and-draw complete mixing reactor operating in batch reaction steps. The biological removal and clarification occur in the same tank.

Mainstream SBRs have reported higher N_2O emissions compared to the other mainstream process groups. EFs range between 0.89% of the influent N-load for an SBR that receives the anoxic selector effluent and operating under feeding (intermittent aeration), aerobic (intermittent aeration), settling and decanting sequences (Duan *et al.*, 2020) and 5.6% of the influent TN-load for an SBR operating under aerated feeding, aerobic and anoxic settling and decanting sequences (1 h each) (Sun *et al.*, 2013). The average EF from SBR reactors is 3.6% of the influent N-load (median: 3.65% of the influent N-load).

Overall, elevated emissions are attributed to (i) NH_4^+ accumulation leading to high AOR during the aerobic SBR phases and to increased production of intermediates (e.g., NH_2OH, NO^-, etc.), (ii) long aerated cycles, (iii) transitions from anoxic to aerobic phases possibly triggering increased N_2O production, (iv) rapid changes in the NH_4^+ and NO_2^- concentrations within the cycle, (v) accumulation of dissolved N_2O during anoxic settling and decanting that is stripped in the subsequent aerobic phase and (vi) accumulation of NO_2^-.

Intermittent aeration and short aerated periods have been suggested to reduce the NO_2^- accumulation in SBR systems and subsequently N_2O emissions. Duan *et al.* (2020), however, showed that elevated DO concentrations (up to 8 mg/L) during intermittent aeration can also be responsible for elevated emissions in the SBR systems and should be avoided. The authors used a multi-pathway N_2O model (Peng *et al.*, 2016) to design a mitigation strategy that was implemented in the studied system. They showed that continuous aeration at DO equal to ~0.5 mg/L that favours simultaneous nitrification-denitrification (SND) can be an effective operational strategy for SBR reactors. The SND operation mode resulted in 35% reduction of the N_2O emissions compared to intermittent over-aeration. The reduction was due to the reduction of DO concentration during feeding and aerated phases that can enhance denitrification during aerated periods and minimize NO_2^- accumulation.

Additionally, in SBR reactors the supply of an external carbon source during denitrification can secure sufficient COD provision and better utilization of influent COD for denitrification (promoting complete heterotrophic denitrification). This allows the system to consume N_2O during denitrification and avoid stripping of residual liquid N_2O in the subsequent aerated phases, thus, reducing N_2O emissions. A cycle configuration with a sequence of aerobic phases (adjusted on site) followed by short non-aerated periods has been proposed as an effective control mechanism to reduce N_2O generation (Rodriguez-Caballero *et al.*, 2015).

In SBR reactors, elevated N_2O emissions are attributed to the NH_2OH pathway when elevated DO is observed during feeding and when high NH_4^+ concentrations are observed without simultaneous NO_2^- increase in the aerated phases. The nitrifier denitrification pathway is the main N_2O triggering mechanism when low DO concentrations in aerobic phases are linked with the N_2O generation and when certain NO_2^- accumulation under aerobic conditions is observed in the reactor. In cases where N_2O generation continues when the aeration finishes, both the nitrifier denitrification and heterotrophic denitrification can contribute to the N_2O formation in the reactor. Finally, the correlation between N_2O emission and influent COD/N, indicates that the incomplete heterotrophic denitrification is mainly responsible for the N_2O generation.

6.2.1.3 Oxidation ditch (OD)

An OD is a modified activated sludge biological treatment process; the removal of biodegradable organics is achieved by applying long SRTs. ODs are considered to approach complete mixing systems, but they can also operate in plug-flow mode.

The N_2O emissions of OD reactor types range from 0.03% of the N-load for an OD reactor favouring simultaneous nitrification denitrification (Ahn *et al.*, 2010) to 2.8% of the N-load for a system consisting of an anaerobic/anoxic/oxic plug-flow reactor followed by two parallel Carrousel reactors (Daelman *et al.*, 2015). The median EF is equal to 0.2% of the influent N-load (average equal to 0.3% of the N-load).

Overall, relatively low emissions have been reported in OD systems; this is attributed to the strong dilution effect (relatively long hydraulic retention time), to the abundance of AOB and nitrite oxidizing bacteria (NOB), and to the more uniform DO profile in the OD process especially when SND is promoted (Li *et al.*, 2016). Abundance of NOB and denitrifiers has been reported in OD systems as contributing to the consumption of NO_2^- during nitrification. The latter reduced NO_2^- accumulation and facilitated complete heterotrophic denitrification (Sun *et al.*, 2015). It is important to note, though, that the majority of the OD reactors have been monitored with gas hoods. The use of floating hoods to monitor GHG emissions in OD systems when aerated with surface aerators has been criticized due to the turbulence commonly observed at the surface affecting the capturing of the emissions in the hood (Ye *et al.*, 2014).

Elevated emissions have been linked to NH_4^+ concentration peaks. In a simulation study, Ni *et al.* (2013) observed that more than 90% of the N_2O emissions were attributed to aerated zones with DO > 2 mg/L and NH_4-N concentration peaks (up to ~9 mg/L). Inadequate anoxic zones, inhibiting complete denitrification have been also reported in OD systems. OD systems with surface aerators are prone to developing zones with reduced DO, inhibiting complete nitrification, that results in nitrite accumulation and increased N_2O emissions.

A similar N_2O emissions pattern has been reported in two OD reactors operating under different control and design (Chen *et al.*, 2019; Daelman *et al.*, 2015). Both systems were monitored over a long term; an increasing trend in N_2O emissions coincided with increase in water temperature whereas, low emissions were observed under lower water temperature. Further studies are required to understand the exact triggering mechanisms at decreasing temperatures and investigate if this N_2O pattern is process-specific.

All N_2O generation pathways have been reported in OD reactors. Incomplete heterotrophic denitrification has been attributed to the competition of the denitrification steps and the preference of the heterotrophic denitrifiers to reduce NO_3^- instead of N_2O under electron donor limitation (Pan *et al.*, 2013). Additionally, heterotrophic denitrification and nitrifier denitrification are the main N_2O triggering mechanisms at insufficient anoxic conditions. Under these conditions NO_2^- accumulation is expected. The NH_2OH oxidation pathway will be triggered in periods with influent NH_4^+ concentration peaks, high ammonia oxidation rate and elevated DO concentrations. Vasilaki *et al.* (2018), showed that the relationships between N_2O emissions and other variables monitored in an OD (i.e. NH_4^+, NO_3^-, DO) are dynamic and affected by seasonal variations. The preferred N_2O pathways were found to be dependent on time and operational conditions.

6.2.1.4 Conventional activated sludge systems

CAS systems consist of aerobic reactors (1-step feed or multiple-step feed) without anoxic compartments. They are characterized by a median EF equal to ~0.4% of the influent N-load (average equal to 0.71%). The NH_4^+ removal is between 38% and 53%. The EF in CAS systems ranges from 0.05% of the N-load (translated to 9% of the NH_4-N removed) (Chen *et al.*, 2019) to 1.7% of the N-load (Gruber *et al.*, 2020).

Peak loads and recirculation of the anaerobic supernatant can be responsible for the N_2O fluxes observed in CAS systems, whereas high aeration rates have been reported, enhancing N_2O stripping (Chen *et al.*, 2016). Additionally, the spatial variation of nutrients in step-fed CAS systems can result in incomplete denitrification and affect the AOR during nitrification (due to uneven substrate-biomass distribution in all feeding points), hence, increasing the total N_2O emissions (Pan *et al.*, 2016). The treatment of the anaerobic supernatant in mainstream CAS systems has been reported to trigger significant N_2O emissions. Gruber *et al.* (2020), monitored the N_2O emissions in two parallel CAS systems and found that elevated emissions were observed solely in the reactor treating the anaerobic supernatant. N_2O emissions can be reduced by up to 80% when influent N-loads are reduced by 30%.

Tumendelger *et al.* (2014) reported that the NH_2OH oxidation pathway was responsible for up to 90% of the N_2O formation under high DO (~2.5 mg/L at the middle and close to the outlet of the aerobic tank) in a CAS system (site preference (SP) isotopic analysis). Both AOB pathways contributed almost equally to N_2O emissions generation at DO levels of ~1.5 mg/L, whereas nitrifier denitrification dominated at DOs lower than 1.5 mg/L. Overall, in activated sludge systems the reduction of aeration rates can decrease the N_2O fluxes stripped and the control of DO has been proposed as a key measure to mitigate N_2O emissions. Additionally, the addition of an anoxic zone to avoid the concurrence of decreased DO and NO_2^- accumulation can have a positive impact on the N_2O generation.

6.2.2 Processes treating high strength (high nitrogen loading) streams

Sidestream processes, such as the partial-nitritation-anammox and nitritation-denitritation are emerging for the low-cost treatment of high-strength municipal wastewater streams (Lackner *et al.*, 2014; Zhou *et al.*, 2018). In the nitritation-denitritation process, ammonium is firstly oxidized to nitrite (nitritation) and then it is reduced to nitrogen gas (denitritation) under anoxic conditions. In the partial-nitritation-anammox process, ammonium is partially oxidized to nitrite and then ammonia and nitrite are converted to nitrogen gas and nitrate under oxygen-free conditions by anaerobic ammonium oxidizers (anammox).

N_2O monitoring studies have been performed in less than 15 sidestream processes. There is a need to improve the understanding of N_2O generation in sidestream processes. For instance, the N_2O emissions were equal to 7.6% of the NH_4-N load in a SCENA process and contributed up to 97% of the operational carbon footprint of the process (Vasilaki *et al.*, 2020). Additionally, the seasonal variation (~1 year) of N_2O emissions in sidestream reactors has not been assessed.

The average EF from full-scale nitritation and partial-nitritation reactors is equal to 4.3% of the influent N-load. One-stage granular anammox reactors have an average EF of 1.1% of the influent N-load. Zhuang *et al.* (2020) showed that in a high-rate anammox granular sludge reactor, N_2O emissions were mainly generated in anammox flocs (~10% total biomass) compared to anammox granules. They reported that the N_2O reduction in flocs was inhibited due to the accumulation of NO. Anammox bacteria concentrations were higher in granules and scavenged NO that was inhibiting the N_2O reduction. In comparison, emissions in lab and pilot-scale single-stage granular anammox reactors ranged from 0.1 to 12.19% of influent N-load (Wan *et al.*, 2019). Therefore, additional studies are required to establish reliable ranges of EFs in sidestream processes and gain insights into the mitigation of N_2O emissions. Low emissions have been also reported in moving bed biofilm reactor (MBBR) annamox technologies. Christensson *et al.* (2013) reported that ~0.75% of the N-reduced were emitted as N_2O at a full-scale deammonification MBBR. Process disturbances and a DO concentration lower than 1 mg/L can increase the N_2O emissions. The authors concluded that stable operation at DO equal to 1 mg/L can result in average daily N_2O of 0.06% of N-reduced. In pilot-scale MBBR-anammox and integrated fixed-film activated sludge (IFAS) – anammox systems Liu *et al.* (2014) reported N_2O EFs equal to 0.52% and 1.7% of the total Kjeldahl nitrogen (TKN) load, respectively.

In the sidestream reactors, the rate of aeration and the DO concentration can significantly impact both the N_2O emissions generation and the N_2O mass fluxes stripped in the atmosphere (Harris *et al.*, 2015; Rathnayake *et al.*, 2015). The influence of the aeration regime on the N_2O generation varies; this can be partially due to the different configurations. For example, Mampaey *et al.* (2016) and Stenström *et al.* (2014) reported an increase of N_2O emissions with lower DO concentrations in a PN-anammox system and a sidestream nitrification-denitrification SBR, respectively. Vasilaki *et al.* (2020) observed increased dissolved N_2O concentration peaks at DO levels lower than 1 mg/L in a SCENA SBR system. The authors reported a Spearman's correlation coefficient between dissolved N_2O concentration and DO equal to −0.7. On the other hand, Kampschreur *et al.* (2009a) could not identify a relationship between the N_2O increase and the higher aeration flowrate during a prolonged aeration experiment in a single-stage nitritation-anammox reactor. As a general remark, it is suggested to have DO concentrations higher than 1 mg/L.

In one-stage PN-anammox reactors, elevated N_2O emissions have been reported during shifts from low to high aeration and linked with high NH_4^+ concentrations and high AOR. Additionally, in nitritation-denitritation SBRs the aerobic dissolved N_2O concentration has been correlated with the decrease of the average aerobic conductivity rate (Spearman's correlation coefficient equal to 0.7) and the changes of conductivity between sequential cycles. Higher emissions have been also linked with high ammonia removal efficiencies (Vasilaki *et al.*, 2020). This means that elevated emissions are due to AOR or higher than average NO_2^- accumulation. N_2O emissions have also increased due to the stripping of the accumulated N_2O in the previous anoxic cycle (accumulated due to incomplete denitritation). In that case, step-feeding, control of initial NH_4^+ concentrations and aeration duration can mitigate the N_2O peaks.

In anammox reactors, a non-negligible generation of N_2O emissions has been reported. Kampschreur *et al.* (2008) observed an EF equal to 0.6% of the influent N-load for the anammox compartment of a full-scale two-stage PN-anammox system treating anaerobic supernatant. Given that N_2 is recognized as the end-product of the anammox process (Jetten *et al.*, 2005), the authors assumed that AOB from the nitritation compartment infiltrated the anammox reactor. Yan *et al.* (2019) observed, via laboratory experiments, that the increase of the COD/N ratio from 0 to 1 can decrease the N_2O generation by 16.7% in a CANON process coupled with denitrification. Therefore, low carbon dosage can be a mitigation strategy for the CANON process or anammox reactors infiltrated with AOB from the nitritation compartment in two-stage PN-anammox processes.

It must be noted, though, that N_2O generation depends not only on a single operational variable but also on the combined effect of several variables (temperature, NH_4^+, NO_2^-, DO, aeration rate). This is supported by Wan *et al.* (2019) who found that higher temperatures resulted in increased N_2O

emissions in the presence of COD and in decreased N_2O emissions in the absence of COD in a one-stage PN-anammox reactor. The latter was attributed to increased anammox activity and reduction of NO_2^- accumulation at higher temperature.

N_2O emissions elevated during shifts from low to high aeration. Under these operational conditions the NH_2OH pathway has been reported as a main generation mechanism (Castro-Barros *et al.*, 2015). At elevated NH_4^+ or DO in the reactor, N_2O production by nitrifier denitrification is enhanced, while NH_2OH oxidation is relatively unimportant (Harris *et al.*, 2015). Both NH_2OH oxidation and nitrifier denitrification can be the main contributors to N_2O accumulation across a range of conditions with varying concentrations of NH_4^+, O_2, and NO_2^-. Harris *et al.* (2015) concluded that when N_2O emissions are relatively low under optimal reactor operation the current understanding of N_2O production and isotopic fractionation is incomplete and needs further investigation.

6.3 CH_4 EMISSIONS FROM FULL-SCALE WWTPs

Compared with N_2O, CH_4 emissions from full-scale WWTPs is less investigated, while it contributes significantly to the overall plant carbon footprint. The results of full-scale CH_4 quantification studies are summarized in Table 6.2. Emissions of CH_4 in WWTPs mainly originate from the influent, anaerobic wastewater treatment and anaerobic sludge handling processes. CH_4 emissions thus vary greatly with different WWTP configurations. For WWTPs without anaerobic sludge handling processes, the majority of the CH_4 may be traced back to the dissolved CH_4 in the influent, which was likely formed in sewer networks. For WWTPs with anaerobic sludge handling processes, anaerobic sludge treatment and handling facilities may contribute the most to CH_4 emissions in plants. When anaerobic treatment is applied in WWTPs for wastewater COD removal, its CH_4 emissions might substantially increase the overall plant carbon footprint.

6.3.1 WWTPs without anaerobic sludge handling

In WWTPs without anaerobic sludge treatment, the largest CH_4 emission source is often the aerobic tank and headworks (especially aerated grit chamber) via the stripping of CH_4 dissolved in the influent. The biological generation of CH_4 requires strict anaerobic conditions. Due to the short residence time, and periodical exposure to oxygen and nitrate or nitrite, it is often not believed that CH_4 can be produced from the headworks or from the aerobic/anoxic wastewater treatment processes (Ribera-Guardia *et al.*, 2019). Instead, it is more likely to be generated in pressurized sewer mains (see next section). By measuring liquid and gas CH_4 concentration, mass balance analyses have been performed in some studies (Daelman *et al.*, 2013; Noyola *et al.*, 2018; Yan *et al.*, 2014), suggesting dissolved CH_4 in the influent could be the main source of CH_4 emissions in WWTPs without anaerobic sludge treatments. In two studied WWTPs in China without sludge stabilization processes, Yan *et al.* (2014) observed 80–98% of total CH_4 was emitted from the wastewater treatment line, and the remaining from headworks. With mass balance analysis, it was concluded that the majority of the CH_4 emissions originated from the CH_4 dissolved in the influent. Similar observations were reported by Daelman *et al.* (2013). In two Dutch WWTPs without anaerobic sludge digestion, 86% and 77% of the total methane emissions stemmed from the influent. Nevertheless, in some cases, CH_4 may be generated during the wastewater treatment processes. A WWTP in Japan without anaerobic sludge digestion saw its CH_4 mainly (86.4%) emitted from the aerobic tank. Considering the relatively small amount of CH_4 in the influent, the CH_4 emitted is likely formed during the wastewater treatment processes under anaerobic conditions (Masuda *et al.*, 2015). Wang *et al.* (2011) also reported CH_4 formation during the wastewater treatment processes, emitting a significant amount of CH_4.

6.3.2 WWTPs with anaerobic sludge handling

Anaerobic sludge digestion is a commonly practised technology for sludge stabilization. During anaerobic sludge digestion, biodegradable organic matters are degraded in the absence of oxygen, to

Table 6.2 A summary of CH_4 quantification results in full-scale WWTPs.

WWTP	Wastewater treatment process	Influent organic strength (mgO_2/L)	Sludge treatment process	Contribution from Headworks (%)	Contribution from secondary treatment (%)	Contribution from sludge management (%)	Emission factor (kg CH_4/kg $COD_{influent}$)	Contribution of total CH_4 emissions to the overall carbon footprint	Source
Durham WWTP, USA	Aerobic treatment	BOD: 250	Aerobic sludge digestion	36	51	13	0.16% (BOD)	Not measured	Czepiel *et al.* (1993)
Jinan WWTP, China	Anaerobic/ anoxic/ oxic (AAO) process	COD: 200	Sludge dewatering, drying	9.3	80% (40% from anaerobic tank, and 40% from aerobic tank)	9.5	0.08%	Not measured	Wang *et al.* (2011)
Papendrecht WWTP, Netherlands	Anaerobic tank followed by anoxic/ oxic carousel reactors	Not available	Sludge dewatering	50	47	3	0.87%	17%	Daelman *et al.* (2013); Daelman *et al.* (2012)
Kortenoord WWTP, Netherland	Anaerobic tank followed by anoxic/ oxic carousel reactors	Not available	Sludge dewatering	45	45	10	0.53%	13%	Daelman *et al.* (2013); Daelman *et al.* (2012)
Kralingseveer WWTP, Netherlands	Anoxic/ aerobic plug flow followed by carousel reactors with surface aerators	COD:339	Anaerobic sludge digester; digestate stored for up to 5 days	<31	Not provided	72 (50% from dewatered sludge storage and buffer tank)	0.8–1.2% (seasonal variation)	5–36%	Daelman *et al.* (2013); Daelman *et al.* (2012)
Granollers WWTP, Spain	Anoxic/oxic plug–flow reactors	COD:730	Anaerobic sludge digestion	Not provided	Not provided	Not provided	0.016%	Not measured	Rodriguez-Caballero *et al.* (2014)

(Continued)

Table 6.2 A summary of CH_4 quantification results in full-scale WWTPs. (*Continued*).

WWTP	Wastewater treatment process	Influent organic strength (mgO_2/L)	Sludge treatment process	Contribution from Headworks (%)	Contribution from secondary treatment (%)	Contribution from sludge management (%)	Emission factor (kg CH_4/kg $COD_{influent}$)	Contribution of total CH_4 emissions to the overall carbon footprint	Source
Beijing WWTP1, China	Oxidation ditch process	COD: 306–689	Sludge thickening, drying and storage	1%	98% (mainly from influent stripping)	Likely negligible	0.17–0.39%	19%	Yan *et al.* (2014)
Beijing WWTP2, China	Reversed AAO process	COD: 353–687	Sludge thickening, drying	11%	89% (mainly from influent stripping)	Likely negligible	0.10–0.19%	15.8%	Yan *et al.* (2014)
Beijing WWTP2, China	AAO process	COD: 353–687	Sludge thickening, drying	19.8%	80% (mainly from influent stripping)	Likely negligible	0.06–0.11%	6.1%	Yan *et al.* (2014)
Sendai WWTP, Japan	Pseudo Anoxic-oxic process	COD: 110	Sludge dewatering, storage	8.2	86.4	<5.4	1.0%	8.3%	Masuda *et al.* (2015)
La Roca del Vallès WWTP, Spain	SBR for COD and N removal	COD: 600	No sludge stabilization	N/A	N/A	N/A	0.02%	Not measured	Rodriguez-Caballero *et al.* (2015)
Akiu WWTP, Japan	Oxidation ditch	BOD: 130	Sludge thickening, storage		97.7 (Grit chamber + OD)	<1	1.3% (BOD)	<4%	Masuda *et al.* (2018)
Hirosegawa WWTP, Japan	Anoxic-oxic process	BOD: 210	Sludge thickening, storage	<75	23.5	Not clear	0.98% (BOD)	<5%	Masuda *et al.* (2018)
Kamiyagari WWTP, Japan	Pseudo anoxic-oxic process	BOD: 150	Sludge thickening, storage	68.1	22.6	<5.6	0.3% (BOD)	<5%	Masuda *et al.* (2018)
Girona WWTP, Spain	Modified Ludzack-Ettinger (MLE) configuration	COD: 410	Anaerobic digestion	N/A	N/A	N/A	0.28–0.49%	45–57%	Ribera-Guardia *et al.* (2019)

CH_4-rich biogas, which can be captured for energy recovery. Undesirable leaks of the generated CH_4 could contribute significantly to the plant overall carbon footprint. In WWTPs with anaerobic sludge digestion, its related CH_4 emissions could contribute the majority of the total CH_4 emissions. Daelman *et al.* (2012) found $72 \pm 23\%$ of the total CH_4 emissions originated from the anaerobic sludge handling facilities: the gravitational thickener for the primary sludge, the centrifuge, the buffer tank for the effluent of the digester, the storage tank that contains the dewatered sludge and methane leakage from the gas engines. Recent studies focusing on methane losses from 23 biogas plants, including those from WWTP facilities, found an average CH_4 emission rate of 10.4 $kgCH_4/h$ with an average loss of 4.6% of the produced CH_4 (Scheutz and Fredenslund, 2019; Tauber *et al.*, 2019). Importantly, Pan *et al.* (2016) identified that the anaerobic sludge drying lagoon could also produce a large amount of CH_4. During a long-term sludge drying process, the degradable organics are converted to CH_4 under anaerobic conditions. Without capturing the produced biogas, the CH_4 emissions from a long-term sludge drying lagoon would represent a quarter to two-thirds of the overall GHG emissions from the investigated WWTP.

6.3.3 WWTPs with anaerobic wastewater treatment technologies

While most WWTPs rely on anoxic/aerobic technologies for COD removal, anaerobic technologies (e.g., upflow anaerobic sludge blanket reactor and anaerobic lagoon) are also applied in WWTPs for COD removal. The anaerobic COD removal wastewater treatment processes often lead to substantial CH_4 emissions. During anaerobic wastewater treatment, biodegradable organics are converted to CH_4. Methane is regarded poorly soluble in water with a relatively high Henry's Law constant. It was previously believed that dissolved methane was saturated at equilibrium with the gas phase methane concentration. However, studies have found dissolved methane is often supersaturated in bulk liquid, and can be several times higher than the predicted equilibrium concentration (Hartley and Lant, 2006). The ratio of the actual dissolved methane concentration to the calculated value from Henry's Law is used to describe the extent of methane supersaturation. For anaerobic treatment systems receiving municipal wastewater, the degree of methane supersaturation measured in many studies falls in the range of 1.34 to 6.9, with a median value of 1.64 (Crone *et al.*, 2016; Hartley and Lant, 2006). Inadequate liquid-to-gas mass transfer of methane due to the lack of mixing and low liquid velocities inherent to the reactor design, results in the observed supersaturation of methane (Crone *et al.*, 2016).

The relatively high dissolved CH_4 concentration in the anaerobic treatment effluent leads to substantial release of CH_4 in downstream processes. Existing quantification studies are mostly conducted in lab-scale and pilot-scale reactors. According to the data summarized by Crone *et al.* (2016), nearly half (49%) of the total CH_4 generated during the anaerobic wastewater treatment is lost in the effluent, which is subject to release in downstream processes. The aerobic activated sludge process is reportedly able to remove 80% of the dissolved CH_4 (Daelman *et al.*, 2012). With COD removal efficiency of anaerobic treatment technologies in the range of 55–80%, the dissolved CH_4 in the anaerobic treatment effluent could lead to CH_4 emissions of about 1.4–2% of the influent COD ($kgCH_4/kgCOD_{influent}$). In comparison, for WWTPs without anaerobic wastewater treatment, the total CH_4 emissions account for 0.02–1.2% of the influent COD (Table 6.2). The anaerobic wastewater treatment process could produce CH_4 emissions higher than an entire WWTP implementing anoxic/aerobic wastewater treatment processes. The CH_4 emissions resulting from the anaerobic wastewater treatment process is still one of the major obstacles for its wide application.

It is clear that CH_4 emissions represent a significant portion of the overall carbon footprint in WWTPs while rarely being the dominant one. The contribution of CH_4 emissions varied mostly from 4% to 19% of the overall carbon footprint (Table 6.2). In cases when N_2O emissions are particularly low, the CH_4 emissions could be the dominant source (45–57%) of overall GHG emissions, as reported by Ribera-Guardia *et al.* (2019). Overall, CH_4 emissions from WWTPs should be monitored, especially in facilities where anaerobic treatment is implemented.

6.4 GHG EMISSIONS FROM SEWER NETWORKS

6.4.1 Reported CH_4 emissions from sewer networks

Anaerobic conditions in sewer pipes together with the high biodegradable COD concentration in the sewage favour the accumulation of methane as the end-product of the methanogenic archaea present in the sewer networks. There are not many studies focusing on the quantification of the overall CH_4 emissions from full-scale sewer systems, probably due to the complexity of the monitoring and the limited accessibility of some parts of the network. To date, overall methane emission data is only available for single pipe rising main and gravity sewers, calculated through the dissolved methane concentration data and following the methods explained in Chapter 4.

The overall methane emission potential of the monitored rising main sewers varies substantially, ranging from 0.04 to 0.32 kg CO_2-equivalent/m^3 with an average value of 0.18 kg CO_2-equivalent/m^3 of wastewater transported. Table 6.3 summarizes the studies reporting CH_4 emissions from sewer networks in the literature.

The majority of the methane formed in rising mains will be eventually stripped to the atmosphere via ventilation in gravity sewers or at WWTPs during the treatment of wastewater, mainly because methane oxidation in sewers is expected to be a slow process (Valentine & Reeburgh, 2000). Therefore, rising main data can be used to calculate potential overall emission rates from sewer systems.

In some other studies, the quantification of overall CH_4 emissions has been carried out by direct measurement of methane emission rate from a discharge manhole (Shah *et al.*, 2011). However, this methodology is expected to underestimate emissions as CH_4 could also be emitted at other locations in the network.

6.4.2 Reported N_2O emissions from sewer networks

Studies providing N_2O emission data from sewer networks are sparse, with very few studies published to date. In 2014, Short *et al.* reported the dissolved N_2O concentrations from the inlet of three WWTPs in Australia during an 8 month monitoring campaign. They found that average levels in the raw wastewater were relatively consistent among the three WWTPs monitored at around 7–10 μg N-N_2O/L. Combining these results with wastewater parameters they were able to calculate presumptive per capita N_2O emission factors, resulting in 1.39–1.84 g N_2O/person year and 0.009–0.02 kg N-N_2O/kg TN.

Another study conducted in the sewer network of the Cincinnati municipality (Fries *et al.*, 2018) reported that its wastewater collection system was a non-point source of N_2O. Based on their results, they estimated approximately an average rate of 151.2 ± 326 g N_2O/d for the whole city.

As the authors from both studies mentioned, all these numbers should be taken with caution as further investigations are needed to better understand the magnitude of sewer N_2O emissions.

6.5 MITIGATION STRATEGIES APPLIED IN FULL-SCALE SYSTEMS

6.5.1 GHG mitigation in WWTPs

There is no standardized methodology for the establishment of N_2O mitigation strategies in full-scale systems. Table 6.1 summarizes the main mitigation strategies that have been proposed or tested in full-scale wastewater treatment processes.

Testing different operational modes is regarded as one of the most effective ways to identify measures for emission mitigation. Several studies have modified the aeration intensity and/or strategy, and optimized the DO set-point and cycle duration to investigate the effect on N_2O emissions in full-scale BNR processes (Castro-Barros *et al.*, 2015; Duan *et al.*, 2020; Kampschreur *et al.*, 2009a, 2009b; Mampaey *et al.*, 2016; Rodriguez-Caballero *et al.*, 2015). For instance, Mampaey *et al.* (2016) achieved a reduction in the N_2O emissions of 56% when the cycles in a one-stage granular SHARON reactor were shortened by 1 h. Rodriguez-Caballero *et al.* (2015) tested different operational conditions in a full-scale SBR. They suggested an optimum control strategy for the minimization of N_2O

Table 6.3 Dissolved methane concentrations and methane emission in rising mains (adapted from Liu *et al.* 2015b).

Country	Name	Type	Length (m)	Diameter (mm)	A/V (m^{-1})	HRT (h) average (min–max)	WW T (°C)	Dissolved CH_4 (mg/L) average (min–max)	Daily flow (m^3/day)	Production ($kgCH_4$/day)	Overall emissions ($kgCO_2eq/m^3$)	References
Australia	UC09	Rising main	828	150	26.7	2.5 (3.1–4.6)	27.7	5.3 (4.4–6.1)	200	1.1	0.11	Foley *et al.* (2011)
Australia	C16	Rising main	1100	300	13.3	2.6 (3.9–11.0)	22.5	15.3 (11.3–3.0)	707	9.8	0.32	Foley *et al.* (2011)
Australia	C16	Rising main	1100	300	13.3	2.6 (3.9–11.0)	23.5	5.2 (3.4–6.6)	707	2.6	0.11	Foley *et al.* (2009)
Australia	C27	Rising main	4400	525	7.6	9.1	24.6	9.1 (5.0–15.0)	2840	24.6	0.19	Liu *et al.* (2015b)
Australia	C27	Rising main	4400	525	7.6	9.1	20.3	7.1 (3.5–12.0)	2840	19.0	0.15	Liu *et al.* (2015b)
Australia	PerthB	Rising main	15 000	900	4.4	–	–	4.8	11 000	52.8	0.10	Liu *et al.* (2015b)
Thailand	RV	Gravity	1000	1000	–	27.9 (22–31.4)	33.3	10.1 (8.0–13.7)	–	–	–	Chaosakul *et al.* (2014)
Thailand	RV	Gravity	1000	1000		7.8 (0–12)	30.2	4.6 (0.1–11.4)	–	–	–	Chaosakul *et al.* (2014)
Australia	CO16 PS	Pumping station	–	–	–	–	23.5	1.5 (1.0–1.92)	707	–	–	Foley *et al.* (2009)
Australia	OR3 PS	Pumping station	–	–	–	–	–	0.51	2000	–	–	Liu *et al.* (2015b)
Spain	Radin	Rising main	2930	500	8.0	4.7 (2.9–6.8)	16.0	1.95 (1.3–2.7)	4210	8.2	0.04	Personal communication
Spain	Collet	Rising main	4800	556	3.6	12.4 (7.3–15.5)	25.4	3.0 (1.9–4.01)	2700	8.1	0.06	Personal communication

emissions based on the application of short aerobic-anoxic cycles (20-min aerobic phase and short duration of anoxic stage).

Activated sludge models have been also applied to identify potential N_2O mitigation strategies in BNR systems. Ni *et al.* (2015) developed a mechanistic model utilizing the data from a two-step plug-flow reactor (Pan *et al.*, 2016) showing that the biomass specific N-loading rate is linked with the elevated N_2O emissions observed in the second step of the process. Different operational conditions were tested with the model demonstrating that lower N_2O emissions (<1% of the N-load) can be achieved if 30% of the total return activated sludge (RAS) stream is recirculated to the second step of the plug-flow reactor (Table 6.1). However, it is unknown whether the suggested mitigation strategy was demonstrated in the system. Similarly, Zaborowska *et al.* (2019) used multiple-pathway activated sludge modelling to investigate N_2O mitigation strategies in an A^2/O reactor. They showed that DO concentrations between 1 and 2 mg/L and mixed liquor recirculation rates above 500% could minimize N_2O emissions and energy consumption during aeration without compromising TN removal in the studied A^2/O reactor. Duan *et al.* (2020) used a multiple-pathway model to test different N_2O mitigation strategies in an SBR reactor. Based on the simulation results, they modified the aeration control of the system. They showed that SND operation mode can result in 35% reduction of the N_2O emissions compared to intermittent over-aeration.

Overall, the main techniques for mitigating the N_2O emissions in wastewater treatment processes target (i) the reduction of the diurnal variation of NH_4^+ loads and avoidance of NH_4^+ peaks and NH_4^+ and NO_2^- build-up (i.e. integration of equalization tanks, recycling steps, optimization of anaerobic supernatant feeding), (ii) the increase of the MLVSS concentration to lower the specific N-loading (i.e. optimization of the RAS or SRT increase), (iii) the facilitation of complete reactions by providing sufficient electron donors (COD) during denitrification (i.e. supply of additional carbon source to ensure complete denitrification) and electron acceptors (O_2) during nitrification, and (iv) the facilitation of N_2O consumption during denitrification (i.e. increasing anoxic duration, lowering DO to enhance SND).

Reports on mitigation of methane from WWTPs are very scarce. Some technologies have been proposed for the removal of dissolved methane from anaerobic effluents, one for the most effective being the application of a degassing membrane (Bandara *et al.*, 2011). However, their application is very limited and no studies for their application in full-scale WWTPs have been found.

Sludge storage also contributes significantly to the fugitive methane emissions from WWTPs as digested sludge has a significant residual methane potential (Daelman *et al.*, 2012). The authors proposed the use of the ventilation air from the buffer tank as combustion air in the gas engines of the cogeneration plant, receiving the biogas produced in the digesters. This would result in less diluted methane streams going to the cogeneration plant, but this should be adapted to handle methane concentrations that exceed the lower explosive limit of methane in air.

Finally, it is important to highlight the need for good housekeeping and regular maintenance of the anaerobic digestion facilities present in WWTPs for sludge digestion, to avoid fugitive CH_4 emissions from these reactors.

6.5.2 GHG mitigation from sewers

CH_4 is the main GHG emitted from sewers and it is usually biogenically formed together with hydrogen sulfide under anaerobic conditions (Chapter 5). The wastewater industry uses several chemical-dosing approaches to mitigate sulfide emissions including the addition of nitrate, oxygen, ferric salts, hydroxide (pH elevation) and free nitrous acid (FNA) (Zhang *et al.*, 2008). But those can also suppress CH_4 formation from sewers because the methanogens are slow growers and are very sensitive to environmental conditions as compared with sulfate reducing bacteria (SRB) (Guisasola *et al.*, 2008). Also, in contrast to SRB, methanogens usually inhabit the deeper zone of sewer biofilms or sediments and are usually protected due to limited penetration of the dosed chemical. Thus, for effective control of methanogens, a higher dosage of chemicals may be needed to achieve full penetration during the

Table 6.4 Summary of the CH_4 mitigation studies conducted in full-scale sewer networks.

Chemical	Dosing levels	Dosing plan	CH_4 reduction (%)	CH_4 production recovery	Reference
Nitrate	17 kg $N\text{-}NO_3^-$/ML	One shock	13	100% in 2 days	Shah *et al.* (2011)
Nitrate	50 kg $N\text{-}NO_3^-$/ML	One shock	27	100% in 2 days	Shah *et al.* (2011)
Hydroxide	pH 11.5	Shock for 6 h	97	3% in 15 days	Gutierrez *et al.* (2014)

initial dosing period, when overall bacterial activity is high. However, continuous dosing, as required for sulfide control with most chemicals, may not be necessary. Table 6.4 summarizes the mitigation studies conducted in full-scale sewer networks.

Today, the current practice of selecting chemicals and design of dosing locations/rates is still mainly based on an individual's experience (Ganigue *et al.*, 2011; Liu *et al.*, 2015a). Constant, flow-paced and profiled dosing rates are currently applied during chemical dosing, again based on experience. Instead, the approach should be based on specific features of the sewer in question. In this respect, the SeweX model (Sharma *et al.*, 2008) consists of an empowering tool in supporting decision-making. Concentrations of methane, sulfide and flows show significant temporal and spatial dynamics in sewers. The rudimentary current methods could be ineffective in methane control, resulting in over-dosing of chemicals during periods with low methane and sulfide production, and conversely underdosing during other periods.

6.6 CONCLUDING REMARKS

Currently, operational strategies at WWTPs do not consider the mitigation of GHG emissions. New objectives, New objectives, including environmental and carbon neutrality targets, in the water industry require approaches to dynamically integrate new parameters (i.e. GHG emissions sensors, energy meters) into the process monitoring, control and decision-making.

Process-based N_2O EF benchmarking is challenging due to (i) differences in the N_2O generation triggered by the site-specific operational characteristics, environmental conditions and control parameters, and (ii) the sensitivity of the quantified EF to differences in monitoring strategies and duration of monitoring campaigns. The quantification of reliable annual EFs requires sampling campaigns lasting at least 1 year. Additional campaigns are required for specific groups of processes (i.e., processes treating high strength streams, biofilm technologies) that have received less attention until now.

Guidelines for N_2O mitigation measures for different process groups have been developed. Further research is required to develop practical approaches to help utilities to quantify, understand and report the N_2O EF and develop dynamically evolving mitigation measures based on the operational conditions. Future research can explore the possibility of coupling artificial intelligence (AI) techniques with multiple-pathway process models for full-scale applications, to facilitate the fast and adaptable online implementation of model predictive control and forecasting decision support tools.

GHG monitoring campaigns carried out in WWTPs should include the monitoring of fugitive CH_4 emissions, which contribute significantly to the overall plant carbon footprint. CH_4 emissions mainly originate from the influent, anaerobic sludge handling processes and anaerobic wastewater treatment in WWTPs. For WWTPs without anaerobic sludge handling processes, the CH_4 emissions can mainly be traced back to the CH_4 dissolved in the influent. The implementation of anaerobic sludge handling processes may contribute the most to CH_4 emissions in WWTPs. When anaerobic treatment is applied in WWTPs for wastewater COD removal, its CH_4 emissions might substantially increase the overall plant carbon footprint.

Finally, more attention should be paid to fugitive GHG emissions from sewer networks. Several studies suggest CH_4 emissions could be important in some parts of the sewer networks, with most of the monitoring campaigns being conducted in pressurized sewer mains. However, very little information is reported for full-scale gravity sewers and very scarce data is available for N_2O emissions from sewer networks.

ACKNOWLEDGEMENTS

Maite Pijuan acknowledges the support from the Economy and Knowledge Department of the Catalan Government through a Consolidated Research Group (ICRA-TECH – 2017 SGR 1318) – Catalan Institute for Water Research and the Spanish Government through the Salvador de Madariaga mobility program (PRX19/00051). Vasileia Vasilaki and Evina Katsou would like to acknowledge the Horizon 2020 research and innovation program, SMART-Plant under grant agreement No 690323. Haoran Duan acknowledges the support of the Australian Research Council (ARC) through project DP180103369.

REFERENCES

Aboobakar A., Cartmell E., Stephenson T., Jones M., Vale P. and Dotro G. (2013). Nitrous oxide emissions and dissolved oxygen profiling in a full-scale nitrifying activated sludge treatment plant. *Water Research*, **47**, 524–534. https://doi.org/10.1016/j.watres.2012.10.004

Adouani N., Limousy L., Lendormi T. and Sire O. (2015). N_2O and NO emissions during wastewater denitrification step: influence of temperature on the biological process. Comptes Rendus Chim. International Chemical Engineering Congress (ICEC) 2013: From fundamentals to applied chemistry and biochemistry, **18**, 15–22. https://doi.org/10.1016/j.crci.2014.11.005

Ahn J. H., Kim S., Park H., Rahm B., Pagilla K. and Chandran K. (2010). N_2O emissions from activated sludge processes, 2008–2009: results of a National Monitoring Survey in the United States. *Environmental Science and Technology*, **44**, 4505–4511. https://doi.org/10.1021/es903845y

Andalib M., Taher E., Donohue J., Ledwell S., Andersen M. H. and Sangrey K. (2017). Correlation between nitrous oxide (N_2O) emission and carbon to nitrogen (COD/N) ratio in denitrification process: a mitigation strategy to decrease greenhouse gas emission and cost of operation. *Water Science and Technology*, **77**, 426–438. https://doi.org/10.2166/wst.2017.558

Ballard S., Porro J. and Trommsdorff C. (2018). The Roadmap to A Low Carbon Urban Water Utility: An International Guide to the WaCCliM Approach. IWA Publishing, Place of publication not identified, London.

Bandara W. M., Satoh H., Sasakawa M., Nakahara Y., Takahashi M. and Okabe S. (2011). Removal of residual dissolved methane gas in an upflow anaerobic sludge blanket reactor treating low-strength wastewater at low temperature with degassing membrane. *Water Research*, **45**(11), 3533–3540.

Baresel C., Andersson S., Yang J. and Andersen M. H. (2016). Comparison of nitrous oxide (N_2O) emissions calculations at a Swedish wastewater treatment plant based on water concentrations versus off-gas concentrations. Advances in Climate Change Research, Including special topic on atmospheric black carbon and its effects on cryosphere, **7**, 185–191. https://doi.org/10.1016/j.accre.2016.09.001

Bellandi G., Porro J., Senesi E., Caretti C., Caffaz S., Weijers S., Nopens I. and Gori R. (2018). Multi-point monitoring of nitrous oxide emissions in three full-scale conventional activated sludge tanks in Europe. *Water Science and Technology*, **77**, 880–890. https://doi.org/10.2166/wst.2017.560

Blomberg K., Kosse P., Mikola A., Kuokkanen A., Fred T., Heinonen M., Mulas M., Lübken M., Wichern M. and Vahala R. (2018). Development of an extended ASM3 model for predicting the nitrous oxide emissions in a full-scale wastewater treatment plant. *Environmental Science and Technology*, **52**, 5803–5811. https://doi.org/10.1021/acs.est.8b00386

Bollon J., Filali A., Fayolle Y., Guerin S., Rocher V. and Gillot S. (2016). N_2O emissions from full-scale nitrifying biofilters. *Water Research*, **102**, 41–51. https://doi.org/10.1016/j.watres.2016.05.091

Brotto A. C., Kligerman D. C., Andrade S. A., Ribeiro R. P., Oliveira J. L. M., Chandran K. and de Mello W. Z. (2015). Factors controlling nitrous oxide emissions from a full-scale activated sludge system in the tropics. *Environmental Science and Pollution Research*, **22**, 11840–11849. https://doi.org/10.1007/s11356-015-4467-x

Cadwallader A. and VanBriesen J. M. (2017). Incorporating uncertainty into future estimates of nitrous oxide emissions from wastewater treatment. *Journal of Environmental Engineering*, **143**, 04017029. https://doi.org/10.1061/(ASCE)EE.1943-7870.0001231

Castellano-Hinojosa A., Maza-Márquez P., Melero-Rubio Y., González-López J. and Rodelas B. (2018). Linking nitrous oxide emissions to population dynamics of nitrifying and denitrifying prokaryotes in four full-scale wastewater treatment plants. *Chemosphere*, **200**, 57–66. https://doi.org/10.1016/j.chemosphere.2018.02.102

Castro-Barros C. M., Daelman M. R. J., Mampaey K. E., van Loosdrecht M. C. M. and Volcke E. I. P. (2015). Effect of aeration regime on N_2O emission from partial nitritation-anammox in a full-scale granular sludge reactor. *Water Research*, **68**, 793–803. https://doi.org/10.1016/j.watres.2014.10.056

Chaosakul T., Koottatep T. and Polprasert C. (2014). A model for methane production in sewers. *Journal of Environmental Science and Health, Part A*, **49**, 1316–1321. https://doi.org/10.1080/10934529.2014.910071

Chen W.-H., Yang J.-H., Yuan C.-S. and Yang Y.-H. (2016). Toward better understanding and feasibility of controlling greenhouse gas emissions from treatment of industrial wastewater with activated sludge. *Environmental Science and Pollution Research*, **23**, 20449–20461. https://doi.org/10.1007/s11356-016-7183-2

Chen X., Mielczarek A. T., Habicht K., Andersen M. H., Thornberg D. and Sin G. (2019). Assessment of full-scale N_2O emission characteristics and testing of control concepts in an activated sludge wastewater treatment plant with alternating aerobic and anoxic phases. *Environmental Science and Technology*, **53**, 12485–12494. https://doi.org/10.1021/acs.est.9b04889

Christensson M., Ekström S., Chan A. A., Le Vaillant E. and Lemaire R. (2013). Experience from start-ups of the first ANITA Mox plants. *Water Science and Technology*, **67**, 2677–2684. https://doi.org/10.2166/wst.2013.156

Crone B. C., Garland J. L., Sorial G. A. and Vane L. M. (2016). Significance of dissolved methane in effluents of anaerobically treated low strength wastewater and potential for recovery as an energy product: A review. *Water Research*, **104**, 520–531.

Czepiel P. M., Crill P. M. and Harriss R. C. (1993). Methane emissions from municipal wastewater treatment processes. *Environmental Science & Technology*, **27**(12), 2472–2477.

Daelman M. R. J., van Voorthuizen E. M., Van Dongen L., Volcke E. I. P. and Van Loosdrecht M. C. M. (2012). Methane emission during municipal wastewater treatment. *Water Research*, **46**, 3657–3.

Daelman M. R. J., van Voorthuizen E. M., Van Dongen L., Volcke E. I. P. and Van Loosdrecht M. C. M. (2013). Methane and nitrous oxide emissions from municipal wastewater treatment–results from a long-term study. *Water Science and Technology*, **67**, 2350–2355, https://doi.org/10.2166/wst.2013.109

Daelman M. R. J., van Voorthuizen E. M., van Dongen U. G. J. M., Volcke E. I. P. and van Loosdrecht M. C. M. (2015). Seasonal and diurnal variability of N_2O emissions from a full-scale municipal wastewater treatment plant. *Science of the Total Environment*, **536**, 1–11. https://doi.org/10.1016/j.scitotenv.2015.06.122

Desloover J., De Clippeleir H., Boeckx P., Du Laing G., Colsen J., Verstraete W. and Vlaeminck S. E. (2011). Floc-based sequential partial nitritation and anammox at full scale with contrasting N_2O emissions. *Water Research*, **45**, 2811–2821. https://doi.org/10.1016/j.watres.2011.02.028

Domingo-Félez C., Calderó-Pascual M., Sin G., Plósz B. G. and Smets B. F. (2017). Calibration of the comprehensive NDHA-N_2O dynamics model for nitrifier-enriched biomass using targeted respirometric assays. *Water Research*, **126**, 29–39. https://doi.org/10.1016/j.watres.2017.09.013

Duan H., van den Akker B., Thwaites B. J., Peng L., Herman C., Pan Y., Ni B.-J., Watt S., Yuan Z. and Ye L. (2020). Mitigating nitrous oxide emissions at a full-scale wastewater treatment plant. *Water Research*, **185**, 116196. https://doi.org/10.1016/j.watres.2020.116196

Ekström S. E. M., Vangsgaard A. K., Lemaire R., Pérez B. V., Benedetti L., Jensen M. M., Plósz B. G., Thornberg D. and Smets B. F. (2017). Simple Control Strategy for Mitigating N_2O Emissions in Phase Isolated Full-Scale WWTPs. In: Proceedings of 12th IWA Specialized Conference on Instrumentation, Control and Automation. Presented at the 12th IWA Specialized Conference on Instrumentation, Control and Automation. IWA Publishing.

Filali A., Fayolle Y., Peu P., Philippe L., Nauleau F. and Gillot S. (2013). Aeration Control in a Full-Scale Activated Sludge Wastewater Treatment Plant: Impact on Performances, Energy Consumption and N_2O Emission. Presented at the 11ème Conférence IWA sur l'instrumentation, le contrôle et l'automatisation. ICA2013, p. 4.

Flowers J. J., Cadkin T. A. and McMahon K. D. (2013). Seasonal bacterial community dynamics in a full-scale enhanced biological phosphorus removal plant. *Water Research*, **47**, 7019–7031. https://doi.org/10.1016/j.watres.2013.07.054

Foley J., Yuan Z. and Lant P. (2009). Dissolved methane in rising main sewer systems: field measurements and simple model development for estimating greenhouse gas emissions. *Water Science and Technology*, **60**, 2963–2971. https://doi.org/10.2166/wst.2009.718

Foley J., de Haas D., Yuan Z. and Lant P. (2010). Nitrous oxide generation in full-scale biological nutrient removal wastewater treatment plants. *Water Research*, **44**, 831–844. https://doi.org/10.1016/j.watres.2009.10.033

Foley J., Yuan Z., Senante E., Chandran K., Willis J., van Loosdrecht M. and van Voorthuizen E. (2011). Global Water Research Coalition.

Foley J., Yuan Z., Keller J., Senante E., Chandran K., Willis J., Shah A., Loosdrecht M. C. M. and Voorthuizen E. (2015). N_2O and CH4 Emission from Wastewater Collection and Treatment Systems: State of the Science Report and Technical Report, IWA Publishing.

Fries A. E., Schifman L. A., Shuster W. D. and Townsend-Small A. (2018). Street-level emissions of methane and nitrous oxide from the wastewater collection system in Cincinnati, Ohio. *Environmental Pollution*, **236**, 247–256.

Ganigue R., Gutierrez O., Rootsey R. and Yuan Z. (2011). Chemical dosing for sulfide control in Australia: an industry survey. *Water Research*, **45**, 6564–6574. https://doi.org/10.1016/j.watres.2011.09.054

Gruber W., Villez K., Kipf M., Wunderlin P., Siegrist H., Vogt L. and Joss A. (2020). N_2O emission in full-scale wastewater treatment: proposing a refined monitoring strategy. *Science of the Total Environment*, **699**, 134157. https://doi.org/10.1016/j.scitotenv.2019.134157

Gruber W., Niederdorfer R., Ringwald J., Morgenroth E., Bürgmann H. and Joss A. (2021). Linking seasonal N_2O emissions and nitrification failures to microbial dynamics in a SBR wastewater treatment plant. *Water Research X*, **11**, 100098. https://doi.org/10.1016/j.wroa.2021.100098

Guisasola A., de Haas D., Keller J. and Yuan Z. (2008). Methane formation in sewer systems. *Water Research*, **42**, 1421–1430, https://doi.org/10.1016/j.watres.2007.10.014

Gustavsson D. J. I. and la Cour Jansen J. (2011). Dynamics of nitrogen oxides emission from a full-scale sludge liquor treatment plant with nitritation. *Water Science and Technology: A Journal of the International Association on Water Pollution Research*, **63**, 2838–2845, https://doi.org/10.2166/wst.2011.487

Gutierrez O., Sudarjanto G., Ren G., Ganigué R., Jiang G. and Yuan Z. (2014). Assessment of pH shock as a method for controlling sulfide and methane formation in pressure main sewer systems. *Water Research*, **48**, 569–578. https://doi.org/10.1016/j.watres.2013.10.021

Haimi H., Mulas M., Corona F. and Vahala R. (2013). Data-derived soft-sensors for biological wastewater treatment plants: an overview. *Environmental Modelling & Software*, **47**, 88–107. https://doi.org/10.1016/j.envsoft.2013.05.009

Hartley K. and Lant P. (2006). Eliminating non-renewable CO2 emissions from sewage treatment: An anaerobic migrating bed reactor pilot plant study. *Biotechnology and Bioengineering*, **95**(3), 384–398.

Harris E., Joss A., Emmenegger L., Kipf M., Wolf B., Mohn J. and Wunderlin P. (2015). Isotopic evidence for nitrous oxide production pathways in a partial nitritation-anammox reactor. *Water Research*, **83**, 258–270. https://doi.org/10.1016/j.watres.2015.06.040

IPCC (2006). 2006 IPCC Guidelines for National Greenhouse Gas Inventories. Intergovernmental Panel on Climate Change.

IPCC (2013). The Physical Science Basis. Contribution of Working Group I to the Fifth Assessment Report of the Intergovernmental Panel on Climate Change. Cambridge University Press, USA.

Jetten M. S. M., Cirpus I., Kartal B., van Niftrik L., van de Pas-Schoonen K. T., Sliekers O., Haaijer S., van der Star W., Schmid M., van de Vossenberg J., Schmidt I., Harhangi H., van Loosdrecht M., Gijs Kuenen J., Op den Camp H. and Strous M. (2005). 1994–2004: 10 years of research on the anaerobic oxidation of ammonium. *Biochemical Society Transactions*, **33**, 119–123. https://doi.org/10.1042/BST0330119

Joss A., Salzgeber D., Eugster J., König R., Rottermann K., Burger S., Fabijan P., Leumann S., Mohn J. and Siegrist H. (2009). Full-scale nitrogen removal from digester liquid with partial nitritation and anammox in one SBR. *Environmental Science and Technology*, **43**, 5301–5306. https://doi.org/10.1021/es900107w

Kampschreur M. J., van der Star W. R. L., Wielders H. A., Mulder J. W., Jetten M. S. M. and van Loosdrecht M. C. M. (2008). Dynamics of nitric oxide and nitrous oxide emission during full-scale reject water treatment. *Water Research*, **42**, 812–826. https://doi.org/10.1016/j.watres.2007.08.022

Kampschreur M. J., Poldermans R., Kleerebezem R., van der Star W. R. L., Haarhuis R., Abma W. R., Jetten M. S. M. and van Loosdrecht M. C. M. (2009a). Emission of nitrous oxide and nitric oxide from a full-scale single-stage nitritation-anammox reactor. *Water Science and Technology*, **60**, 3211–3217. https://doi.org/10.2166/wst.2009.608

Kampschreur M. J., Temmink H., Kleerebezem R., Jetten M. S. M. and Loosdrecht M. C. M. (2009b). Nitrous oxide emission during wastewater treatment. *Water Research*, **43**, 4093–4103. https://doi.org/10.1016/j.watres.2009.03.001

Kosonen H., Heinonen M., Mikola A., Haimi H., Mulas M., Corona F. and Vahala R. (2016). Nitrous oxide production at a fully covered wastewater treatment plant: results of a long-term online monitoring campaign. *Environmental Science and Technology*, **50**, 5547–5554. https://doi.org/10.1021/acs.est.5b04466

Lackner S., Gilbert E. M., Vlaeminck S. E., Joss A., Horn H. and van Loosdrecht M. C. (2014). Full-scale partial nitritation/anammox experiences–an application survey. *Water Research*, **55**, 292–303, https://doi.org/10.1016/j.watres.2014.02.032

Law Y., Ye L., Pan Y. and Yuan Z. (2012). Nitrous oxide emissions from wastewater treatment processes. *Philosophical Transactions of the Royal Society B*, **367**, 1265–1277. https://doi.org/10.1098/rstb.2011.0317

Li H., Peng D., Liu W., Wei J., Wang Z. and Wang B. (2016). N_2O generation and emission from two biological nitrogen removal plants in China. *Desalination and Water Treatment*, **57**, 11800–11806. https://doi.org/10.1080/19443994.2015.1046145

Liu M., Smal N., Barry J., Morton R., Tang C.-C., Friess P. L., Bell J. and Zhao H., 2014. Pilot-scale evaluation of ANITATM mox for centrate nitrogen removal at the joint water pollution control plant. *Proceedings of the Water Environment Federation*, **2014**, 4460–4482, https://doi.org/10.2175/193864714815941199

Liu Y., Ni B. J., Sharma K. R. and Yuan Z. (2015a). Methane emission from sewers. *Science of the Total Environment*, **524–525**, 40–51. https://doi.org/10.1016/j.scitotenv.2015.04.029

Liu Y., Sharma K. R., Fluggen M., O'Halloran K., Murthy S. and Yuan Z. (2015b). Online dissolved methane and total dissolved sulfide measurement in sewers. *Water Research*, **68**, 109–118. https://doi.org/10.1016/j.watres.2014.09.047

Mampaey K. E., De Kreuk M. K., van Dongen U. G. J. M., van Loosdrecht M. C. M. and Volcke E. I. P. (2016). Identifying N_2O formation and emissions from a full-scale partial nitritation reactor. *Water Research*, **88**, 575–585. https://doi.org/10.1016/j.watres.2015.10.047

Mampaey K. E., Spérandio M., van Loosdrecht M. C. M. and Volcke E. I. P. (2019). Dynamic simulation of N_2O emissions from a full-scale partial nitritation reactor. *Biochemical Engineering Journal*, **152**, 107356. https://doi.org/10.1016/j.bej.2019.107356

Mannina G., Ekama G., Caniani D., Cosenza A., Esposito G., Gori R., Garrido-Baserba M., Rosso D. and Olsson G. (2016). Greenhouse gases from wastewater treatment–a review of modelling tools. *Science of the Total Environment*, **551**, 254–270, https://doi.org/10.1016/j.scitotenv.2016.01.163

Masuda S., Suzuki S., Sano I., Li Y.-Y. and Nishimura O. (2015). The seasonal variation of emission of greenhouse gases from a full-scale sewage treatment plant. *Chemosphere*, **140**, 167–173.

Massara T. M., Malamis S., Guisasola A., Baeza J. A., Noutsopoulos C. and Katsou E. (2017). A review on nitrous oxide (N_2O) emissions during biological nutrient removal from municipal wastewater and sludge reject water. *Science of the Total Environment*, **596**, 106–123, https://doi.org/10.1016/j.scitotenv.2017.03.191

Masuda S., Otomo S., Maruo C. and Nishimura O. (2018). Contribution of dissolved N_2O in total N_2O emission from sewage treatment plant. *Chemosphere*, **212**, 821–827. https://doi.org/10.1016/j.chemosphere.2018.08.089

Mello W. Z., Ribeiro R. P., Brotto A. C., Kligerman D. C., Piccoli A. de S. and Oliveira J. L. M. (2013). Nitrous oxide emissions from an intermittent aeration activated sludge system of an urban wastewater treatment plant. *Química Nova*, **36**, 16–20. https://doi.org/10.1590/S0100-40422013000100004

Ni B.-J., Ye L., Law Y., Byers C. and Yuan Z. (2013). Mathematical modeling of nitrous oxide (N_2O) emissions from full-scale wastewater treatment plants. *Environmental Science and Technology*, **47**, 7795–7803. https://doi.org/10.1021/es4005398

Ni B.-J., Pan Y., van den Akker B., Ye L. and Yuan Z. (2015). Full-scale modeling explaining large spatial variations of nitrous oxide fluxes in a step-feed plug-flow wastewater treatment reactor. *Environmental Science and Technology*, **49**, 9176–9184. https://doi.org/10.1021/acs.est.5b02038

Noyola A., Paredes M. G., Güereca L. P., Molina L. T. and Zavala M. (2018). Methane correction factors for estimating emissions from aerobic wastewater treatment facilities based on field data in Mexico and on literature review. *Science of The Total Environment*, **639**, 84–91.

Pan Y., Ni B.-J., Bond P. L., Ye L. and Yuan Z. (2013). Electron competition among nitrogen oxides reduction during methanol-utilizing denitrification in wastewater treatment. *Water Research*, **47**, 3273–3281. https://doi.org/10.1016/j.watres.2013.02.054

Pan Y., Akker B., Ye L., Ni B.-J., Watts S., Reid K. and Yuan Z. (2016). Unravelling the spatial variation of nitrous oxide emissions from a step-feed plug-flow full scale wastewater treatment plant. *Scientific Reports*, **6**. 1–10. https://doi.org/10.1038/srep20792

Peng L., Ni B.-J., Ye L. and Yuan Z. (2015). The combined effect of dissolved oxygen and nitrite on N_2O production by ammonia oxidizing bacteria in an enriched nitrifying sludge. *Water Research*, **73**, 29–36, https://doi.org/10.1016/j.watres.2015.01.021

Peng L., Ni B.-J., Law Y. and Yuan Z. (2016). Modeling N_2O production by ammonia oxidizing bacteria at varying inorganic carbon concentrations by coupling the catabolic and anabolic processes. *Chemical Engineering Science*, **144**, 386–394. https://doi.org/10.1016/j.ces.2016.01.033

Rathnayake R. M., Oshiki M., Ishii S., Segawa T., Satoh H. and Okabe S. (2015). Effects of dissolved oxygen and pH on nitrous oxide production rates in autotrophic partial nitrification granules. *Bioresource Technology*, **197**, 15–22, https://doi.org/10.1016/j.biortech.2015.08.054

Reino C., van Loosdrecht M. C. M., Carrera J. and Pérez J. (2017). Effect of temperature on N_2O emissions from a highly enriched nitrifying granular sludge performing partial nitritation of a low-strength wastewater. *Chemosphere*, **185**, 336–343. https://doi.org/10.1016/j.chemosphere.2017.07.017

Ren Y. g., Wang J. h., Li H. f., Zhang J., Qi P. y. and Hu Z. (2013). Nitrous oxide and methane emissions from different treatment processes in full-scale municipal wastewater treatment plants. *Environmental Technology*, **34**, 2917–2927. https://doi.org/10.1080/09593330.2012.696717

Ribeiro R. P., Bueno R. F., Piveli R. P., Kligerman D. C., Mello W. Z. and Oliveira J. L. M. (2017). The response of nitrous oxide emissions to different operating conditions in activated sludge wastewater treatment plants in southeastern Brazil. *Water Science and Technology*, **76**, 2337–2349. https://doi.org/10.2166/wst.2017.399

Ribera-Guardia A., Bosch Ll., Corominas Ll. and Pijuan M. (2019). Nitrous oxide and methane emissions from a plug-flow full scale bioreactor and assessment of its carbon footprint. *Journal of Cleaner Production*, **212**, 162–172. https://doi.org/10.1016/j.jclepro.2018.11.286

Rodriguez-Caballero A., Aymerich I., Poch M. and Pijuan M. (2014). Evaluation of process conditions triggering emissions of green-house gases from a biological wastewater treatment system. *Science of the Total Environment*, **493**, 384–391. https://doi.org/10.1016/j.scitotenv.2014.06.015

Rodriguez-Caballero A., Aymerich I., Marques R., Poch M. and Pijuan M. (2015). Minimizing N_2O emissions and carbon footprint on a full-scale activated sludge sequencing batch reactor. *Water Research*, **71**, 1–10. https://doi.org/10.1016/j.watres.2014.12.032

Sabba F., Terada A., Wells G., Smets B. F. and Nerenberg R. (2018). Nitrous oxide emissions from biofilm processes for wastewater treatment. *Applied Microbiology and Biotechnology*, **102**, 9815–9829. https://doi.org/10.1007/s00253-018-9332-7

Samuelsson J., Delre A., Tumlin S., Hadi S., Offerle B. and Scheutz C. (2018). Optical technologies applied alongside on-site and remote approaches for climate gas emission quantification at a wastewater treatment plant. *Water Research*, **131**, 299–309. https://doi.org/10.1016/j.watres.2017.12.018

Schaubroeck T., De Clippeleir H., Weissenbacher N., Dewulf J., Boeckx P., Vlaeminck S. E. and Wett B. (2015). Environmental sustainability of an energy self-sufficient sewage treatment plant: improvements through DEMON and co-digestion. *Water Research*, **74**, 166–179. https://doi.org/10.1016/j.watres.2015.02.013

Scheutz C. and Fredenslund A. M. (2019). Total methane emission rates and losses from 23 biogas plants. *Waste Management*, **97**, 38–46.

Schulthess R. von, Wild D. and Gujer W. (1994). Nitric and nitrous oxides from denitrifying activated sludge at low oxygen concentration. *Water Science and Technology*, **30**, 123–132, https://doi.org/10.2166/wst.1994.0259

Shah A., Willis J. and Fillmore L. (2011). Quantifying Methane Evolution from Sewers: Results from WERF/ Dekalb Phase 2 Continuous Monitoring at Honey Creek Pumping Station and Force Main. Proceedings of the Water Environment Federation, pp. 475–485. https://doi.org/10.2175/193864711802836841

Sharma K. R., Yuan Z., de Haas D., Hamilton G., Corrie S. and Keller J. (2008). Dynamics and dynamic modelling of H_2S production in sewer systems. *Water Research*, **42**, 2527–2538. https://doi.org/10.1016/j.watres.2008.02.013

Spinelli M., Eusebi A. L., Vasilaki V., Katsou E., Frison N., Cingolani D. and Fatone F. (2018). Critical analyses of nitrous oxide emissions in a full scale activated sludge system treating low carbon-to-nitrogen ratio wastewater. *Journal of Cleaner Production*, **190**, 517–524. https://doi.org/10.1016/j.jclepro.2018.04.178

Stenström F., Tjus K. and Jansen J. la C. (2014). Oxygen-induced dynamics of nitrous oxide in water and off-gas during the treatment of digester supernatant. *Water Science and Technology*, **69**, 84–91. https://doi.org/10.2166/wst.2013.558

Sun S., Cheng X. and Sun D. (2013). Emission of N_2O from a full-scale sequencing batch reactor wastewater treatment plant: characteristics and influencing factors. *International Biodeterioration & Biodegradation*, **85**, 545–549. https://doi.org/10.1016/j.ibiod.2013.03.034

Sun S., Bao Z. and Sun D. (2015). Study on emission characteristics and reduction strategy of nitrous oxide during wastewater treatment by different processes. *Environmental Science and Pollution Research*, **22**, 4222–4229. https://doi.org/10.1007/s11356-014-3654-5

Sun S., Bao Z., Li R., Sun D., Geng H., Huang X., Lin J., Zhang P., Ma R., Fang L., Zhang X. and Zhao X. (2017). Reduction and prediction of N_2O emission from an anoxic/oxic wastewater treatment plant upon DO control and model simulation. *Bioresource Technology*, **244**, 800–809. https://doi.org/10.1016/j.biortech.2017.08.054

Tallec G., Garnier J., Billen G. and Gousailles M. (2006). Nitrous oxide emissions from secondary activated sludge in nitrifying conditions of urban wastewater treatment plants: effect of oxygenation level. *Water Research*, **40**, 2972–2980. https://doi.org/10.1016/j.watres.2006.05.037

Tauber J., Parravicini V., Svardal K. and Krampe J. (2019). Quantifying methane emissions from anaerobic digesters. *Water Science and Technology*, **80**(9), 1654–1661.

Townsend-Small A., Pataki D. E., Tseng L. Y., Tsai C.-Y. and Rosso D. (2011). Nitrous oxide emissions from wastewater treatment and water reclamation plants in southern California. *Journal of Environmental Quality*, **40**, 1542–1550. https://doi.org/10.2134/jeq2011.0059

Tumendelger A., Toyoda S. and Yoshida N. (2014). Isotopic analysis of N_2O produced in a conventional wastewater treatment system operated under different aeration conditions. *Rapid Communications in Mass Spectrometry*, **28**, 1883–1892. https://doi.org/10.1002/rcm.6973

Valentine D. L. and Reeburgh W. S. (2000). New perspectives on anaerobic methane oxidation. *Environmental Microbiology*, **2**, 477–484. https://doi.org/10.1046/j.1462-2920.2000.00135.x

Van Hulle S. W. H., Vandeweyer H. J. P., Meesschaert B. D., Vanrolleghem P. A., Dejans P. and Dumoulin A. (2010). Engineering aspects and practical application of autotrophic nitrogen removal from nitrogen rich streams. *Chemical Engineering Journal*, **162**, 1–20. https://doi.org/10.1016/j.cej.2010.05.037

Vasilaki V., Volcke E. I. P., Nandi A. K., van Loosdrecht M. C. M. and Katsou E. (2018). Relating N_2O emissions during biological nitrogen removal with operating conditions using multivariate statistical techniques. *Water Research*, **140**, 387–402. https://doi.org/10.1016/j.watres.2018.04.052

Vasilaki V., Massara T. M., Stanchev P., Fatone F. and Katsou E. (2019). A decade of nitrous oxide (N_2O) monitoring in full-scale wastewater treatment processes: a critical review. *Water Research*, **161**, 392–412. https://doi.org/10.1016/j.watres.2019.04.022

Vasilaki V., Conca V., Frison N., Eusebi A. L., Fatone F. and Katsou E. (2020). A knowledge discovery framework to predict the N_2O emissions in the wastewater sector. *Water Research*, **178**, 115799. https://doi.org/10.1016/j.watres.2020.115799

Wan X., Baeten J. E. and Volcke E. I. P. (2019). Effect of operating conditions on N_2O emissions from one-stage partial nitritation-anammox reactors. *Biochemical Engineering Journal*, **143**, 24–33. https://doi.org/10.1016/j.bej.2018.12.004

Wang J., Zhang J., Wang J., Qi P., Ren Y. and Hu Z. (2011). Nitrous oxide emissions from a typical northern Chinese municipal wastewater treatment plant. *Desalination and Water Treatment*, **32**, 145–152. https://doi.org/10.5004/dwt.2011.2691

Wang Y., Fang H., Zhou D., Han H. and Chen J. (2016a). Characterization of nitrous oxide and nitric oxide emissions from a full-scale biological aerated filter for secondary nitrification. *Chemical Engineering Journal*, **299**, 304–313. https://doi.org/10.1016/j.cej.2016.04.050

Wang Y., Lin X., Zhou D., Ye L., Han H. and Song C. (2016b). Nitric oxide and nitrous oxide emissions from a full-scale activated sludge anaerobic/anoxic/oxic process. *Chemical Engineering Journal*, **289**, 330–340. https://doi.org/10.1016/j.cej.2015.12.074

Weissenbacher N., Takacs I., Murthy S., Fuerhacker M. and Wett B. (2010). Gaseous nitrogen and carbon emissions from a full-scale deammonification plant. *Water Environment Research*, **82**, 169–175. https://doi.org/10.2175/106143009X447867

Wunderlin P., Mohn J., Joss A., Emmenegger L. and Siegrist H. (2012). Mechanisms of N_2O production in biological wastewater treatment under nitrifying and denitrifying conditions. *Water Research*, **46**, 1027–1037. https://doi.org/10.1016/j.watres.2011.11.080

Yan X., Li L. and Liu J. (2014). Characteristics of greenhouse gas emission in three full-scale wastewater treatment processes. *Journal of Environmental Sciences*, **26**, 256–263. https://doi.org/10.1016/S1001-0742(13)60429-5

Yan P., Li K., Guo J.-S., Zhu S.-X., Wang Z.-K. and Fang F. (2019). Toward N_2O emission reduction in a single-stage CANON coupled with denitrification: investigation on nitrite simultaneous production and consumption and nitrogen transformation. *Chemosphere*, **228**, 485–494. https://doi.org/10.1016/j.chemosphere.2019.04.148

Yang X., Wang S. and Zhou L. (2012). Effect of carbon source, C/N ratio, nitrate and dissolved oxygen concentration on nitrite and ammonium production from denitrification process by Pseudomonas stutzeri D6. *Bioresource Technology*, **104**, 65–72. https://doi.org/10.1016/j.biortech.2011.10.026

Ye L., Ni B.-J., Law Y., Byers C. and Yuan Z. (2014). A novel methodology to quantify nitrous oxide emissions from full-scale wastewater treatment systems with surface aerators. *Water Research*, **48**, 257–268. https://doi.org/10.1016/j.watres.2013.09.037

Zaborowska E., Lu X. and Makinia J. (2019). Strategies for mitigating nitrous oxide production and decreasing the carbon footprint of a full-scale combined nitrogen and phosphorus removal activated sludge system. *Water Research*, **162**, 53–63. https://doi.org/10.1016/j.watres.2019.06.057

Zhang L., De Schryver P., De Gusseme B., De Muynck W., Boon N. and Verstraete W. (2008). Chemical and biological technologies for hydrogen sulfide emission control in sewer systems: a review. *Water Research*, **42**, 1–12, https://doi.org/10.1016/j.watres.2007.07.013

Zhou X., Zhang X., Zhang Z. and Liu Y. (2018). Full nitration-denitration versus partial nitration-denitration-anammox for treating high-strength ammonium-rich organic wastewater. *Bioresource Technology*, **261**, 379–384. https://doi.org/10.1016/j.biortech.2018.04.049

Zhuang J., Zhou Y., Liu Y. and Li W. (2020). Flocs are the main source of nitrous oxide in a high-rate anammox granular sludge reactor: insights from metagenomics and fed-batch experiments. *Water Research*, **186**, 116321. https://doi.org/10.1016/j.watres.2020.116321

NOMENCLATURE

A/O	Anoxic/aerobic
A^2/O	Anaerobic/anoxic/aerobic
AMO	Ammonia monooxygenase
Anammox	Anaerobic ammonium oxidation
AOA	Ammonia oxidizing archaea
AOB	Ammonia oxidizing bacteria
AOR	Ammonia oxidation rate
BNR	Biological nutrient removal
CANON	Completely autotrophic nitrogen removal over nitrite
CAS	Conventional activated sludge
CO_2	Carbon dioxide
COD	Chemical oxygen demand
Comammox	Complete ammonium oxidizer
CuO	Copper oxide
dGAO	Denitrifying glycogen accumulating organisms
DO	Dissolved oxygen
dPAO	Denitrifying polyphosphate accumulating organism
EF	Emission factor
FA	Free ammonia
FNA (HNO_2)	Free nitrous acid
GHG	Greenhouse gas
H_2S	Hydrogen sulphide
HRT	Hydraulic retention time
MLE	Modified Ludzack-Ettinger
MLVSS	Mixed liquor volatile suspended solids
N_2	Nitrogen gas
N_2O	Nitrous oxide

N_2O_4	Nitrogen tetroxide
NaR	Nitrate reductase
NH_2OH	Hydroxylamine
NH_3	Ammonia
NH_4^+	Ammonium
NiR	Nitrite reductase
NO	Nitric oxide
NO_2^-	Nitrite
NO_3^-	Nitrate
NOB	Nitrite oxidizing bacteria
NOH	Nitrosyl radical
NoR	Nitric oxide reductase
NoS	Nitrous oxide reductase
OD	Oxidation ditch
PN	Partial nitrification
RT-qPCR	Real time quantitative polymerase chain reaction
SBR	Sequencing batch reactor
SCENA	Short cut enhanced nutrient abatement
SP	Site-preference
WWTPs	Wastewater treatment plants

doi: 10.2166/9781789060461_167

Chapter 7

Modelling N_2O production and emissions

Mathieu Spérandio[1], Longqi Lang[1], Fabrizio Sabba[2], Robert Nerenberg[2], Peter Vanrolleghem[3], Carlos Domingo-Félez[4], Barth F. Smets[4], Haoran Duan[5], Bing-Jie Ni[5] and Zhiguo Yuan[5]

[1]TBI, Université de Toulouse, CNRS, INRAE, INSA, Toulouse, France. E-mail: sperandio@insa-toulouse.fr
[2]Department of Civil and Environmental Engineering and Earth Sciences, University of Notre Dame, Notre Dame, IN 46556, USA
[3]Modeleau, Département de génie civil et de génie des eaux, Université Laval, 1065 av. de la Médecine, Québec, QC G1V 0A6, Canada
[4]Department of Environmental Engineering, Technical University of Denmark, 2800 Kongens Lyngby, Denmark
[5]Advanced Water Management Centre, The University of Queensland, St Lucia, QLD 4072, Australia. E-mail: z.yuan@awmc.uq.edu.au

SUMMARY

Mathematical modelling of N_2O emissions is of great importance for the understanding and reduction of the environmental impact of wastewater treatment systems. This chapter reviews the current status of the modelling of N_2O emissions from wastewater treatment. The existing mathematical models describing all known microbial pathways for N_2O production are reviewed and discussed. These include N_2O production and consumption by heterotrophic denitrifiers, N_2O production by ammonia-oxidizing bacteria (AOB) through the hydroxylamine oxidation pathway and the AOB denitrification pathway and the integration of these pathways in single-pathway N_2O models. The two-pathway models are compared to single-pathway models. The calibration and validation of these models using lab-scale and full-scale experimental data is also reviewed. The mathematical modelling of N_2O production, while still being enhanced by new knowledge development, has reached a maturity that facilitates the estimation of site-specific N_2O emissions and the development of mitigation strategies for wastewater treatment plants taking into account the specific design and operational conditions of the plant.

Keywords: AOB pathways, calibration, heterotrophic denitrification, modelling, N_2O

TERMINOLOGY

Term	Definition
Mathematical model	A system of mathematical equations that describes physical and biological processes. It is a simplified representation of the real process.
Model parameters	Model parameters are model constituents (stoichiometric and kinetic) determined according to model applications. The value of a parameter for the given application is ideally constant.

State variables	State variables represent time-varying concentrations or other properties to be determined by a solver based on their derivatives.
Model calibration	The estimation and adjustment of model parameters to enhance the agreement between model output and experimental data.
Model validation	The comparison of model simulated output with real observations using data not used in model development. The model is validated if the simulation during the validation period lies within acceptable limits around the observations.
Kinetics	Kinetics describe the rate of chemical or biological reactions, by considering factors that influence the rate of reactions. Kinetics are associated with the fundamental mechanisms of the reaction.
Stoichiometric relationship	The quantitative relationship among the amounts of substances consumed or produced in a chemical or biological reaction.
Metabolic pathway	A series of biochemical reactions occurring within microorganisms. The reactants, products, and intermediates of an enzymatic reaction which are known as metabolites, are linked by the metabolic pathway.
Emission factor (N_2O)	The N_2O emission factor is defined as the ratio between N_2O nitrogen emitted and the ammonium nitrogen converted.

7.1 INTRODUCTION

Mathematical models have been widely applied to the prediction of nitrogen removal in wastewater treatment, and are gaining increasing attention for the prediction of N_2O accumulation and emission during nitrification and denitrification processes (CH2MHill, 2008; Corominas *et al.*, 2012; Guo and Vanrolleghem, 2014; Harper *et al.*, 2015; Hiatt and Grady, 2008; Mannina *et al.*, 2016; Ni *et al.*, 2011; Pocquet *et al.*, 2013). The ability to predict N_2O production by modelling provides an opportunity to include N_2O production as an important consideration in the design, operation and optimization of biological nitrogen removal processes (Ni *et al.*, 2011, 2013a). Furthermore, mathematical modelling should be a more appropriate method for estimating site-specific emissions of N_2O than oversimplified models with fixed N_2O emission factors (Corominas *et al.*, 2012; Guo and Vanrolleghem, 2014; Mampaey *et al.*, 2013; Ni *et al.*, 2011, 2013a; Pocquet *et al.*, 2013). In addition, mathematical modelling provides a method for verifying hypotheses related to the mechanisms for N_2O production, and thus serves as a tool to support the development of mitigation strategies (Duan *et al.*, 2020; Ni *et al.*, 2013b; Vasilaki *et al.*, 2020; Zaborowska *et al.*, 2019).

N_2O modelling has evolved rapidly in the past few years, with models based on various production pathways proposed. These models have been calibrated with data obtained from laboratory reactors and full-scale wastewater treatment plants operated under various conditions. Each of these models has its underlying assumptions and has been calibrated/validated to various degrees based on the understanding of the processes of the distinct model creators. These models displayed various predictive abilities (usually good fit with own data but failure with foreign data). Despite the obvious importance of N_2O modelling, and the increasing number of publications, model comparisons and comprehensive reviews are rare (Mannina *et al.*, 2016, Spérandio *et al.*, 2016). This chapter aims to compare these models and provide guidance for their use. The existing mathematical models describing all known microbial and chemical pathways for N_2O production and consumption, as well as their underlying assumptions, are reviewed, discussed and compared.

This work includes the single-pathway and two-pathway models of ammonia-oxidizing bacteria (AOB), the N_2O models of heterotrophic denitrifiers, and the integrated N_2O models including both AOB and heterotrophic denitrifier activities. An overview of the model evaluations using lab-scale and full-scale experimental data is also presented to provide insights into the applicability of these N_2O models under various conditions.

7.2 N_2O KINETIC MODEL STRUCTURES

N_2O is produced during biological nitrogen removal in wastewater treatment, typically attributed to autotrophic AOB (Chandran *et al.*, 2011; Kampschreur *et al.*, 2009; Tallec *et al.*, 2006) and heterotrophic denitrifiers (Kampschreur *et al.*, 2009; Lu and Chandran, 2010; Pan *et al.*, 2012). There are three main microbial pathways involved in N_2O formation, namely the NH_2OH oxidation, nitrifier (AOB) denitrification, and heterotrophic denitrification pathways (Wunderlin *et al.*, 2012, 2013). The latter pathway is the only known microbial pathway that allows N_2O consumption. Table S1 in the supplementary information (SI) lists the definitions of the all the state variables used in the models described in this chapter.

In addition N_2O might be potentially produced through chemical pathways (Harper *et al.*, 2015; Schreiber *et al.*, 2009). Such processes involve hydroxylamine with different oxidants (HNO_2, Fe^{3+}, O_2) or hydroxylamine disproportionation, or HNO_2 reduction by Fe^{2+}. The kinetic model structure for such a chemical pathway is relatively simple, based on the first order regarding the reactants for instance, as proposed for the reaction between hydroxylamine and free nitrous acid (Harper *et al.*, 2015). Moreover the recent work of Su *et al.* (2019a) demonstrated that these chemical reactions are strongly influenced by pH and become important only at acidic pH ($\leq$5). Consequently abiotic N_2O production contributes little (<3% of total N_2O production) to total N_2O emissions in typical nitritation reactor systems between pH 6.5 and 8. Hence, in this chapter the description will be focused on biological models.

7.2.1 Modelling of N_2O production and consumption by Heterotrophic Denitrifiers

7.2.1.1 Introduction

N_2O is a known intermediate in heterotrophic denitrification (Pan *et al.*, 2012, 2013a; von Schulthess and Gujer, 1996). Heterotrophic denitrification converts the nitrate and/or nitrite generated from autotrophic nitrification to nitrogen gas (N_2) and thus removes nitrogen from wastewater. It consists of four consecutive steps, which produce three obligatory intermediates, namely NO_2^-, NO and N_2O. These steps are individually catalysed by four different denitrification reductases, that is nitrate reductase (Nar), nitrite reductase (Nir), NO reductase (NOR) and N_2O reductase (N_2OR). N_2O is produced by the sequential action of the NO_3^-, NO_2^- and NO reductases.

Many factors could affect the denitrification process and thus impact N_2O emission, such as chemical oxygen demand (COD) to N ratios, the substrate and biomass types, pH levels and temperature, among others (Lu and Chandran, 2010; Pan *et al.*, 2012, 2013a). On the other hand, the four parallel denitrification steps could also exert influence on each other through electron competition, which could result in accumulation of various intermediates including N_2O. The four denitrification steps all require electrons from carbon oxidation, and they could face competition for electrons when the electron supply rate from carbon oxidation does not meet the demand for electrons by the four steps of denitrification combined (Pan *et al.*, 2013a).

To predict denitrification intermediates accumulation, denitrification needs to be modelled as a multiple-step process (von Schulthess and Gujer, 1996). Figure 7.1 gives an overview of the major models. Four-step denitrification models have been proposed and widely applied to predict the accumulation of all denitrification intermediates including N_2O (Hiatt and Grady, 2008; Kampschreur *et al.*, 2007; Ni *et al.*, 2011; Pan *et al.*, 2013b). To date, two distinct concepts have been proposed (Table 7.1), which are represented by the activated sludge model for nitrogen (ASMN) (Hiatt and Grady, 2008) and the activated sludge model with indirect coupling of electrons (ASM-ICE) (Pan *et al.*, 2013b), respectively. Table S2 in the SI lists the kinetic and stoichiometric matrices for the two models, which are fundamentally different in describing the electron allocation among different steps of heterotrophic denitrification (Table 7.1).

7.2.1.2 Activated sludge model for nitrogen (ASMN)

The '*direct coupling approach*', represented by ASMN (Model OHO-A in Table 7.1, Hiatt and Grady, 2008), directly couples the carbon oxidation and nitrogen reduction processes in the model. This

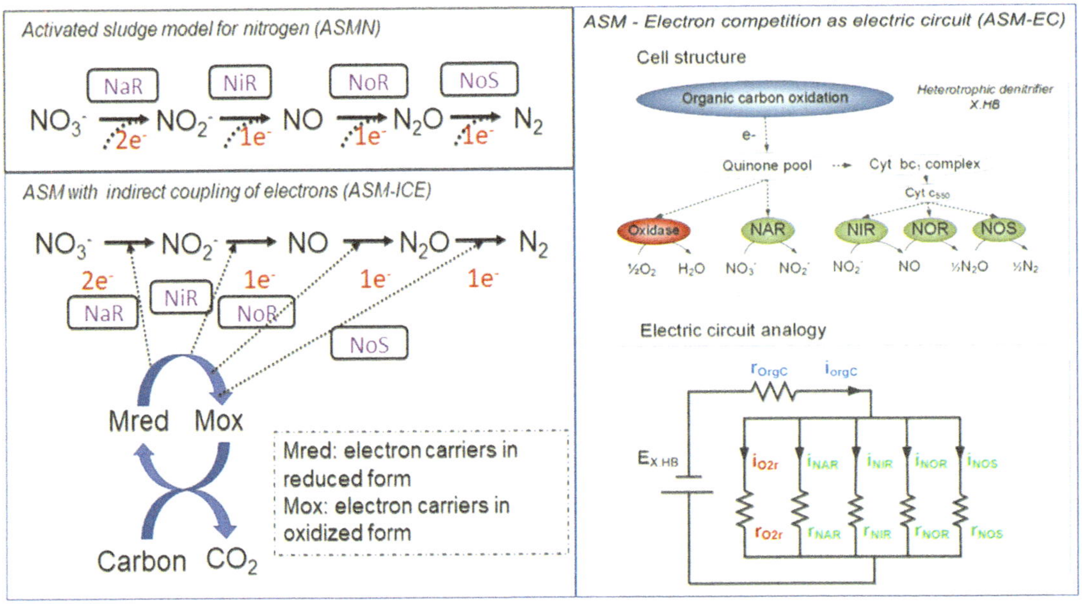

Figure 7.1 Schematic illustration of denitrification models. ASMN: Hiatt and Grady (2008); ASM-ICE: Pan *et al.* (2013b); ASM-EC electric circuit analogy: Domingo-Félez and Smets (2020a).

model describes each of the four steps as a separate and independent oxidation-reduction reaction (Table S2 in SI), with the kinetics of each step modelled according to the nitrogen reduction reaction kinetics and using a stoichiometric relationship obtained through an electron balance. Model OHO-A ignores the fact that the nitrogen oxides reduction and carbon oxidation are carried out by different enzymes with their specific kinetics, and consequently either of the two processes could limit the rate of denitrification. In addition, this coupling approach describes each denitrification step independently with its rate not being affected by other denitrification steps that draw electrons from the same electron supply. Essentially, the carbon oxidation rate is modelled as the sum of the carbon requirements by all denitrification steps, with the underlying assumption that electron supply will always be able to meet the predicted total electron demand.

This model's structure is close to what wastewater modellers are used to, that is kinetics depending on soluble and particulate components, and less on detailed metabolic pathway information. The importance of N_2O accumulation and emission logically depend on respective consumption and production rates. For instance it could be mentioned that this model predicts more N_2O production in the case of organic matter limitation by using a higher Ks value for organic matter in the last reaction (original set of parameter values by Hiatt and Grady, 2008). This is an important point in terms of acceptability and usability in the profession.

7.2.1.3 Activated sludge model with indirect coupling of electrons (ASM-ICE)

The 'indirect coupling approach', proposed by Pan *et al.* (2013b) and named ASM-ICE, decouples the carbon oxidation and nitrogen reduction processes. Electron carriers are introduced as a new component in this model to link carbon oxidation to nitrogen oxides reduction, in which carbon oxidation reduces carriers and nitrogen oxides reduction oxidizes carriers (Model OHO-B in Table 7.1, Pan *et al.*, 2013b). In this way, each step of heterotrophic denitrification can be regulated by

Table 7.1 Key differences among the N_2O models by heterotrophs, single-pathway models by AOB and two-pathway models by AOB

N_2O models		Model components	Stoichiometric	Kinetic	Reference
N_2O models by heterotrophs	Model OHO-A	Without electron carriers.	Coupling carbon oxidation and nitrogen reduction (4 processes).	Without electron competition concept.	Hiatt and Grady (2008)
	Model OHO-B	With electron carriers.	Decoupling carbon oxidation and nitrogen reduction (5 processes).	With electron competition concept.	Pan *et al.* (2013b)
Single-pathway models by AOB	Model A – AOB denitrification	Using SNH_4 and SNO_2; With SNH_2OH.	Two-step NH_4^+ oxidation; Two-step NO_2^- reduction; Cell growth during NH_2OH oxidation.	Two different oxygen affinity constants; Oxygen inhibition on NO_2^- and NO reductions; Anoxic reduction factor.	Ni *et al.* (2011)
	Model A1 – AOB denitrification	Using SNH_3 and $SHNO_2$; With SNH_2OH.	Same as Model A.	Two different oxygen affinity constants; Without oxygen inhibition; NH_3 inhibition on NH_3 oxidation; Anoxic reduction factor.	Pocquet *et al.* (2013)
	Model B – AOB denitrification	Using SNH_3 and $SHNO_2$; Without SNH_2OH.	One-step NH_4^+ oxidation; Two-step NO_2- reduction; Cell growth during all 3 processes.	Only one oxygen affinity constant; Without oxygen inhibition; Anoxic reduction factor.	Mampaey *et al.* (2013)
	Model B1 – AOB denitrification	Same as Model B.	Same as Model B.	Only one oxygen affinity constant; NH_3 and HNO_2 inhibitions on NH_3 oxidation; Haldane function for oxygen limitation/inhibition; Anoxic reduction factor.	Guo and Vanrolleghem (2014)
	Model C – NH_2OH pathway (via NOH)	Using SNH_4 and SNO_2; With SNOH.	Three-step NH_4^+ oxidation via NOH; Cell growth during NH_2OH oxidation.	Two different oxygen affinity constants; NOH breakdown to produce N_2O.	Law *et al.* (2012)
	Model D – NH_2OH pathway (via NO)	Using SNH_4 and SNO_2; With SNO.	Three-step NH_4^+ oxidation via NO; Cell growth during NH_2OH oxidation.	Two different oxygen affinity constants; NO reduction to produce N_2O; Without oxygen inhibition.	Ni *et al.* (2013b)

(*Continued*)

Table 7.1 Key differences among the N_2O models by heterotrophs, single-pathway models by AOB and two-pathway models by AOB (*Continued*).

N_2O models		Model components	Stoichiometric	Kinetic	Reference
Two-pathway models by AOB	Model E	Using SNH_4 and SNO_2; With electron carriers.	Three-step NH_4^+ oxidation; One-step NO_2^- reduction; Without cell growth.	Applying electron competition concept; Without oxygen inhibition; Without anoxic reduction factor.	Ni *et al.* (2014)
	Model F	Mostly same as Model E; With SCO_2; With energy carriers.	Mostly same as Model E; With energy carriers involved; With cell growth considered.	Mostly same as Model E; With energy carriers involved; With effect of inorganic carbon considered.	Peng *et al.* (2015a)
	Model G	Using SNH_3 and $SHNO_2$ With SNH_2OH.	Two-step NH_3 oxidation; One-step HNO_2 reduction; Cell growth during NH_2OH oxidation.	With oxygen inhibition of denitrification; With reduction factor; Three different oxygen affinity constants.	Pocquet *et al.* (2016)
	Model H	Using SNH_3 and $SHNO_2$ With SNH_2OH.	Two-step NH_3 oxidation; Two-step HNO_2 reduction; Cell growth during NH_2OH oxidation; O_2-independent NH_2OH oxidation.	With oxygen inhibition of denitrification; With reduction factor; Two different oxygen affinity constants; Two different NH_2OH affinity constants.	Domingo-Félez and Smets (2016)

both the nitrogen reduction and the carbon oxidation processes. The possibility of either the carbon oxidation or electron transfer being a limiting step in denitrification is thus considered in the model. In heterotrophic denitrifiers, competition for electrons may occur between the four reduction steps when the electron supply rate from the oxidation process cannot meet the demand for electrons from the four reduction steps (Pan *et al.*, 2013b), which plays an important role in the accumulation and emission of N_2O (Pan *et al.*, 2013a). The electron competition between the four denitrifying steps can be modelled by assigning different values to the affinity constants responsible for Processes 2, 3, 4 and 5 with respect to Mred, which are provided by Process 1. Model OHO-B can be used as a practical tool for predicting N_2O accumulation during denitrification, with the complex biochemical reactions and electron transfer processes involved in biological denitrification by different microbial species being lumped into one oxidation and four reduction reactions that are linked through electron carriers.

7.2.1.4 Activated sludge model – electron competition (ASM-EC)

Almeida *et al.* (1997) proposed that the electron flow through the respiratory chain can be modelled similarly to electron flow across resistors in an electric circuit. A model structure describing four-step denitrification, aerobic respiration, and organic carbon oxidation was proposed using the analogy between electron competition during respiration and electron distribution in a multi-resistor electric circuit (Domingo-Félez and Smets, 2020a). A potential is created by the presence of heterotrophic denitrifying bacteria that mediate electron transfer between an electron donor and an electron acceptor pair. The competition for electrons between multiple denitrifying enzymes is analogous to the electron flow through parallel resistors. Reaction rates are analogous to the current intensity through a resistor (Ohm's law). Individual resistances vary with the substrate concentrations, and the electron current released through electron donor (i.e., organic substrates) oxidation will be distributed between reduction rates. Following conservation of potential and conservation of charge, the current through any resistor can be calculated. The ASM–EC model can substitute the process rates describing denitrification in ASM-type models and includes fewer parameters than ASMN or ASM-ICE.

7.2.2 Modelling N_2O production by AOB

7.2.2.1 Introduction

AOB are chemolithotrophs that oxidize ammonia (NH_3) to nitrite (NO_2-) via hydroxylamine (NH_2OH) as their predominant energy-generating metabolism (Arp and Stein, 2003; Arp *et al.*, 2007) (Figure 7.2a). The first step is catalysed by ammonia monooxygenase (AMO) where NH_3 is oxidized to NH_2OH with the reduction of molecular oxygen (O_2). In the second step, NH_2OH is oxidized to NO_2- by hydroxylamine oxidoreductase (HAO), with O_2 as the primary electron acceptor. However, AOB contain a periplasmic copper-containing nitrite reductase (NirK) and a nitric oxide reductase (Nor) (Chandran *et al.*, 2011; Hooper *et al.*, 1997) (as shown in Figure 7.2a). NirK could speed up NH_2OH oxidation by channelling electrons from the cytochrome pool to NO_2- (to form NO) and thus play a facilitative role in NH_3 oxidation itself (Chandran *et al.*, 2011; Hooper *et al.*, 1997). AOB also possess the inventory to alternatively convert NO into N_2O, using a haem–copper nitric oxide reductase, sNOR (Chandran *et al.*, 2011).

Although N_2O is not an obligate intermediate in NH_3 oxidation, N_2O can be produced by AOB through two major pathways according to the current understanding (Figure 7.2a): (i) N_2O as a by-product of incomplete oxidation of NH_2OH to NO_2-, typically referred to as the NH_2OH oxidation pathway (Chandran *et al.*, 2011; Law *et al.*, 2012; Poughon *et al.*, 2000; Stein, 2011a), and (ii) N_2O as the final product of AOB denitrification with NO_2- as the terminal electron acceptor and NO as an intermediate, the so-called nitrifier or AOB denitrification pathway (Chandran *et al.*, 2011; Ni *et al.*, 2013b; Stein, 2011b).

It is generally accepted that NO_2- and NO reduction for N_2O production is carried out by AOB under oxygen limiting or completely anoxic conditions (Kampschreur *et al.*, 2009; Law *et al.*, 2013). Increased N_2O production under high NO_2- concentrations has been suggested to be due to AOB

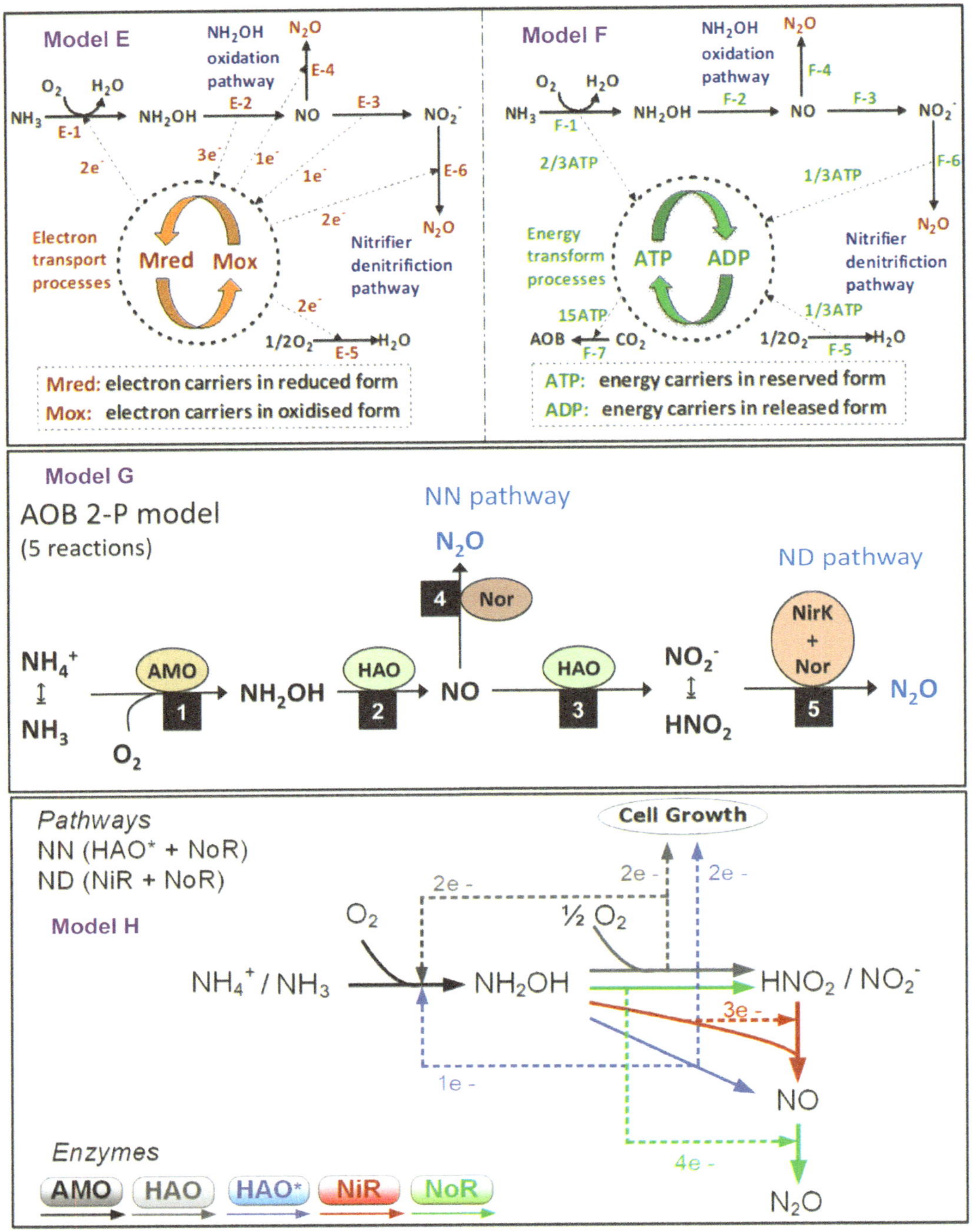

Figure 7.2 General description of the two-pathway AOB models (E: Ni *et al.*, 2014, F: Peng *et al.*, 2015a, G: Pocquet *et al.*, 2016; H: Domingo-Félez and Smets, 2016). NN: hydroxylamine pathway, ND: nitrifier denitrification pathway.

denitrification (Yang *et al.*, 2009; Yu *et al.*, 2010). On the other hand, there is also evidence supporting N_2O production from NH_2OH oxidation by AOB. The higher NH_3 oxidation rate could result in the accumulation of NH_2OH and other reaction intermediates such as NO or NOH (Law *et al.*, 2012), which in turn result in N_2O formation with detailed reactions yet to be fully elucidated (Chandran *et al.*, 2011; Stein, 2011a).

As the fundamental metabolic pathways for N_2O production by AOB are now coming to light (Castro-Barros *et al.*, 2015; Harris *et al.*, 2015; Kampschreur *et al.*, 2007; Okabe *et al.*, 2011; Perez-Garcia *et al.*, 2014; Schreiber *et al.*, 2009; Stein, 2011a; Yu *et al.*, 2010), several mechanistic models have been proposed for N_2O production by AOB in mixed culture based on one or two of the known N_2O production pathways of AOB, that is AOB denitrification and NH_2OH oxidation pathways. To date, two categories of N_2O production models by AOB in mixed culture have been proposed, which are represented by single-pathway models and two-pathway models.

7.2.2.2 Single-pathway models

Six different single-pathway model structures available in literature are presented in Table S3 in the SI, detailed with their kinetic and stoichiometric matrices. Table 7.1 presents the key differences among the model structures of these single-pathway models by AOB.

Model A (Ni *et al.*, 2011) and Model B (Mampaey *et al.*, 2013) are based on the AOB denitrification pathway. In Model A (Table 7.1, Ni *et al.*, 2011), AOB denitrification with NO_2- as the terminal electron acceptor produces NO and subsequently N_2O by consuming NH_2OH as the electron donor. Similarly, in Model B (Table 7.1, Mampaey *et al.*, 2013), AOB denitrification occurs in parallel with ammonium oxidation, reducing NO_2- to NO and then to N_2O with ammonium as the electron donor. The key difference between these two models is that in Model A, dissolved oxygen (DO) is assumed to inhibit nitrite and NO reduction by AOB, while in Model B, this inhibition is absent. A further minor difference is that ammonia oxidation is modelled as a two-step (ammonia to hydroxylamine and then to nitrite) process in Model A, but as a one-step process (ammonia to nitrite) in Model B.

Model A1 (Pocquet *et al.*, 2013) and Model B1 (Guo and Vanrolleghem, 2014) are also based on the AOB denitrification pathway, and are the two modified versions of Models A and B which describe N_2O production in several studies (Guo and Vanrolleghem, 2014; Pocquet *et al.*, 2013). In Model A1 (Table 7.1, Pocquet *et al.*, 2013), the oxygen inhibition of the AOB denitrification pathway was removed. In addition, free ammonia (FA) and free nitrous acid (FNA) were considered as the substrate for the AOB reactions, in order to explicitly consider the effect of pH variation. In Model B1 (Table 7.1, Guo and Vanrolleghem, 2014), oxygen limitation and inhibition was added through a Haldane function in the kinetics of both nitrite reduction and NO reduction processes (Guo and Vanrolleghem, 2014). Inhibition by FA was also considered in Model A1 and inhibition by both FA and FNA were included in Model B1.

Model C (Law *et al.*, 2012) and Model D (Ni *et al.*, 2013b) are based on the NH_2OH oxidation pathway. Model C assumes that N_2O production is due to the chemical decomposition of the unstable NOH, an intermediate of NH_2OH oxidation (Law *et al.*, 2012). In contrast, Model D assumes that the reduction of NO, produced from the oxidation of NH_2OH, resulted in N_2O production by consuming NH_2OH as the electron donor. Model D (Table 7.1, Ni *et al.*, 2013b) assumes that DO has no inhibitory effect on NO reduction (Yu *et al.*, 2010), as in Model B.

7.2.2.3 Two-pathway models

The two N_2O production pathways of AOB (NN: hydroxylamine pathway, ND: nitrifiers denitrification) have been integrated into different two-pathway models. Table 7.1 and Figure 7.2 compare the key differences between these four models (E–H). Two of them (E–F) are based on the decoupling approach with electron carriers (Ni *et al.*, 2014; Peng *et al.*, 2015a) whereas the two others (G–H) are based on the coupling approach (Domingo-Félez and Smets, 2016; Pocquet *et al.*, 2016).

In Model E (Table 7.1, Ni *et al.*, 2014), the complex biochemical reactions and electron transfer processes involved in AOB metabolism are lumped into three oxidation and three reduction reactions (Figure 7.2). Electron carriers are introduced as a new component in the model to link the electron transfer from oxidation to reduction. By decoupling the oxidation (E-1 to E-3 in Figure 7.2) and reduction (E-4 to E-6 in Figure 7.2) reactions through the use of electron carriers, the electron distribution between O2, NO_2- and NO as electron sinks is modelled by assigning different kinetic values to Processes E-4, E-5 and E-6 with respect to electron carriers. These electron carriers are regenerated by Processes E-2 and E-3. In this way, the model can predict the relative contribution of the two pathways to total N_2O production by AOB, as well as the shifts in the dominating pathway at various DO and nitrite level conditions.

Model F (Peng *et al.*, 2015a) is based on the decoupling approach with both electron and energy (adenosine triphosphate) balances, which are proposed by the extension of Model E to describe the dependency of N_2O production by AOB on the inorganic carbon (IC) concentration (Peng *et al.*, 2015a). In addition to the electron carriers that link electron transfer from oxidation to reduction, adenosine triphosphate (ATP)/adenosine diphosphate (ADP) are also introduced as components in the model (Table 7.1) to link energy generation to IC fixation for biomass growth (Figure 7.2). The energy distribution between ammonia oxidation, NO_2- reduction and oxygen reduction as energy source (ATP) is modelled through assigning different kinetic values to Processes F-1, F-5 and F-6 with respect to ADP, which is consumed by Process F-7 with IC as substrate for AOB growth. In this way, the possible effect of IC on AOB growth, and subsequently the N_2O production from different pathways by AOB, can be explicitly described when the IC concentration in the bioreactor varies temporally or spatially, with N_2O production increasing with an increase in IC concentrations.

In Model G (Table 7.1, Pocquet *et al.*, 2016) the two pathways are combined based on a direct coupling approach. The model includes five enzymatic reactions (Figure 7.2). As in models E and F, NO is considered as an intermediary compound during oxidation of hydroxylamine into nitrite and N_2O is supposed to be produced by both the reduction of NO (hydroxylamine pathway) and the reduction of nitrite (AOB denitrification). Free ammonia and free nitrous acid are considered as substrate for nitrification and denitrification, respectively. As in model E and F, the NO intermediary is not considered in the denitrification pathway which avoids the NO loop (i.e., production via nitrite reduction and re-oxidation into nitrite). The inhibition of AOB denitrification by oxygen is considered by a modified Haldane equation as in Guo and Vanrolleghem (2014).

In Model H (Table 7.1, Domingo-Félez and Smets, 2016) the two pathways are also combined based on a direct coupling approach and it includes five enzymatic reactions (Figure 7.2). Free ammonia and free nitrous acid are considered the substrates for nitritation and denitritation, respectively. Hydroxylamine is oxidized aerobically producing nitrite and is independent of oxygen presence producing NO (hydroxylamine pathway). The inhibition of AOB denitrification by oxygen is considered by an inverse Michaelis-Menten-like equation and produces NO (AOB denitrification). Hence, NO is an intermediate of the two pathways with different dependencies on oxygen and free nitrous acid concentrations. A single autotrophic N_2O-producing process accounts for the combined NO reduction.

7.3 MODEL INTEGRATION, USE AND CALIBRATION

7.3.1 Integrated N_2O models

N_2O can be generally produced by both AOB and heterotrophic denitrifiers in WWTPs and consumed by heterotrophic denitrifiers (Guo and Vanrolleghem, 2014; Kampschreur *et al.*, 2009; Law *et al.*, 2012). Therefore, integrated N_2O models incorporating N_2O production/consumption by both AOB and heterotrophic denitrifiers would contribute to more powerful models that predict the N_2O dynamics more accurately in WWTPs, which could also be a useful tool for the development of N_2O mitigation strategies.

Two approaches have been reported to integrate the N_2O production/consumption by both AOB and heterotrophic denitrifiers into a comprehensive N_2O model: (i) ASM-type models that combine one of the single-pathway models of AOB (e.g., Models A–D, Table S3) with ASMN of heterotrophic denitrifiers (Model OHO-A, Table S2) (Guo and Vanrolleghem, 2014; Ni *et al.*, 2011; Pocquet *et al.*, 2013; Spérandio *et al.*, 2016), and (ii) electron balance-based models that integrate the electron carrier-based two-pathway model of AOB (Model E, Table S4) and ASMN (Model OHO-A, Table S2) (Ni *et al.*, 2015). Both modelling approaches have been successfully applied to describe N_2O emissions from mixed culture nitrification-denitrification systems and to identify the relative contributions of AOB and heterotrophic denitrifiers to total N_2O production (Ni *et al.*, 2011, 2013b, 2015; Spérandio *et al.*, 2016). A third potential approach to integrate the N_2O production/consumption by both AOB and heterotrophic denitrifiers could be a full electron balance-based model integrating the electron carrier-based two-pathway model of AOB (Model E, Table S4) and the electron carrier-based model of heterotrophs (Model OHO-B, Table S2), this requires future testing though. Model H integrates a two-pathway AOB model, ASMN of heterotrophic denitrifiers and abiotic reactions considering free nitrous acid and hydroxylamine.

It should be noted that the possible consumption of N_2O by heterotrophic denitrification as an N_2O sink may occur and reduce overall N_2O production in an integrated model under the conditions of high COD to N ratio and/or low DO level.

7.3.2 Model Evaluation against experimental data

The N_2O models have to be tested to predict N_2O emission data from experiments in order for the models to become useful tools in practical applications. During recent years, many measurement campaigns have been performed. All available N_2O models have been evaluated with experimental data collected from different systems to reveal their performance under various process conditions and shed light on the conditions under which each of the models would be suitable for application.

7.3.2.1 Heterotrophic denitrification

For denitrifying N_2O models, Model OHO-A was found generally able to reproduce the nitrate, nitrite and N_2O profiles when only one nitrogen oxide species was added (Ni *et al.*, 2011; Pan *et al.*, 2015), but Model OHO-A failed to reproduce the results when two or more nitrogen oxide species were added together. In contrast, Model OHO-B was shown to be able to describe general COD consumption, nitrate reduction and nitrite accumulation by an enriched denitrifying culture (Pan *et al.*, 2015), and the influence of nitrite and N_2O addition on nitrate reduction, as well as the experimental results when one or more nitrogen oxide species were added (Pan *et al.*, 2015). Therefore the decoupling approach of Model OHO-B might be essential to describe complex conditions with addition of multiple nitrogen oxide species, but in the many situations for which only nitrate and nitrite are provided Model OHO-A is still applicable. For both models it can be noted that an independent calibration of each of the successive steps of denitrification has rarely been possible. As heterotrophic denitrification is an important N_2O mechanism to be considered for future mitigation strategies, more measurements of intermediates is recommended in future studies to improve the robustness of the models calibration.

7.3.2.2 Single-pathway AOB models

The six single-pathway AOB models (Models A–D, Table 7.1) were evaluated and compared (Ni *et al.*, 2011, 2013a; Spérandio *et al.*, 2016) based on their ability to capture the observed N_2O production results from different experiments (Kim *et al.*, 2010; Law *et al.*, 2012; Spérandio *et al.*, 2016; Yang *et al.*, 2009). Model A could predict well the observed trend of a decrease in N_2O production at high DO concentrations (Yang *et al.*, 2009), whereas Model B was not able to predict such a trend due to the absence of oxygen inhibition on AOB denitrification in Model B (Ni *et al.*, 2013a). Model B could not describe well the N_2O peak that is likely related to the dynamics of NH_2OH (Ni *et al.*, 2013a), which was not included in Models B and B1. Models A, A1, B and B1 have been tested and found to reasonably describe N_2O production data with high nitrite accumulation (Figure 7.3) (Pocquet *et al.*, 2013; Spérandio *et al.*, 2016). In contrast, both Models C and D were not able to capture the observed dependency of N_2O production on nitrite availability (Kim *et al.*, 2010; Spérandio *et al.*, 2016; Yang *et al.*, 2009) due to the fact that the two models are linked to incomplete NH2OH oxidation. However, Models C and D were able to reproduce the experimental observations that the N_2O production increased/decreased with increasing/decreasing DO concentration (Law *et al.*, 2012). The kinetic structure of Model B also ensured that the N_2O production rate was dependent on oxygen availability, resulting in a similar N_2O dynamic trend (increase in the N_2O production rate with an increase in DO concentration). On the contrary, Model A predicted the opposite to such an observation (Law *et al.*, 2012). These results suggest that DO inhibition might be required to describe AOB denitrification and NH_2OH needs to be included as a necessary intermediate. The use of FA and FNA in model structures would be recommended for a better description of the pH effect and possible FNA inhibition. NOH would be preferably used as an N_2O precursor for describing the NH_2OH pathway under extremely high nitrite accumulation conditions, whereas NO could be generally applied as an intermediate for N_2O production from NH_2OH oxidation under common wastewater conditions.

7.3.2.3 Two-pathway AOB models

With respect to the two-pathway models of AOB, Model E has satisfactorily described the N_2O data from several different nitrifying cultures (partial nitritation culture or/and full nitrification culture) and under various DO and NO_2- concentration conditions (Figure 7.4) (Ni *et al.*, 2014; Peng *et al.*, 2014; Sabba *et al.*, 2015). Model F has also predicted well these different nitrifying cultures (partial nitritation and full nitrification culture) but also under various IC conditions (Peng *et al.*, 2015a). Although the electron-based two-pathway models (Models E and F) have been demonstrated to be effective, electron carriers may not necessarily be the only approach to the integration of the two

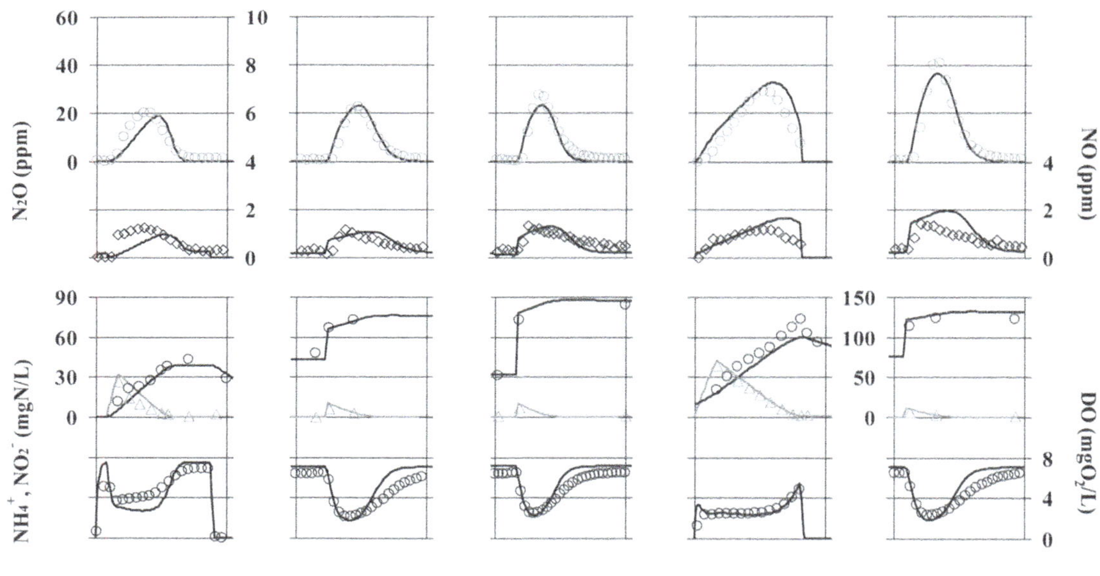

Figure 7.3 Comparison of simulation (A1, Pocquet *et al.*, 2013) and measured data for five batch experiments at different nitrite and ammonium levels. NO in gas phase (◊), N_2O in gas phase (○), ammonium (Δ), nitrite (○) and dissolved oxygen (○). Duration of experiments: 1 h. N_2O emission factors: 1.39%, 2.58%, 3.86%, 1.83%, 4.52% (gN-N_2O/gN-NH_{4^+}), respectively (Spérandio *et al.*, 2016).

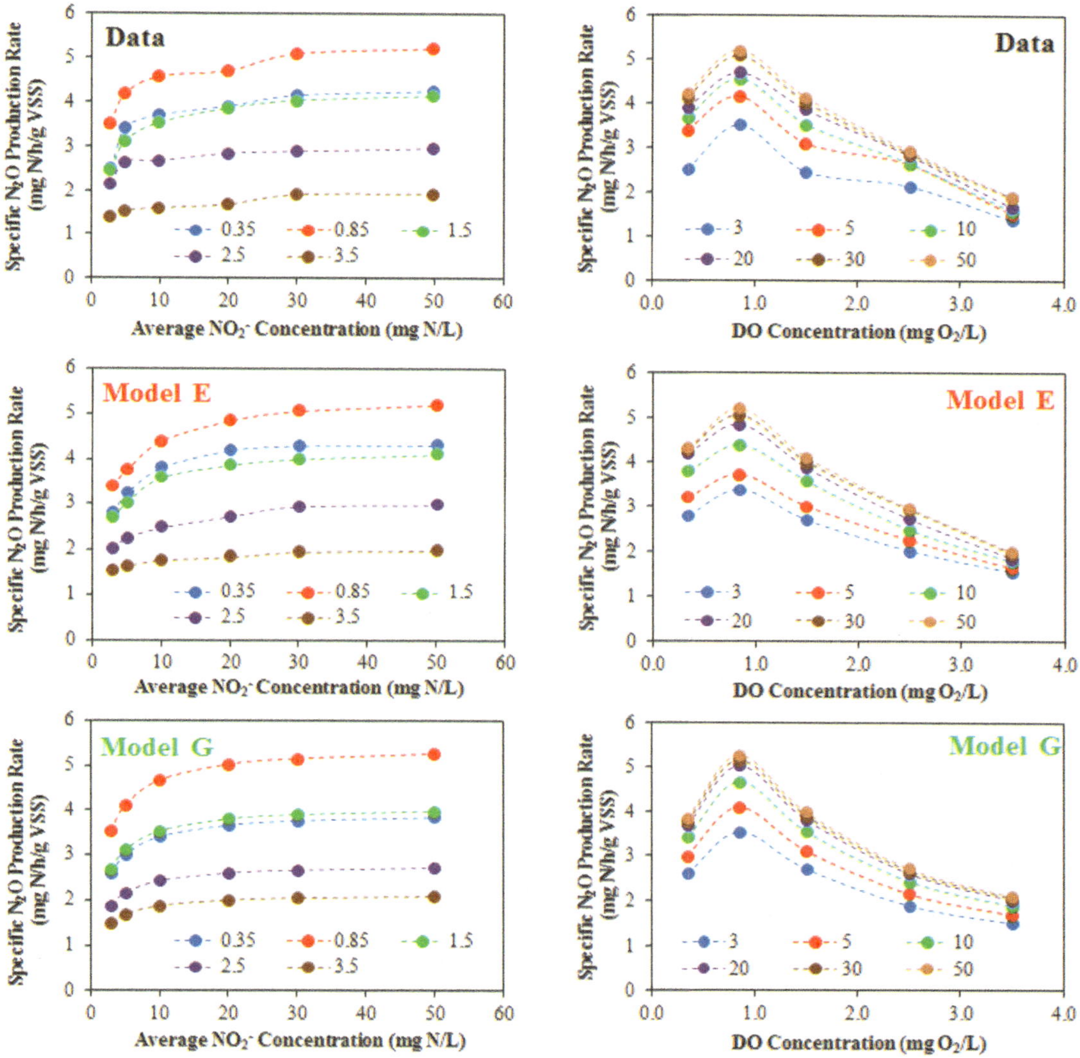

Figure 7.4 Comparison of Model E and Model G predictions (two-pathway models). Experimental and simulated effect of dissolved oxygen and nitrite concentrations on specific N_2O production rate during short-term (batch) experiments (Lang *et al.*, 2017).

pathways into one model. Model G is based on the *coupling approach* without considering electron carriers. It was calibrated with batch experiments and validated with long-term data collected in a sequencing batch reactor performing nitritation and denitrification (Pocquet *et al.*, 2016; Mampaey *et al.*, 2019). A good prediction of the N_2O emissions for varying nitrite concentrations was obtained. Model G is also capable of describing the trends observed for the NO emissions and the variation of the NO/N_2O ratio depending on the pathways' contribution. The combined effect of nitrite (via free nitrous acid) and dissolved oxygen (DO) is also correctly predicted by Model G (Figure 7.4) (Lang *et al.*, 2017).

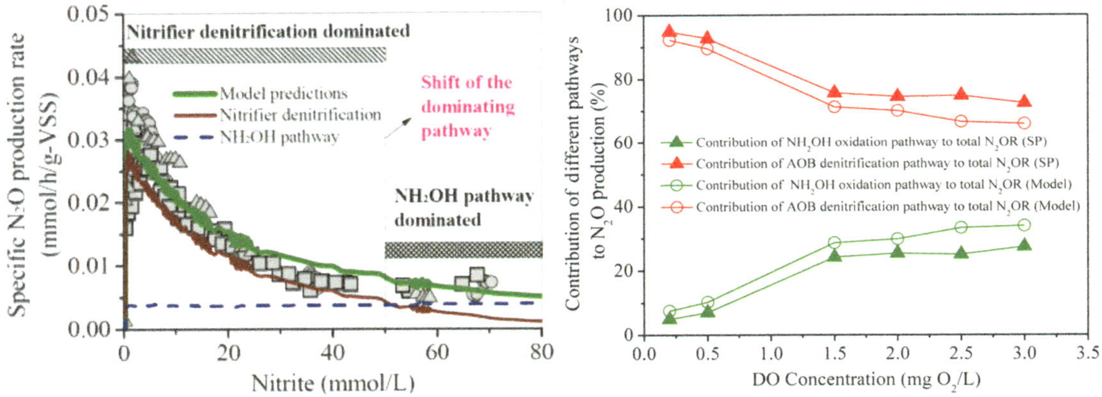

Figure 7.5 Predicted contributions from the nitrifier denitrification pathway and the NH_2OH pathway as well as their shifts using Model E (real data: symbols, model predictions: lines) for a partial nitrification (left panel adapted from Ni *et al.*, 2014) and a full nitrification system (right panel adapted from Peng *et al.*, 2014).

Two-pathway models E and G successfully predicted shifts of the dominating pathway at various DO, nitrite and/or IC levels (see Figure 7.5, Lang *et al.*, 2017), consistent with experimental observations that N_2O was produced from both nitrifier denitrification and NH_2OH oxidation pathways by AOB (Ni *et al.*, 2014; Peng *et al.*, 2014; Wunderlin *et al.*, 2013). The model results suggested that the contribution of AOB denitrification decreased as DO increased, accompanied by a corresponding increase in the contribution by the NH_2OH oxidation pathway. This was verified by site preference (SP) isotopic measurements (Peng *et al.*, 2014). The two-pathway models also successfully predicted the increase of the AOB denitrification pathway with nitrite (at low nitrite concentrations) and the inhibition of AOB denitrification at high nitrite concentrations (see Figure 7.5) (Ni *et al.*, 2014; Pocquet *et al.*, 2016).

Model H was calibrated with AOB-enriched biomass and activated sludge mixed liquor biomass (Domingo-Félez and Smets, 2020b; Domingo-Félez *et al.*, 2017; Su *et al.*, 2019b). Optimal extant respirometry and anaerobic batch experiments that target endogenous and exogenous processes (of both autotrophic ammonium/nitrite oxidation and heterotrophic denitrification), together with the associated net N_2O production were designed and executed. The calibrated model predicts the NO and N_2O dynamics at varying ammonium, nitrite and dissolved oxygen levels in the two independent systems.

7.3.3 Selection of models for N_2O Prediction

7.3.3.1 Single-pathway AOB models versus two-pathway AOB models

The model evaluation results strongly suggest that appropriate selection of available N_2O models is important for accurate N_2O prediction in different engineering nitrogen removal systems under different operational conditions. Figure 7.6 presents a possible guideline for model selection in their further applications.

For N_2O production by heterotrophic denitrifiers, Model OHO-A can be used to predict the overall nitrogen and COD removal performance in a wastewater treatment plant, as in most cases the low level accumulation of denitrification intermediates does not significantly affect the overall nitrogen removal rate. However, in the context of predicting the N_2O production by heterotrophic denitrifiers, Model OHO-B is inadequate due to its structural deficiency in describing the electron competition process in denitrification. Model H enhanced our ability to predict N_2O production by heterotrophic denitrifiers and has the potential to describe all N_2O data under different conditions. However, it

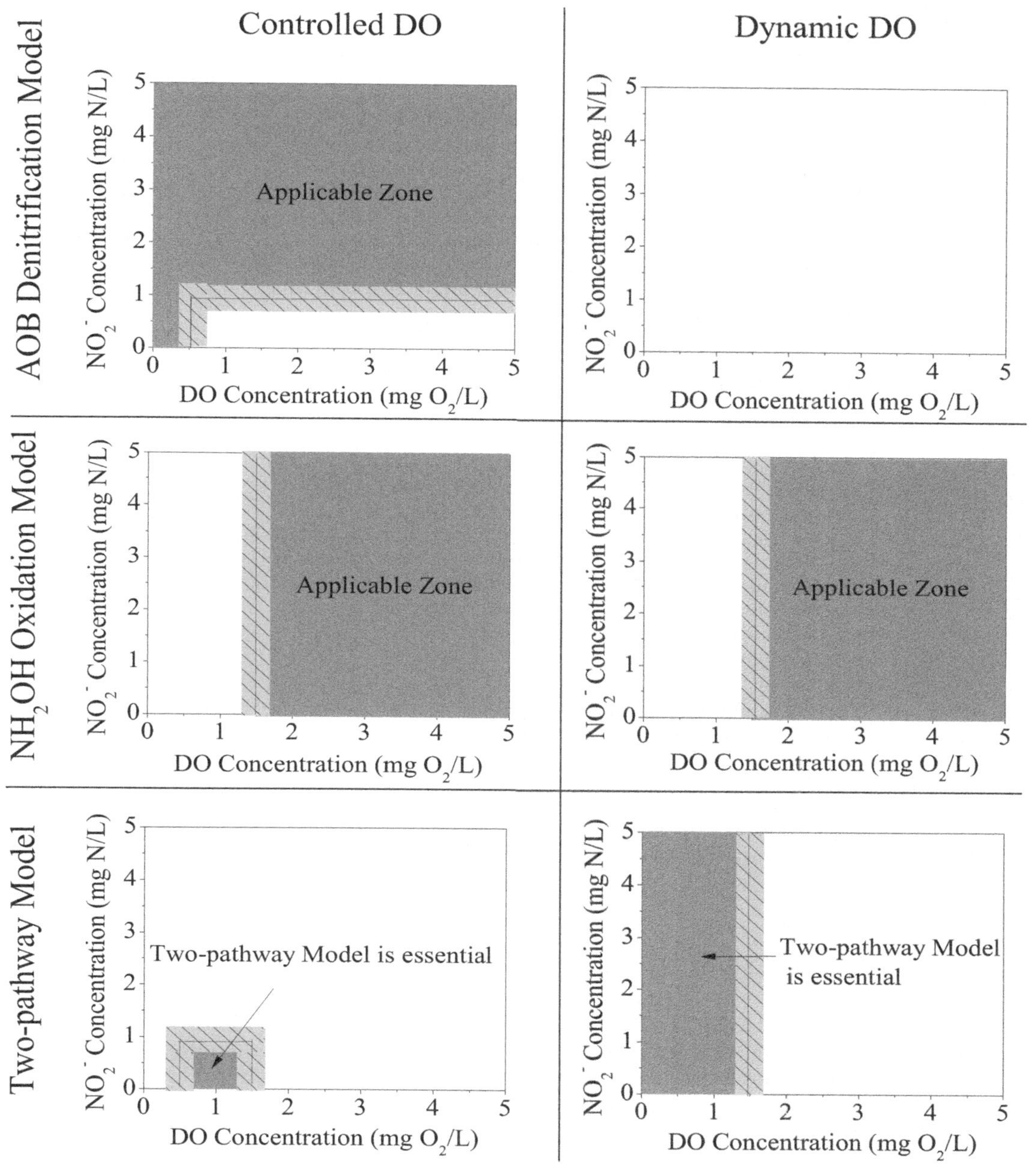

Figure 7.6 Summary of applicable regions for the AOB denitrification model, the NH_2OH oxidation model and the two-pathway model under various DO and NO_2^- concentrations. The applicable regions were insensitive to the variations of key parameters governing N_2O production by the two-pathway model (Peng *et al.*, 2015b).

Table 7.2 Guideline for model selection for predicting N_2O production by AOB and heterotrophic denitrification

N_2O models	Single-pathway models by AOB	Two-pathway models by AOB	N_2O models by heterotrophs
Applicable conditions	• Models A, A1, B and B1 to describe the regulation of N_2O production by nitrite (or FNA) • Model A to predict possible DO inhibition on N_2O production at high DO levels • Models A1, B and B1 to predict possible pH effect and FA/FNA inhibition on N_2O production • Models C and D to describe N_2O emissions at high DO levels and low nitrite accumulation	• Model E to predict N_2O production at varying DO and NO_2- with constant IC • Model F to describe N_2O production under highly dynamic IC condition • Model G to predict NO and N_2O production at varying DO and NO_2- and possible pH and FA/FNA effects on N_2O production	• Model OHO-A to predict the overall nitrogen and COD removal performance with low level accumulation of denitrification intermediates • Model OHO-B to describe N_2O production under different conditions
Inabilities of the models	• Model A not to describe the increase of N_2O production with increasing DO • Models B and B1 not to predict the N_2O production related to the dynamics of NH_2OH • Models C and D not to predict the effect of nitrite accumulation on N_2O production	• Model E and G not to describe N_2O production with dynamic IC • Model E and F not to describe pH effect on N_2 production • Model E and F not to describe NO production	• Model OHO-A not to describe N_2O production with electron competition
Key parameters for calibration	• The half saturation constant for nitrite or FNA ($K_{NO2,AOB}$ or $K_{HNO2,AOB}$ for Models A, A1, B, B1) • The reduction factor for N_2O production (η_{AOB}, for all six single-pathway models)	• E, F: The affinity constants with respect to electrons (e.g., $K_{Mred,3}$, and $K_{Mred,4}$) • E, F: The ratios among the affinity constants to electrons • G, H: The reduction factors, affinity constant for HNO_2 and NH_2OH	• The N_2O production and reduction rates • The relative ratios between electron affinity constants

requires information on both the carbon oxidation reaction kinetics and the nitrogen reduction kinetics.

For N_2O production by AOB, the single-pathway models (Models A–D) have simpler structures (one single pathway involved) and fewer parameters, which is convenient for model calibration (Table 7.2). This makes their use preferential under certain conditions, even though they may not be able to reproduce all N_2O data. The two-pathway models (Models E–G) have the potential to describe all N_2O data with different operational conditions, but may require more efforts in model calibration because of their larger number of parameters. Specifically (Table 7.2), Models A, A1, B and B1 might be used to describe the regulation of N_2O production by nitrite (or FNA) concentrations. Models C and D might be able to describe N_2O emissions from systems under relatively high DO concentrations and low nitrite accumulation that likely favour the NH_2OH oxidation pathway for N_2O production. In addition, according to the analysis by Peng *et al.* (2015b) (Figure 7.6), for the AOB denitrification model to be used (e.g., Model A) it is preferable that the DO concentration in the system is well controlled at a constant level. NH_2OH oxidation models (e.g., Model D) can be applied under high DO conditions.

Under other conditions, the two-pathway models (e.g., Model E or G) should be applied. Model E or Model G could be used under varying DO and NO_2- concentrations, but stable IC conditions are required, while Model F would be preferable under highly dynamic IC conditions.

7.3.3.2 Two-pathway AOB models: direct versus indirect coupling approach

The two different two-pathway models E and G based on indirect coupling or direct coupling approaches, respectively, were compared by Lang *et al.* (2017). Both were calibrated to describe the experimental N_2O emissions collected from 43 kinetic experiments considering a large range for both DO and nitrite concentrations and three different AOB-enriched cultures (Lang *et al.*, 2017) (Figure 7.4).

Both models enabled the prediction that the increase of DO enhances the hydroxylamine pathway contribution while it reduces the contribution of the AOB denitrification pathway (Figure 7.7). This is related to competition between oxygen and nitrite for electron carriers in Model E, whereas it is described by an inhibition term in the AOB denitrification kinetics in Model G. Regarding the nitrite effect, both concepts similarly describe the shift from the hydroxylamine pathway to the AOB denitrification pathway. Considering FNA in Model G also indirectly enables the pH influence to be described.

The choice between these two concepts will depend on simulation objectives and calibration experience. Regarding the last extended model of Peng *et al.* (2016) which also includes the inorganic carbon effect (Model F), the 'indirect coupling' approach is able to reveal some metabolic relations between N_2O production and the cell's anabolism. This mathematical framework constitutes an ideal approach for investigating intracellular and metabolic mechanisms. In comparison, the direct coupling approach is simpler and easily understandable for practitioners as it is based on the conventional ASM approach. Another advantage of Model G is that it considers NO as an external state variable (not an intracellular compound like Model E) which makes it possible to use such data for model calibration (Pocquet *et al.*, 2016).

Finally, actual experience shows that both concepts enable good predictions. Work is now recommended with data from full-scale systems in which the mixed liquor complexity and the combination with other biochemical reactions could reveal stronger differences between these two models.

7.3.4 Key kinetic and stoichiometric parameters for calibration

The calibration approach for N_2O models is based on two successive steps. The first step consists of adjusting the 'conventional' parameters (i.e., growth rate, decay rate, substrate removal rate and the

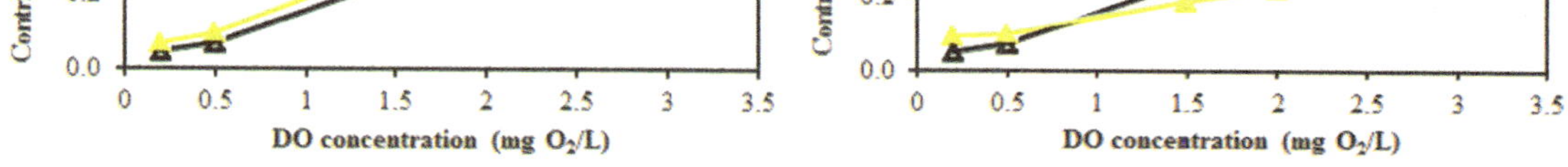

Figure 7.7 Effect of dissolved oxygen on contribution of the two AOB pathways (left: Model E, right: Model G) SP: data from site-preference measurements (Lang *et al.*, 2017).

affinity constant) to obtain a good prediction of the main substances dynamics: ammonium, nitrite, nitrate, oxygen. In a second step, the specific parameters influencing N_2O emissions are calibrated. A final iteration should be performed if the prediction of the main substances is slightly affected by this second calibration step.

Typical values of the model parameters can be found in literature (Ni and Yuan, 2015). Different sets of parameter values obtained after calibration on different sets of data are provided by Spérandio *et al.* (2016) and Lang *et al.* (2017) for the single-pathway AOB models and the two-pathway AOB models, respectively. Continued testing against more experimental data would delineate a range/pattern in parameter values. It should be noted that these parameters were estimated under different conditions of temperature, sludge retention time and feeding composition, and therefore correction factors must be adjusted by, for example, Arrhenius equations (Snip *et al.*, 2014). Furthermore, the parameter values estimated during batch experiments may not be adequate for continuous processes and may not be compatible with the values of other parameters (Ni *et al.*, 2013a; Snip *et al.*, 2014; Spérandio *et al.*, 2016).

Regarding the ASM-ICE of the heterotrophic denitrifiers (Model OHO-B in Table S2), information on both the carbon oxidation reaction kinetics and the nitrogen reduction kinetics is required for its calibration and application (Table 7.2). Due to the lack of understanding of the electron competition process in most of the previous studies, the respective reaction kinetics of the carbon oxidation and nitrogen reduction processes were not well established. For instance, the maximum carbon source oxidation rate ($r_{COD,max}$), which is the key parameter to restrict the overall model predictions of the carbon oxidation (electron supply) rate, is not available in literature and thus needs to be measured or estimated (Pan *et al.*, 2015). Similar to the two-pathway models of AOB, the relative ratios between electron affinity constants ($K_{Mred,1}$, $K_{Mred,2}$, $K_{Mred,3}$, and $K_{Mred,4}$) rather than their absolute values are important for the reaction rate. Therefore, increased efforts are needed to provide more information on these key parameters of the ASM-ICE model for its further implementation (Table 7.2).

For the six single-pathway AOB models (Models A–D in Table S3), the model parameters were obtained after significant calibration efforts, and thus some of the parameters showed wide variation (more than 100%) among case studies during model evaluations (Ni *et al.*, 2011, 2013a; Spérandio *et al.*, 2016). Among them, the half saturation constant for nitrite or FNA ($K_{NO2,AOB}$ or $K_{HNO2,AOB}$ for Models A, A1, B, B1) and the reduction factor for N_2O production (η_{AOB}, for all six single-pathway models) were most variable and very influential on N_2O emissions (Spérandio *et al.*, 2016). Regarding the models based on the AOB denitrification pathway (e.g., Models A, A1, B and B1), the large variation of these two key parameters was related to the range of nitrite (or FNA) concentrations observed in each system (Spérandio *et al.*, 2016), likely due to the adaptation of enzymatic activity (NirK). Regarding the models based on the NH_2OH oxidation pathway (e.g., Models C and D) the large variation of η_{AOB} might be dependent on the possible NO accumulation in each system. High NO accumulation would lead to a low value for η_{AOB} (Spérandio *et al.*, 2016). Thus, calibration will be required for the application of the single-pathway models regarding these key parameters (Table 7.2).

For the electron balance-based two-pathway AOB models (Models E and F in Table S4), the affinity constants with respect to electrons (e.g., $K_{Mred,3}$, and $K_{Mred,4}$) are unique to the two-pathway models and the key parameters governing the N_2O production via the two pathways. The values represent the affinity of the corresponding reduction reaction to electrons, with lower values indicating a higher affinity and thus a higher ability to compete for electrons. For example, the estimated $K_{Mred,3}$ has a value that is about one magnitude smaller than $K_{Mred,4}$ (Ni *et al.*, 2014), indicating that O_2 reduction has a higher ability to compete for electrons than the main electron acceptor during NH_2OH oxidation. Ni *et al.* (2014) revealed that the absolute value of Ctot is not critical for model calibration and predictions, and it is the ratios between parameters K_{Mox}, $K_{Mred,1}$, $K_{Mred,2}$, $K_{Mred,3}$, and $K_{Mred,4}$ and parameter Ctot that affect the model output. Therefore, attention should be paid to these ratios for the calibration and application of the two-pathway models (see Table 7.2).

For the two-pathway model G, the reduction factor for both AOB denitrification and hydroxylamine pathway (η_{AOB_ND}, η_{AOB_NN}) are the two major influential parameters which control the maximal specific N_2O production rates of each pathway (Lang *et al.*, 2017). The affinity constant for FNA also has to be calibrated from one culture to another, especially when working at very different nitrite concentrations. The hydroxylamine pathway contribution is also sensitive to the affinity constant for nitric oxide as determined by NO measurements (Pocquet *et al.*, 2016). In parallel, the AOB denitrification contribution is influenced by the inhibition constant for oxygen which is a key parameter for predicting the effect of lowering aeration on N_2O emissions (Lang *et al.*, 2017).

Model H was calibrated following a global sensitivity analysis and an information-based parameter selection procedure (Domingo-Félez and Smets, 2020b; Domingo-Félez *et al.*, 2017). First, parameters associated with heterotrophic denitrification were fitted. Then, parameters associated with aerobic nitrite and ammonia oxidation were sequentially fitted to DO consumption profiles by isolating individual processes, followed by N_2O production profiles. In the AOB-enriched biomass the reduction factors for the NN pathway and the N_2O-production process were estimated together with the HNO_2 and NH_2OH affinity constants. In the activated sludge mixed liquor biomass three reduction factors were estimated: NO-producing NN and ND pathways, and N_2O production processes. The pH-dependency of AOB-driven N_2O production and heterotrophic N_2O consumption was also described (Domingo-Félez and Smets, 2020b; Su *et al.*, 2019a). The uncertainty associated with parameter estimation results was propagated to validate the model response, and is recommended to be included with best-fit simulations.

7.3.5 Application of N_2O models in biofilm systems

The previous sections provide a basis for modelling the formation and consumption of N_2O by AOB and heterotrophic denitrifying bacteria. In this section, we discuss how these kinetics are applied to biofilm processes. Biofilm treatment processes, such as the moving bed biofilm reactor (MBBR), integrated fixed-film activated sludge (IFAS), biological aerated filters (BAF), denitrifying filters, and granular sludge, are becoming increasingly popular for wastewater treatment. Due to substrate gradients and microbial stratification, the behaviour of biofilms is typically different from suspended growth processes, and this may be especially true for N_2O production and emissions (Law *et al.*, 2012; Schreiber *et al.*, 2009; Sutka *et al.*, 2006). These systems appear to have among the highest N_2O emission rates (e.g., Bollon *et al.*, 2016).

A schematic of a biofilm, with diffusion of substrates and products, is shown in Figure 7.8. In conventional biofilms, both the electron donor and acceptor substrates diffuse from the bulk liquid into the biofilm. Substrates penetrate into the biofilm by diffusion and are consumed within the biofilm by microbially catalysed reactions. Substrates diffusing into the biofilm from the bulk liquid side first pass the liquid diffusion (or boundary) layer. The liquid diffusion layer adds diffusive resistance to substrate transport into the biofilm, decreasing the substrate concentration at the biofilm surface with respect to the bulk liquid.

When modelling N_2O emissions from biofilms, the underlying rate expressions are the same as those described for the suspended growth processes. However, diffusion and microbial stratification within the biofilm can change the observed behaviour. For example, suspended growth bacteria in an aerobic zone of a treatment process are unlikely to have appreciable denitrification. However, biofilms in aerobic zones of a treatment process may have anoxic zones in their interior. This can allow heterotrophic denitrification, including formation and consumption of N_2O and AOB denitrification.

Sharp gradients of O_2 and other substrates within a biofilm, combined with different microbial species in close proximity, allow the diffusion of intermediates to different redox environments or zones with different microbial metabolisms. For example, NH_2OH can be produced by AOB in the outer, aerobic zones of a biofilm, and consumed in inner, anoxic zones where it leads to peaks in N_2O formation due to AOB denitrification. Nitrite oxidizing bacteria (NOB) can enhance this effect by increasing the O_2 gradients within the biofilm (Sabba *et al.* submitted).

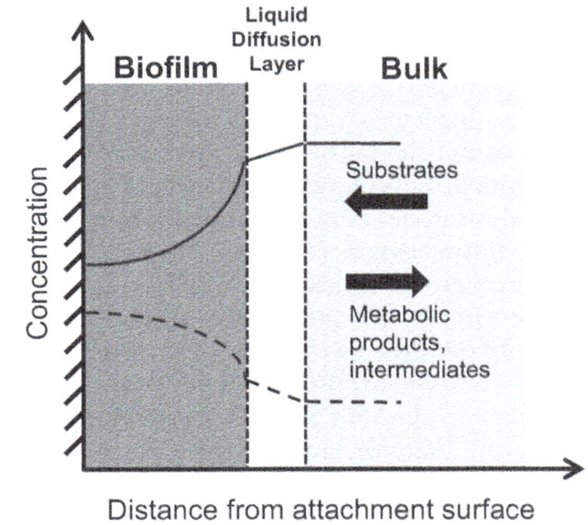

Figure 7.8 Schematic of a biofilm with substrate penetration and metabolic products/intermediates formation.

The importance of intermediate diffusion is illustrated in Figure 7.9, which shows N_2O emissions from a pilot-scale granular sludge reactor for side-stream nitritation (Pijuan *et al.*, 2014). When modelled with very limited NH_2OH diffusion (1% of the actual value), the model could not capture the actual N_2O emissions (Figure 7.9). However, when NH_2OH diffusion was included in the model, the model provided an excellent fit to the data (Sabba *et al.*, 2015).

Another example of interactions within a biofilm is the scavenging of N_2O formed in the outer zone of a biofilm, for example by AOB, by heterotrophic denitrifiers in the deeper, anoxic zones of the biofilm. This can lead to lower net N_2O emissions, although the complexity of biofilms makes this highly dependent on the specific reactor conditions.

Biofilm processes can be more challenging to calibrate than suspended growth processes. Information is needed about the biofilm thickness, density, substrate diffusivities, and microbial community structure. In most cases, a one-dimensional model can capture biofilm behaviour. Special care should be taken when analysing putative suspended-growth processes that may actually display biofilm behaviour. This may be true for processes with large flocs. It also may be true for bench- or pilot-scale systems, as reactor wall area is more significant, relative to reactor volume. This can lead to a greater impact of biofilms growing on walls than in a full-scale system.

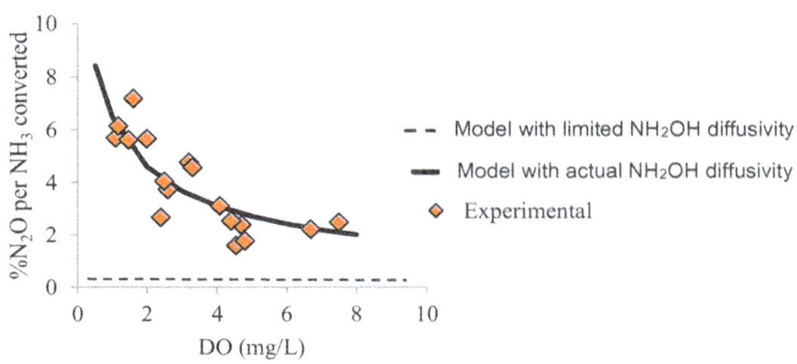

Figure 7.9 Comparison between measured $\%N_2O$ from NH_3 converted and model-calculated values in steady-state conditions (Sabba *et al.*, 2015).

Recently the two pathway AOB model (G) and the multiple step denitrification model (OHO-A) were combined for describing N_2O emissions from a nitrifying biofilter and denitrifying biofilter (Fiat *et al.*, 2019; Zhu *et al.*, 2019) as well as a granular sludge system (Lang *et al.*, 2019). Biofilm structure was described by a one-dimensional model. For a granular sludge partial nitrification anammox (PNA) system, the model was successfully calibrated to experimental data by adjusting the affinity constant for hydroxylamine mainly (Lang *et al.*, 2019). The effect of varying nitrite concentration and air flow rate was correctly predicted. Simulation demonstrated that a part of the N_2O produced by AOB was consumed by heterotrophic denitrification in the biofilm, and N_2O emission was highly influenced by the level of NOB repression. This work clarified the behaviour of N_2O emission under very low oxygen concentration, indicating that nitrifier denitrification was the major contributing pathway.

Regarding the biofilters systems, the N_2O models (A and G) were used in successive reactors to describe longitudinal heterogeneity in the biofilter. The simulations were compared to data monitored on full scale installations (results are described in the next paragraph).

In summary, biofilm processes are significantly more complex than suspended growth processes, and modelling can be a critical tool to understand the mechanisms and predict the N_2O formation and emissions.

7.3.6 Application of N_2O models in full-scale WWTPs

Mathematical modelling of N_2O emissions from full-scale WWTPs was first conducted successfully by using ASM-type models that combine one of the single-pathway models of AOB with ASMN of heterotrophic denitrifiers (Ni *et al.*, 2013b). Ni *et al.* (2013b) applied a model based on the NH_2OH pathway model of AOB (Model D, Table 7.1) and ASMN (Model OHO-A, Table 7.1) to describe the N_2O emissions from full-scale WWTPs. The model described well the dynamic ammonium, nitrite, nitrate, DO and N_2O data collected from both an open oxidation ditch (OD) system with surface aerators and a sequencing batch reactor (SBR) system with bubbling aeration. Ni *et al.* (2013b) also performed additional evaluations on the other three single-pathway N_2O models of AOB (Model A, Model B and Model C in Table 7.1) to evaluate the experimentally observed N_2O data from the two full-scale WWTPs. The results indicated that Model A could not predict the N_2O data from either WWTP (Ni *et al.*, 2013b; Spérandio *et al.*, 2016). Models B and C, on the contrary, obtained very similar good fits between the model-predicted and experimentally observed N_2O data (Ni *et al.*, 2013b, Spérandio *et al.*, 2016).

Dynamic simulations were also confronted to the data collected on the UCT (University Cape Town configuration) process of the Eindhoven plant by using ASM-type models that combine one of the single-pathway models of AOB with ASMN of heterotrophic denitrifiers (Guo and Vanrolleghem, 2014; Spérandio *et al.*, 2016). Model A1 + Model OHO-A, Model B1 + Model OHO-A and Model D + Model OHO-A were all implemented for this plant and calibrated using data collected in a 1-month measurement campaign. The conclusion was that all these models could be calibrated to the same level of fit (Spérandio *et al.*, 2016). They had similar performance and could follow the dynamic variations in the measured N_2O data (see Figure 7.10). In addition, the results showed that there was less N_2O emission under wet-weather conditions compared to dry-weather conditions and all three models showed better simulation performance under dry-weather conditions than wet-weather conditions (Spérandio *et al.*, 2016).

Mathematical modelling of N_2O emissions from full-scale WWTPs was then conducted successfully by using electron balance-based models that integrate the two-pathway model of AOB and the ASMN of heterotrophic denitrifiers (Ni *et al.*, 2015). Ni *et al.* (2015) applied an integrated model incorporating the electron balance-based two-pathway model of AOB (Model E, Table 7.1) and the ASMN of heterotrophic denitrifiers (Model OHO-A, Table 7.1) to describe N_2O emissions from a step-feed full-scale WWTP. The model described well all dynamic ammonium, nitrite, nitrate, DO and N_2O emission data. Modelling results revealed that the AOB denitrification rate decreased and the NH_2OH oxidation rate increased along the path of both steps, with the second step of the full-scale

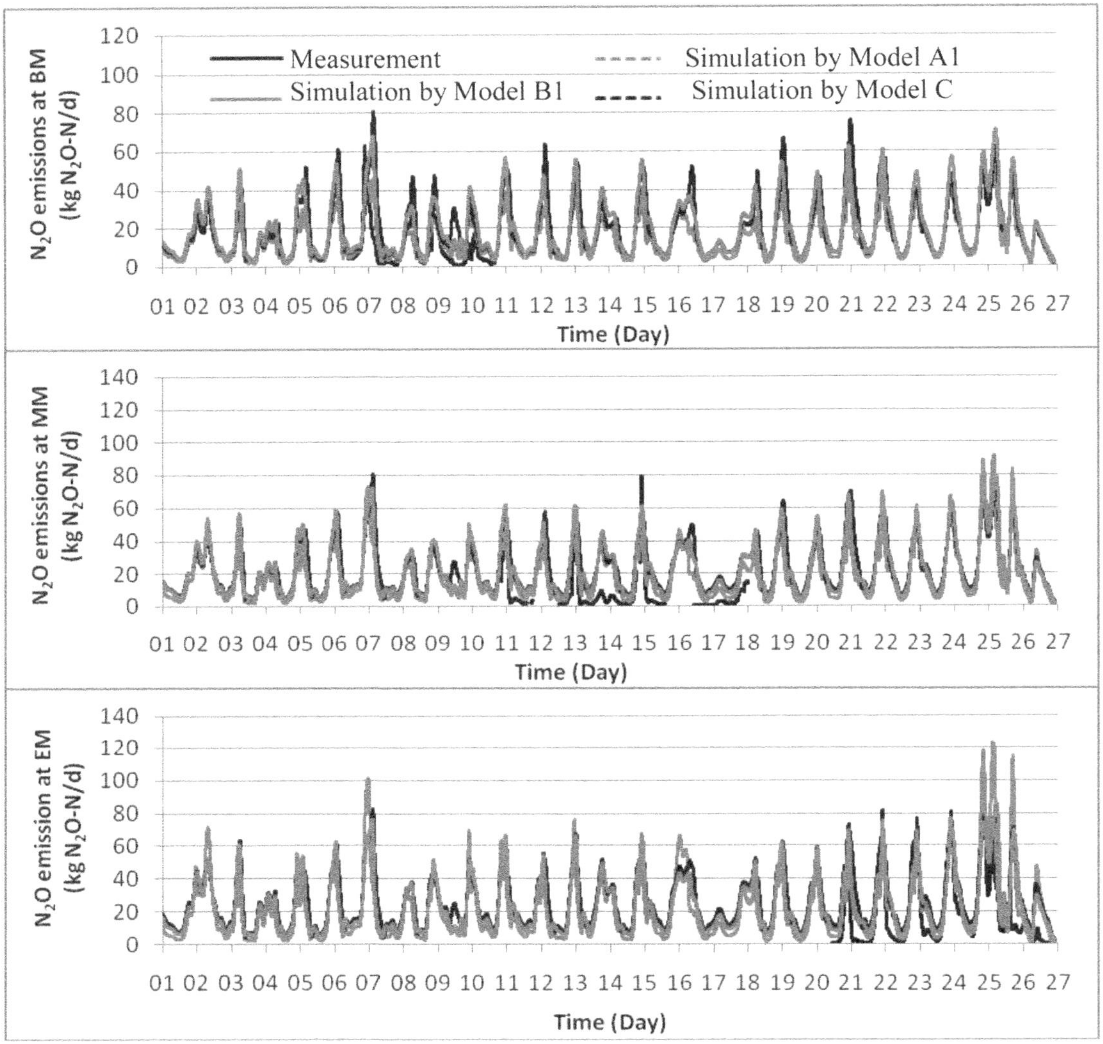

Figure 7.10 Model evaluation results for N_2O emissions using the measurement results at the beginning (BM) (upper panel), the middle (MM) (middle panel) and the end section (EM) (bottom panel) of the summer aeration package on the UCT process at the Eindhoven treatment plant by using ASM-type models that combine one of the single-pathway models of AOB (Models A1, B1 and C) with the ASMN (Model OHO-A) of heterotrophic denitrifiers (Spérandio *et al.*, 2016).

WWTP having much higher N_2O emission than the first step. The integrated N_2O model captured all these trends regarding the shifting/distribution between the different N_2O pathways observed in this full-scale WWTP (see Figure 7.11). A potential strategy to mitigate N_2O emission from this plant was also evaluated using the model. The overall N_2O emission from the step-feed WWTP would be largely mitigated if 30% of the returned activated sludge was returned to the second step with the remaining 70% returning to the first step.

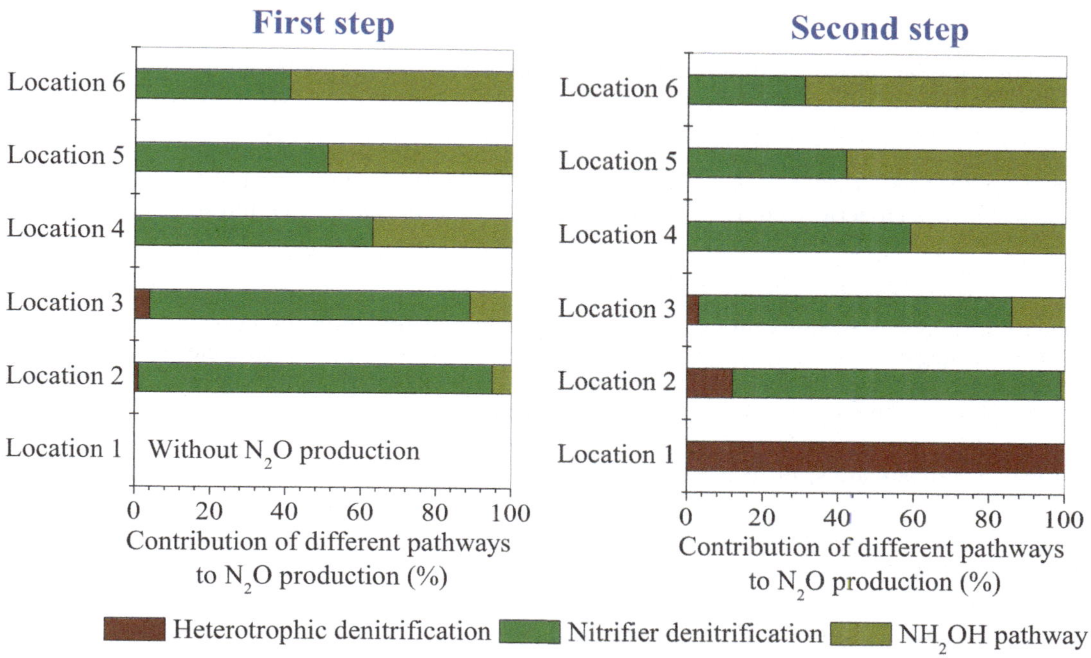

Figure 7.11 Model predicted percentage contributions from the three N_2O pathways to total N_2O productions at six different locations of the first step (left panel) and the second step (right panel) in the step-feed full-scale WWTP, that is, the nitrifier denitrification pathway, the NH_2OH pathway and the heterotrophic denitrification pathway (Ni *et al.*, 2015).

More recently, the electron balance-based model (Model F) has been successfully applied to guide a full-scale N_2O mitigation study (Duan *et al.*, 2020). Full-scale treatment plants have inevitably more complexities than laboratory reactor operations (Ahn *et al.*, 2010a, b; de Haas and Hartley, 2004; Foley *et al.*, 2010; Inventory of U.S. Greenhouse Gas Emissions and Sinks: 1990–2010 (2012); IPCC, 2007; Ye *et al.*, 2014). It was inconclusive whether the model predictions can be reliably applied to guide full-scale mitigations. In this recent work, Model F was calibrated and validated against full-scale results, before being applied to predict the N_2O emissions and nutrient removal performances with different N_2O mitigation measures. The close agreement between the measured emission factor (EF) ($0.58 \pm 0.06\%$) after the implementation of the proposed mitigation strategy, and the EF predicted by the mathematical model (0.55%), showed that the N_2O mathematical model is indeed a useful tool to evaluate N_2O mitigation strategies at full-scale. The model can be a powerful tool for the prediction of N_2O emissions from full-scale WWTPs and development of effective mitigation strategies, although it may require more efforts on model calibration.

Regarding biofilm systems, only a few studies are currently available at full scale. In the work of Fiat *et al.* (2019) and Zhu *et al.* (2019) tertiary nitrifying and denitrifying biofilters were modelled including the main N_2O biological pathways. Simulations were confronted to full-scale data from Seine Aval, the largest wastewater resource recovery facility in Europe. Zhu *et al.* (2019) obtained a satisfying prediction of the emission factor which was higher in the winter period (5.9%) than the value obtained in the summer period (2.9%), in accordance with experimental observations. Fiat *et al.* (2019) demonstrated that the model should include a mass balance on the gaseous phase in each reactor compartment of the BAF in order to correctly describe the N_2O gas-liquid partition and N_2O emissions. Preliminary modifications of the model structure were made to include the gas phase as a compartment of the model, which significantly affected the prediction of nitrification. In particular, considering gas hold-up influenced the prediction of the hydraulic retention time, and thus nitrification performances. Finally, the value of the volumetric oxygen transfer coefficient was adjusted to successfully predict both nitrification and N_2O emissions.

It should be noted that there are still only a limited number of studies presented in literature regarding the real application of N_2O models at full-scale WWTPs although many full-scale measurement campaigns have been performed in different places during recent years. More full-scale applications of the models using these full-scale N_2O data are still needed for the models to be developed into a useful tool for practical applications. In addition, the requirement for good fundamental knowledge on N_2O emission by the modeller/engineer might also hinder the N_2O model applications due to the complicated procedure for model selection and calibration, which consequently limit the development of effective mitigation strategies. Hopefully this chapter will facilitate the selection of suitable N_2O models, the estimation of site-specific N_2O emissions and the development of mitigation strategies for wastewater treatment plants taking into account the specific design and operational conditions of the plant.

7.4 CONCLUSIONS AND PERSPECTIVES

In this chapter, the existing N_2O models available in literature based on the three major N_2O production pathways were reviewed and compared to illuminate their structural differences, their capabilities and inabilities in describing experimental data, and their potential range of applications. The key conclusions are:

Our understanding of fundamental mechanisms related to N_2O production and consumption has strongly progressed in recent decades, leading to the development of N_2O models with different mathematical structures but relatively similar metabolic pathways.

For AOB, the two-pathway models have the potential to describe most of the N_2O data. The combined effect of DO and nitrite is described well by these models, either based on the direct or indirect coupling approach. In comparison, the single-pathway models can be used under several particular conditions depending on the concentrations of oxygen and nitrite which determine the dominating pathway. Despite calibration works still being necessary, recent studies have demonstrated good prediction capabilities for both lab-scale and full-scale observations. The uncertainties around parameters and their propagation on prediction should be considered appropriately.

For heterotrophic denitrifiers, the ASMN-type model was the most used model for predicting the overall nitrogen and COD removal performance in the case where there is only low accumulation of intermediates, whereas new alternatives have been proposed recently and should be considered in the future. In a heterotrophic denitrifier biofilm, the potential for N_2O accumulation together with its consumption in the inner biofilm should be taken into account. The ASM-ICE type model has the potential to describe all N_2O data, but requires more information on reaction kinetics. Full-scale data sets still need to be properly consolidated by adding highly different reactor set-ups, measurement methods, culture history, documentation, and/or interpretations, which would limit the failure of model predictions. Numerous full-scale data sets are starting to be available for suspended growth systems. However, very few studies have identified emissions from biofilm systems.

Future efforts should be devoted to comparing the multiple pathway models to data from real WWTPs to observe the key differences and to enhance their practical applications. Ideally, more information on pathway contributions should be collected in such systems (by means of isotope techniques, or NO:N_2O ratio variations).

Although suspended growth models seem to capture N_2O emissions efficiently, biofilm mechanisms of N_2O production need further investigation. Only a few experiences with mathematical modelling of N_2O emission from biofilm systems have been reported and this should also be conducted using more monitoring data from such systems.

Mathematical modelling of N_2O production has reached a maturity that facilitates the estimation of site-specific N_2O emissions and the development of mitigation strategies. Although existing models still have limitations, their application will undoubtedly increase in the near future. Their confrontation to full-scale data should improve the robustness of the parameters and would certainly suggest further model improvement. For instance the coupling of an N_2O model with more detailed description of hydrodynamics and heterogeneities is probably a future need.

Integration of N_2O models with the models describing other sections of the WWTPs into a plant-wide model could be a powerful tool for future optimization works.

ACKNOWLEDGEMENTS

Research on N_2O emissions at INSA-TBI was supported by the French National Research Agency (ANR) through the research project N_2OTrack. Longqi Lang was supported by a scholarship granted by the China Scholarship Council. Zhiguo Yuan is a recipient of the ARC Laureate Fellowship (FL170100086). Research on N_2O dynamics at DTU has been supported by the Danish Agency for Science, Technology and Innovation through the Research Project LaGas (12-132633), the Danish Council for Independent Research through Project N_2Oman (17H01893) and the Danish Ministry of Environment and Food through the MUDP project HEPWAT. Research at Notre Dame on N_2O emissions from biofilm processes has been supported by NSF project CBET0954918 and WERF project U2R10. Additional support was provided by a Bayer Corporation Fellowship to F.S.

REFERENCES

Ahn J. H., Kim S., Park H., Rahm B., Pagilla K. and Chandran K. (2010a). N_2O emissions from activated sludge processes, 2008–2009: results of a national monitoring survey in the United States. *Environmental Science & Technology*, **44**, 4505–4511, https://doi.org/10.1021/es903845y

Ahn J. H., Kim S., Park H., Katehis D., Pagilla K. and Chandran K. (2010b). Spatial and temporal variability in atmospheric nitrous oxide generation and emission from full-scale biological nitrogen removal and non-BNR processes. *Water Environment Research*, **82**, 2362–2372, https://doi.org/10.2175/106143010X12681059116897

Almeida J. S., Reis M. A. M. and Carrondo M. J. T. (1997). A unifying kinetic model of denitrification. *Journal of Theoretical Biology*, **186**(2), 241–249, https://doi.org/10.1006/jtbi.1996.0352

Arp D. J. and Stein L. Y. (2003). Metabolism of inorganic N compounds by ammonia-oxidizing bacteria. *Critical Reviews in Biochemistry and Molecular Biology*, **38**, 471–495, https://doi.org/10.1080/10409230390267446

Arp D. J., Chain P. S. G. and Klotz M. G. (2007). The impact of genome analyses on our understanding of ammonia-oxidizing bacteria. *Annual Review of Microbiology*, **61**, 503–528, https://doi.org/10.1146/annurev.micro.61.080706.093449

Bollon J., Filali A., Fayolle Y., Guerin S., Rocher V. and Gillot S. (2016). N_2O emissions from full-scale nitrifying biofilters. *Water Research*, **102**, 41–51, https://doi.org/10.1016/j.watres.2016.05.091

Castro-Barros C. M., Daelman M. R. J., Mampaey K. E., van Loosdrecht M. C. M. and Volcke E. I. P. (2015). Effect of aeration regime on N_2O emission from partial nitritation-anammox in a full-scale granular sludge reactor. *Water Research*, **68**, 793–803, https://doi.org/10.1016/j.watres.2014.10.056

Chandran K., Stein L. Y., Klotz M. G. and van Loosdrecht M. C. M. (2011). Nitrous oxide production by lithotrophic ammonia-oxidizing bacteria and implications for engineered nitrogen-removal systems. *Biochemical Society Transactions*, **39**, 1832–1837, https://doi.org/10.1042/BST20110717

Corominas L., Flores-Alsina X., Snip L. and Vanrolleghem P. A. (2012). Comparison of different modelling approaches to better understand and minimize greenhouse gas emissions from wastewater treatment plants. *Biotechnology and Bioengineering*, **109**, 2854–2863, https://doi.org/10.1002/bit.24544

CH2MHill. (2008). Discussion paper for a wastewater treatment plant sector greenhouse gas emissions reporting protocol, Final report prepared for California Wastewater Climate Change Group, Oakland, California.

de Haas D. and Hartley K. J. (2004). Greenhouse gas emissions from BNR plant – do we have the right focus? Proceedings: Sewage Management – Risk Assessment and Triple Bottom Line. Queensland Environmental Protection Agency, April 2004, Cairns, pp. 5–7.

Domingo-Félez C. and Smets B. F. (2016). A consilience model to describe N_2O production during biological N removal. *Environmental Science: Water Research & Technology*, **2**, 923–930, https://doi.org/10.1039/C6EW00179C

Domingo-Félez C. and Smets B. F. (2020a). Modelling denitrification as an electric circuit accurately captures electron competition between individual reductive steps: the activated sludge model–electron competition model. *Environmental Science & Technology*, **54**(12), 7330–7338, https://doi.org/10.1021/acs.est.0c01095

Domingo-Félez C. and Smets B. F. (2020b). Modelling N_2O dynamics of activated sludge biomass: Uncertainty analysis and pathway contributions. *Chemical Engineering Journal*, **379**, 122311, https://doi.org/10.1016/j.cej.2019.122311

Domingo-Félez C., Calderó-Pascual M., Sin G., Plósz B. G. and Smets B. F. (2017). Calibration of the comprehensive NDHA-N_2O dynamics model for nitrifier-enriched biomass using targeted respirometric assays. *Water Research*, **126**, 29–39, https://doi.org/10.1016/j.watres.2017.09.013

Duan H., van den Akker B., Thwaites B. J., Peng L., Herman C., Pan Y., Ni B. J., Watt S., Yuan Z. and Ye L. (2020). Mitigating nitrous oxide emissions at a full-scale wastewater treatment plant. Water Research, 116196, https://doi.org/10.1016/j.watres.2020.116196

Fiat J., Filali A., Fayolle Y., Bernier J., Rocher V., Spérandio M. and Gillot S. (2019). Considering the plug-flow behavior of the gas phase in nitrifying BAF models significantly improves the prediction of N_2O emissions. *Water Research*, **156**, 337–346, https://doi.org/10.1016/j.watres.2019.03.047

Foley J., de Haas D., Yuan Z. and Lant P. (2010). Nitrous oxide generation in full-scale biological nutrient removal wastewater treatment plants. *Water Research*, **44**, 831–844, https://doi.org/10.1016/j.watres.2009.10.033

Guo L. and Vanrolleghem P. A. (2014). Calibration and validation of an activated sludge model for greenhouse gases No. 1 (ASMG1) – prediction of temperature dependent N_2O emission dynamics. *Bioprocess and Biosystems Engineering*, **37**, 151–163, https://doi.org/10.1007/s00449-013-0978-3

Harper Jr, W. F., Takeuchi Y., Riya S., Hosomi M. and Terada A. (2015). Novel abiotic reactions increase nitrous oxide production during partial nitrification: modelling and experiments. *Chemical Engineering Journal*, **281**, 1017–1023, https://doi.org/10.1016/j.cej.2015.06.109

Harris E., Joss A., Emmenegger L., Kipf M., Wolf B., Mohn J. and Wunderlin P. (2015). Isotopic evidence for nitrous oxide production pathways in a partial nitritation-anammox reactor. *Water Research*, **83**, 258–270, https://doi.org/10.1016/j.watres.2015.06.040

Hiatt W. C. and Grady Jr, C. P. L. (2008). An updated process model for carbon oxidation, nitrification, and denitrification. *Water Environment Research*, **80**, 2145–2156, https://doi.org/10.2175/106143008X304776

Hooper A. B., Vannelli T., Bergmann D. J., and Arciero D. M. (1997). Enzymology of the oxidation of ammonia to nitrite by bacteria. *Antonie Van Leeuwenhoek*, **71**, 59–67, https://doi.org/10.1023/A:1000133919203

Inventory of U.S. Greenhouse Gas Emissions and Sinks: 1990–2010. (2012). EPA 430-R-12-001. U. S. Environmental Protection Agency, Washington, DC.

IPCC. (2007). Climate Change 2007: The Physical Science Basis. In: Contribution of Working Group I to the Fourth Assessment Report of the Intergovernmental Panel on Climate Change, S. Solomon, D. Qin, M. Manning, Z. Chen, M. Marquis, K. B. Averyt, M. Tignor and H. L. Miller (eds.), Cambridge University Press, Cambridge, United Kingdom and New York, NY, USA.

Kampschreur M. J., Picioreanu C., Tan N., Kleerebezem R., Jetten M. S. M. and van Loosdrecht M. C. M. (2007). Unraveling the source of nitric oxide emission during nitrification. *Water Environment Research*, **79**, 2499–2509, https://doi.org/10.2175/106143007X220815

Kampschreur M. J., Temmink H., Kleerebezem R., Jetten M. S. M. and van Loosdrecht M. C. M. (2009). Nitrous oxide emission during wastewater treatment. *Water Research*, **43**, 4093–4103, https://doi.org/10.1016/j.watres.2009.03.001

Kim S. W., Miyahara M., Fushinobu S., Wakagi T. and Shoun H. (2010). Nitrous oxide emission from nitrifying activated sludge dependent on denitrification by ammonia-oxidizing bacteria. *Bioresource Technology*, **101**, 3958–3963, https://doi.org/10.1016/j.biortech.2010.01.030

Lang L., Pocquet M., Ni B.J., Yuan Z. and Spérandio M. (2017). Comparison of different 2-pathway models for describing the combined effect of DO and nitrite on the nitrous oxide production by ammonia-oxidizing bacteria. *Water Science and Technology*, **75**(3), 491–500, https://doi.org/10.2166/wst.2016.389

Lang L., Piveteau S., Azimi S., Rocher V. and Spérandio M. (2019). Modelling N_2O emission from PNA process under oxygen limitation: model calibration and prospects. Proceeding of Watermatex 2019 Conference, 1–4 September 2019, Copenhagen, Danemark.

Law Y., Ni B. J., Lant P. and Yuan Z. (2012). Nitrous oxide (N_2O) production by an enriched culture of ammonia oxidising bacteria depends on its ammonia oxidation rate. *Water Research*, **46**, 3409–3419, https://doi.org/10.1016/j.watres.2012.03.043

Law Y., Lant P. A. and Yuan Z. (2013). The confounding effect of nitrite on N_2O production by an enriched ammonia-oxidising culture. *Environmental Science & Technology*, **47**, 7186–7194, https://doi.org/10.1021/es4009689

Lu H. and Chandran K. (2010). Factors promoting emissions of nitrous oxide and nitric oxide from denitrifying sequencing batch reactors operated with methanol and ethanol as electron donors. *Biotechnology and Bioengineering*, **106**, 390–398, https://doi.org/10.1002/bit.22704

Mannina G., Ekama G., Caniani D., Cosenza A., Esposito G., Gori R., Garrido-Baserba M., Rosso D. and Olsson G. (2016). Greenhouse gases from wastewater treatment – a review of modelling tools. *Science of the Total Environment*, **551–552**, 254–270, https://doi.org/10.1016/j.scitotenv.2016.01.163

Mampaey K. E., Beuckels B., Kampschreur M. J., Kleerebezem R., Van Loosdrecht M. C. M. and Volcke E. I. P. (2013). Modelling nitrous and nitric oxide emissions by autotrophic ammonia-oxidizing bacteria. *Environmental Technology*, **34**, 1555–1566, https://doi.org/10.1080/09593330.2012.758666

Mampaey K., Spérandio M., van Loosdrecht M. C. M. and Volcke E. I. P. (2019). Dynamic simulation of N_2O emissions from a full-scale partial nitritation reactor. *Biochemical Engineering Journal*, **152**, 107356, https://doi.org/10.1016/j.bej.2019.107356

Ni B. J. and Yuan Z. (2015). Recent advances in mathematical modelling of nitrous oxides emissions from wastewater treatment processes. *Water Research*, **87**, 336–346, https://doi.org/10.1016/j.watres.2015.09.049

Ni B. J., Ruscalleda M., Pellicer-Nacher C. and Smets B. F. (2011). Modelling nitrous oxide production during biological nitrogen removal via nitrification and denitrification: extensions to the general ASM models. *Environmental Science & Technology*, **45**, 7768–7776, https://doi.org/10.1021/es201489n

Ni B. J., Yuan Z., Chandran K., Vanrolleghem P. A. and Murthy S. (2013a). Evaluating four mathematical models for nitrous oxide production by autotrophic ammonia-oxidizing bacteria. *Biotechnology and Bioengineering*, **110**, 153–163, https://doi.org/10.1002/bit.24620

Ni B. J., Ye L., Law Y., Byers C. and Yuan Z. (2013b). Mathematical modelling of nitrous oxide (N_2O) emissions from full-scale wastewater treatment plants. *Environmental Science & Technology*, **47**, 7795–7803, https://doi.org/10.1021/es4005398

Ni B. J., Peng L., Law Y., Guo J. and Yuan Z. (2014). Modelling of nitrous oxide production by autotrophic ammonia-oxidizing bacteria with multiple production pathways. *Environmental Science & Technology*, **48**, 3916–392, https://doi.org/10.1021/es405592h

Ni B. J., Pan Y., van den Akker B., Ye L. and Yuan Z. (2015). Full-scale modelling explaining large spatial variations of nitrous oxide fluxes in a step-feed plug-flow wastewater treatment reactor. *Environmental Science & Technology*, **49**, 9176–9184, https://doi.org/10.1021/acs.est.5b02038

Okabe S., Oshiki M., Takahashi Y. and Satoh H. (2011). N_2O emission from a partial nitrification–anammox process and identification of a key biological process of N_2O emission from anammox granules. *Water Research*, **45**, 6461–6470, https://doi.org/10.1016/j.watres.2011.09.040

Pan Y., Ye L., Ni B. J. and Yuan Z. (2012). Effect of pH on N_2O reduction and accumulation during denitrification by methanol utilizing denitrifiers. *Water Research*, **46**(15), 4832–4840, https://doi.org/10.1016/j.watres.2012.06.003

Pan Y., Ni B. J., Bond P. L., Ye L. and Yuan Z. (2013a). Electron competition among nitrogen oxides reduction during methanol-utilizing denitrification in wastewater treatment. *Water Research*, **47**(10), 3273–3281, https://doi.org/10.1016/j.watres.2013.02.054

Pan Y., Ni B. J. and Yuan Z. (2013b). Modelling electron competition among nitrogen oxides reduction and N_2O accumulation in denitrification. *Environmental Science & Technology*, **47**, 11083–11091, https://doi.org/10.1021/es402348n

Pan Y., Ni B. J., Lu H., Chandran K., Richardson D. and Yuan Z. (2015). Evaluating two concepts for the modelling of intermediates accumulation during biological denitrification in wastewater treatment. *Water Research*, **71**, 21–31, https://doi.org/10.1016/j.watres.2014.12.029

Peng L., Ni B. J., Erler D., Ye L. and Yuan Z. (2014). The effect of dissolved oxygen on N_2O production by ammonia-oxidizing bacteria in an enriched nitrifying sludge. *Water Research*, **66**, 12–21, https://doi.org/10.1016/j.watres.2014.08.009

Peng L., Ni B. J., Law Y. and Yuan Z. (2015a). Modelling of N_2O production by ammonia oxidizing bacteria: integration of catabolism and anabolism. The 9th IWA Symposium on Systems Analysis and Integrated Assessment (Watermatex 2015), 14–17 June, Gold Coast, Australia.

Peng L., Ni B. J., Ye L. and Yuan Z. (2015b). Selection of mathematical models for N_2O production by ammonia oxidizing bacteria under varying dissolved oxygen and nitrite concentrations. *Chemical Engineering Journal*, **281**, 661–668, https://doi.org/10.1016/j.cej.2015.07.015

Peng L., Ni B. J., Law Y. and Yuan Z. (2016). Modeling N_2O production by ammonia oxidizing bacteria at varying inorganic carbon concentrations by coupling the catabolic and anabolic processes. *Chemical Engineering Science*, **144**, 386–394.

Perez-Garcia O., Villas-Boas S. G., Swift S., Chandran K. and Singhal N. (2014). Clarifying the regulation of NO/N_2O production in Nitrosomonas europaea during anoxic-oxic transition via flux balance analysis of a metabolic network model. *Water Research*, **60**, 267–277, https://doi.org/10.1016/j.watres.2014.04.049

Pijuan M., Tora J., Rodriguez-Caballero A., Cesar E., Carrera J. and Perez J. (2014). Effect of process parameters and operational mode on nitrous oxide emissions from a nitritation reactor treating reject wastewater. *Water Research*, **49**, 23–33, https://doi.org/10.1016/j.watres.2013.11.009

Pocquet M., Queinnec I. and Spérandio M. (2013). Adaptation and identification of models for nitrous oxide (N_2O) production by autotrophic nitrite reduction. Proceedings 11th IWA Conference on Instrumentation, Control and Automation (ICA2013), 18–20 September, Narbonne, France.

Pocquet M., Wu Z., Queinnec I. and Spérandio M. (2016). A two pathway model for N_2O emissions by ammonium oxidizing bacteria supported by the NO/N_2O variation. *Water Research*, **88**, 948–959.

Poughon L., Dussap C. G. and Gros J. B. (2000). Energy model and metabolic flux analysis for autotrophic nitrifiers. *Biotechnology and Bioengineering*, **72**, 416–433, https://doi.org/10.1002/1097-0290(20000220)72:4<416::AID-BIT1004>3.0.CO;2-D

Sabba F., Picioreanu C., Pérez J. and Nerenberg R. (2015). Hydroxylamine diffusion can enhance N_2O emissions in nitrifying biofilms: a modelling study. *Environmental Science & Technology*, **49**, 1486–1494, https://doi.org/10.1021/es5046919

Schreiber F., Loeffler B., Polerecky L., Kuypers M. M. M. and de Beer D. (2009). Mechanisms of transient nitric oxide and nitrous oxide production in a complex biofilm. *The ISME Journal*, **3**, 1301–1313, https://doi.org/10.1038/ismej.2009.55

Snip L. J. P., Boiocchi R., Flores-Alsina X., Jeppsson U. and Gernaey K. V. (2014). Challenges encountered when expanding activated sludge models: a case study based on N_2O production. *Water Science and Technology*, **70**(7), 1251–1260, https://doi.org/10.2166/wst.2014.347

Spérandio M., Pocquet M., Guo L., Vanrolleghem P., Ni B. J. and Yuan Z. (2016). Evaluation of different nitrous oxide production models with four continuous long-term wastewater treatment process data series. *Bioprocess and Biosystems Engineering*, **39**, 493–510, https://doi.org/10.1007/s00449-015-1532-2

Stein L. Y. (2011a). Surveying N_2O-producing pathways in bacteria. *Methods in Enzymology*, **486**, 131–152, https://doi.org/10.1016/B978-0-12-381294-0.00006-7

Stein L. Y. (2011b). Heterotrophic nitrification and nitrifier denitrification. In: Nitrification, B. B. Ward, D. J. Arp and M. G. Klotz (eds.), American Society for Microbiology Press, Washington DC, pp. 95–114.

Su Q., Domingo-Félez C., Zhang Z., Blum J. M., Jensen M. M. and Smets B. F. (2019a). The effect of pH on N_2O production in intermittently-fed nitritation reactors. *Water Research*, **156**, 223–231, https://doi.org/10.1016/j.watres.2019.03.015

Su Q., Domingo-Félez C., Jensen M. M. and Smets B. F. (2019b) Abiotic nitrous oxide (N_2O) production is strongly pH dependent, but contributes little to overall N_2O emissions in biological nitrogen removal systems. *Environmental Science & Technology*, **53**(7), 3508–3516, https://doi.org/10.1021/acs.est.8b06193

Sutka R., Ostrom N., Ostrom P., Breznak J., Gandhi H., Pitt A. and Li F. (2006). Distinguishing nitrous oxide production from nitrification and denitrification on the basis of isotopomer abundances. *Applied Environmental Microbiology*, **72**(1), 638–644, https://doi.org/10.1128/AEM.72.1.638-644.2006

Tallec G., Garnier J., Billen G. and Gousailles M. (2006). Nitrous oxide emissions from secondary activated sludge in nitrifying conditions of urban wastewater treatment plants: effect of oxygenation level. *Water Research*, **40**, 2972–2980, https://doi.org/10.1016/j.watres.2006.05.037

Vasilaki V., Conca V., Frison N., Eusebi A. L., Fatone F. and Katsou, E. (2020). A knowledge discovery framework to predict the N_2O emissions in the wastewater sector. *Water Research*, **178**, 115799, https://doi.org/10.1016/j.watres.2020.115799

von Schulthess R. and Gujer W. (1996). Release of nitrous oxide (N_2O) from denitrifying activated sludge: Verification and application of a mathematical model. *Water Research*, **30**(3), 521–530, https://doi.org/10.1016/0043-1354(95)00204-9

Wunderlin P., Mohn J., Joss A., Emmenegger L. and Siegrist H. (2012). Mechanisms of N_2O production in biological wastewater treatment under nitrifying and denitrifying conditions. *Water Research*, **46**, 1027–1037, https://doi.org/10.1016/j.watres.2011.11.080

Wunderlin P., Lehmann M. F., Siegrist H., Tuzson B., Joss A., Emmenegger L. and Mohn J. (2013). Isotope signatures of N_2O in a mixed microbial population system: constraints on N_2O producing pathways in wastewater treatment. *Environmental Science & Technology*, **47**, 1339–1348, https://doi.org/10.1021/es402971p

Yang Q., Liu X., Peng C., Wang S., Sun H. and Peng Y. (2009). N_2O Production during nitrogen removal via nitrite from domestic wastewater: main sources and control method. *Environmental Science & Technology*, **43**, 9400–9406, https://doi.org/10.1021/es9019113

Ye L., Ni B.-J., Law Y., Byers C. and Yuan Z. (2014). A novel methodology to quantify nitrous oxide emissions from full-scale wastewater treatment systems with surface aerators. *Water Research*, **48**, 257–268, https://doi.org/10.1016/j.watres.2013.09.037

Yu R., Kampschreur M. J., van Loosdrecht M. C. M. and Chandran K. (2010). Mechanisms and specific directionality of autotrophic nitrous oxide and nitric oxide generation during transient anoxia. *Environmental Science & Technology*, **44**, 1313–1319, https://doi.org/10.1021/es902794a

Zaborowska E., Lu X. and Makinia J. (2019). Strategies for mitigating nitrous oxide production and decreasing the carbon footprint of a full-scale combined nitrogen and phosphorus removal activated sludge system. *Water Research*, **162**, 53–63, https://doi.org/10.1016/j.watres.2019.06.057

Zhu J., Bernier J., Patry B., Azimi S., Pauss A., Rocher V. and Vanrolleghem P. A. (2019). Comprehensive modelling of full-scale nitrifying and post-denitrifying biofilters. Proceedings WEF Nutrient Removal and Recovery Symposium 2019 – 21st Century Vision, 23–25 July 2019, Minneapolis, MN, USA.

NOMENCLATURE

ADP	Adenosine diphosphate
AMO	Ammonia monooxygenase
AOB	Ammonia-oxidizing bacteria
ASM	Activated sludge model
ASM-EC	Activated sludge model – electron competition
ASM-ICE	Activated sludge model with indirect coupling of electrons
ASMN	Activated sludge model for nitrogen
ATP	Adenosine triphosphate
COD	Chemical oxygen demand
DO	Dissolved oxygen
EF	Emission factor
FA	Free ammonia
FNA	Free nitrous acid
HAO	Hydroxylamine oxidoreductase
IC	Inorganic carbon
Mox	Electron carriers in oxidized form
Mred	Electron carriers in reduced form
N_2OR	N_2O reductase
Nar	Nitrate reductase
ND	Nitrifiers denitrification

Nir	Nitrite/nitric oxide oxidoreductase
NirK	Nitrite reductase
NN	Hydroxylamine pathway
NOB	Nitrite oxidizing bacteria
NOR	NO reductase
OHO	Ordinary heterotrophic organisms
SP	Site-preference
sNOR	Haem–copper nitric oxide reductase
WWTP	Wastewater treatment plant

Table S1 in the supplementary information (SI) lists the definitions of the all the state variables used in the models described in this chapter. Please see doi: 10.2166/9781789060461_S1

doi: 10.2166/9781789060461_197

Chapter 8

Modelling of methane production and emissions

Keshab Sharma[1], Oriol Gutierrez[2], Zhiguo Yuan[1], Matthijs R. J. Daelman[3], Mark C. M. van Loosdrecht[3] and Eveline I. P. Volcke[4]

[1]Advanced Water Management Centre, The University of Queensland, St. Lucia, Queensland, Australia. E-mail: keshab@awmc.uq.edu.au; zhiguo@awmc.uq.edu.au
[2]Catalan Institute of Water Research (ICRA), Girona, Spain. E-mail: ogutierrez@icra.cat
[3]Department of Biotechnology, Delft University of Technology, Delft, the Netherlands. E-mail: matthijs.daelman@gmail.com; M.C.M.vanLoosdrecht@tudelft.nl
[4]Department of Green Chemistry and Technology, Ghent University, Belgium. E-mail: Eveline.Volcke@UGent.be

SUMMARY

This chapter provides a review of the models available for estimating the production and emission of methane from wastewater collection and treatment systems. The details of a number of mechanistic models as well as the simplified empirical models have been summarized. Their limitations have been identified and general methods for calibration and validation have been presented.

Keywords: Activated sludge, emission, methane, model, oxidation, production, sewer

TERMINOLOGY

Term	Definition
Greenhouse gas	Gas that absorbs and emits radiant energy within the thermal infrared range.
Collection system	A system of sewer pipes that collects wastewater from different sources and delivers it to a wastewater treatment plant
SeweX	A dynamic model for simulating hydrogen sulfide and methane generation in a sewer system
Sulfate reducing bacteria (SRB)	A group of bacteria found in anaerobic biofilm, which can perform anaerobic respiration utilizing sulfate as the terminal electron acceptor and reducing it to hydrogen sulfide. Organic carbon is generally used as the electron donor.
Methanogens	A group of microorganisms (archaea) that produce methane as a metabolic by-product under anaerobic conditions.

Methanotrophs	Prokaryotes that metabolize methane as their source of carbon and energy. They can be either bacteria or archaea and can grow aerobically or anaerobically. These require single-carbon compounds to survive.
Model calibration	A process of adjustment of the model parameters to obtain a model representation of the processes of interest that satisfies prescribed criteria.
Model validation	A process by which model outputs are systematically compared to independent real-world observations to judge the quantitative and qualitative correspondence with reality.
Anaerobic digestion	A biochemical process through which microorganisms break down organic matter in the absence of oxygen generating methane-rich biogas.
Empirical model	A model based on statistical relationships between the output and inputs, which are developed using experimental data.
Dissolved methane	Methane (CH_4) gas present in dissolved form in the water phase.
Activated sludge process	A wastewater treatment process for treating sewage or industrial wastewaters using aeration and biological flocs composed of bacteria and protozoa.

8.1 INTRODUCTION

Modelling of methane production in wastewater systems emerged from the modelling of anaerobic digestion, in which the production of methane gas has been the major focus. The methane model for anaerobic digestion has been widely reported in literature. However, due to the continuous evolution of sewer models during the past 3 decades and renewed interest in greenhouse gas (GHG) emissions from the collection systems, there has been significant development in methane modelling for sewer systems. It is not only the production, but also the consumption of methane, which serves as a sink for methane, that has attracted the interest of many researchers in recent years. This has led to the development of models for methane removal in aerobic systems, primarily the aerobic methane oxidation. This chapter summarizes the models for methane production and removal in an urban wastewater system.

8.2 CH_4 MODELLING FOR COLLECTION SYSTEM

Due to the operational complexity of sewer systems and dynamic nature of methane production as well as emissions, it is not practical to estimate overall CH_4 emissions from large sewer networks through either online or offline measurements presented in the earlier chapters. Mathematical modelling of the methanogenic activity is a viable option for predicting the methane production and emission in sewer networks. A mathematical model also serves as a powerful tool for the water industry, supporting operational optimization and the development of mitigation strategies for GHG emission control from their collection systems. To date, a number of different models for predicting methane production in sewers have been developed. These models are described in the following sections.

8.2.1 Mechanistic model for CH_4 production in sewer biofilms

Guisasola *et al.* (2009) developed a mechanistic model for CH_4 production in sewer biofilms, which has been incorporated in the sewer model presented in Sharma *et al.* (2008) to account for the methanogenic activity. The sewer model, which is now known as the SeweX model (Cesca *et al.*, 2015; Nguyen *et al.*, 2015), is a dynamic sewer model, describing in-sewer biological, chemical, and physical processes. It predicts both the temporal and spatial variations of wastewater characteristics, including sulfate, sulfide and methane, using sewer network configuration, pipe geometry, sewage characteristics and hydraulic data as the inputs. SeweX is the first sewer model capable of predicting the spatial and temporal variation in dissolved and gas phase methane concentrations in a sewer system.

The processes included in the sewer CH_4 model are listed in Table 8.1, while a schematic presentation of these processes is shown in Figure 8.1. The following processes that are responsible for methane production in sewers are included in the SeweX model.

1. Acidogenesis
2. Acetogenesis
3. Acetoclastic methanogenesis
4. Hydrogenotrophic methanogenesis
5. Acetate-based sulfidogenesis
6. Hydrogenotrophic sulfidogenesis
7. Propionate-based sulfidogenesis

The Monod type kinetic expressions are used for the biofilm-catalysed processes and higher values of saturation constants are employed to account for substrate diffusion limitations in the biofilm. Some of the key features of this model are:

1. The sewer biofilm is considered the main contributor to sulfide and methane production.
2. Fermentation is modelled considering the acetate, propionate, and hydrogen as the products.
3. Acetoclastic methanogenesis is the predominant mechanism for methane production.
4. Glucose has been used to represent the fermentable substrates in the biochemical reactions as in Anaerobic Digestion Model No. 1 (ADM1, Batstone *et al.*, 2002).
5. Given the fact that direct propionate utilization by methanogens is not possible and propionate in real sewage is at a low concentration, propionate is considered as an electron donor only for sulfate reduction, not for methane generation.
6. The fermentative bacteria are likely to outcompete sulfate reducing bacteria (SRB) for the fermentable substrates (e.g., sugars or other carbohydrates). For this reason, sulfate reduction using these substrates is not considered in the model and the use of these substrates by SRB is accounted for by considering the use of the fermentation products from these substrates.

The details of the stoichiometric and kinetic parameters included in the SeweX model, which describe the interactions between sulfate reducing bacteria, fermentative bacteria (FB) and methanogenic archaea (MA) can be found in Guisasola *et al.* (2009).

The SeweX model with parameters initially calibrated using the data collected from lab-scale experiments (Guisasola *et al.*, 2009), was subsequently validated using manually sampled, offline methane data from two sewer sites, one in Australia and another in Spain. Figure 8.2 shows a comparison of measured CH_4 data (offline) with the model predicted results for a sewer system in Australia (Guisasola *et al.*, 2009), while Figure 8.3 shows a similar comparison for a sewer system in Spain. Figure 8.4 compares a long-term field CH_4 measurement (online) data from another sewer system in Australia (Liu *et al.*, 2015b) with the CH_4 results predicted using the calibrated model. These comparisons clearly demonstrate the validity of the sewer CH_4 model discussed above.

Although the model predictions and field data showed very good correlations in the above presented cases, more online field measurement data are needed for further calibration and validation of the methane related kinetics, especially under a wide range of sewer conditions.

8.2.2 Methane oxidation under aerobic environment

Despite there being a strong possibility of methane oxidation under aerobic conditions in a gravity sewer by methanotrophs, there is no information on this available in the literature. The lack of sufficient information suggests that there has been no attempt made so far to model the methane oxidation in gravity sewers. Modelling efforts have been focused only on the anaerobic sewer biofilm in the rising main, which is the source of methane in a sewer system.

Table 8.1 Stoichiometry and kinetics of the model describing the interactions among MA, FB and SRB (Guisasola *et al.*, 2009).

Process	CH_4 (Methane)	$C_2H_4O_2$ (Acetate)	$C_6H_{12}O_6$ (Glucose)	$C_3H_6O_2$ (Propionate)	CO_2	H_2	H_2O	H_2SO_4	H_2S	Kinetics
	mol/L									
Hydrogenotrophic methanogenesis	1				−1	−4	2			$k_{CH_4,H_2} \cdot \frac{S_{H_2}}{K_{H_2,MA} + S_{H_2}} \cdot \frac{A}{V} \cdot \alpha^{T-20}$
Acetoclastic methanogenesis	1	−1			1					$k_{CH_4,S_{AC}} \cdot \frac{S_{AC}}{K_{S_{AC},MA} + S_{AC}} \cdot \frac{A}{V} \cdot \alpha^{T-20}$
Acetogenesis		2	−1		2	4	−2			$q_{ACETOG} \cdot \frac{S_F}{K_F + S_F} \cdot \frac{A}{V} \cdot \alpha^{T-20}$
Acidogenesis		2	−3	4	2		2			$q_{ACIDOG} \cdot \frac{S_F}{K_F + S_F} \cdot \frac{A}{V} \cdot \alpha^{T-20}$
Hydrogenotrophic sulfidogenesis						−4	4	−1	1	$k_{H_2S,H_2} \cdot \frac{S_{H_2}}{K_{H_2,SRB} + S_{H_2}} \cdot \frac{S_{SO_4}}{K_{SO_4} + S_{SO_4}} \cdot \frac{A}{V} \cdot \alpha^{T-20}$
Acetate-based sulfidogenesis		−1			2		2	−1	1	$k_{H_2S,S_{AC}} \cdot \frac{S_{AC}}{K_{AC,SRB} + S_{AC}} \cdot \frac{S_{SO_4}}{K_{SO_4} + S_{SO_4}} \cdot \frac{A}{V} \cdot \alpha^{T-20}$
Propionate-based sulfidogenesis		1		−1	1		2	−3/4	3/4	$k_{H_2S,S_{PROP}} \cdot \frac{S_{PROP}}{K_{PROP,SRB} + S_{PROP}} \cdot \frac{S_{SO_4}}{K_{SO_4} + S_{SO_4}} \cdot \frac{A}{V} \cdot \alpha^{T-20}$

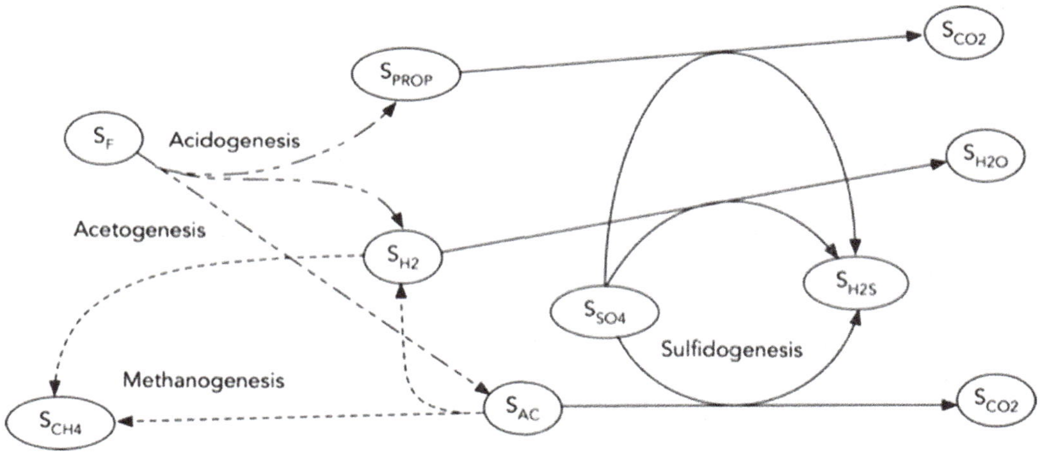

Figure 8.1 Schematic representation of the methane biological model. Sulfate reducing bacteria processes (solid line), fermentative bacteria processes (dash-dotted line) and methanogenic archaea processes (dashed line) (Guisasola *et al.*, 2009).

8.2.3 Methane production in sewer sediments

Liu *et al.* (2015a) developed a detailed, but simple, one-dimensional sediment model to predict methane and sulfide production and microbial distribution in a sewer sediment based on the biological reactions proposed by Guisasola *et al.* (2009). The proposed model is presented in Equation (8.1).

$$r_{CH4} = k \times S_F{}^{0.5} \tag{8.1}$$

where, r_{CH4} is the areal methane production rate (g CH_4/m^2·day); k is the rate constant for methane production expressed as (g CH_4/m)$^{0.5}$/day; and S_F is the bulk fermentable chemical oxygen demand (COD) concentration (mg/L).

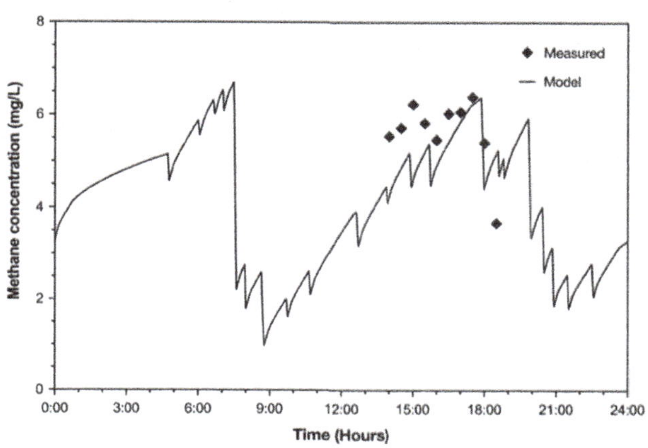

Figure 8.2 SeweX model predictions vs offline CH_4 data collected from a sewer system in Australia (Guisasola *et al.*, 2009).

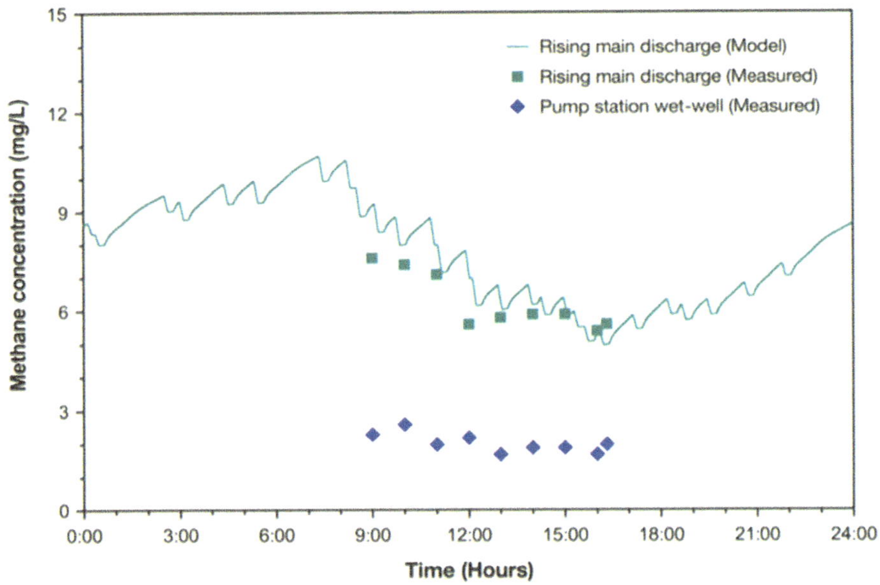

Figure 8.3 SeweX model predictions vs measured CH_4 data (off-line) for a sewer system in Spain.

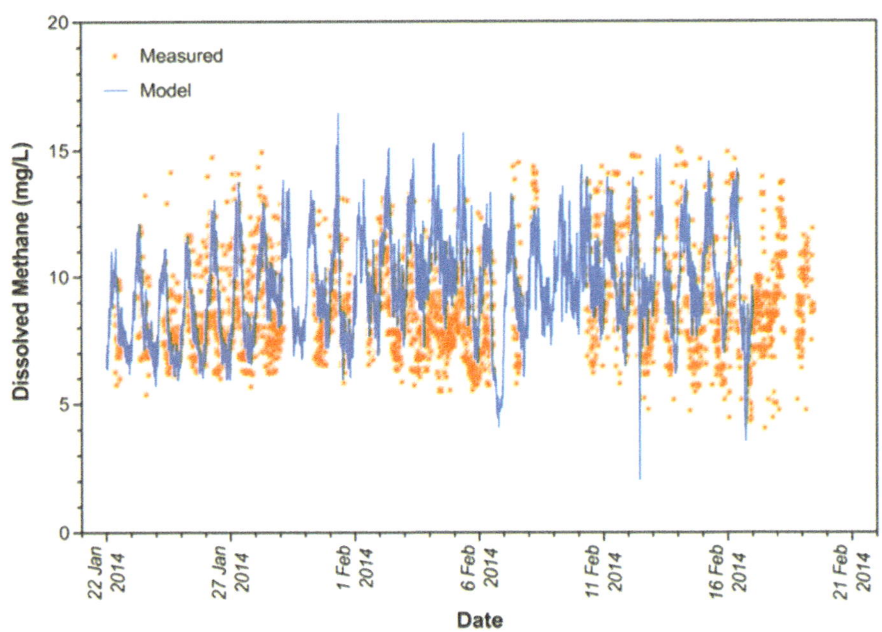

Figure 8.4 Calibrated seweX model predictions vs measured CH_4 data (on-line) for a sewer system in Australia.

The parameter k was calibrated using the least squares method after comparing the model predicted methane production rate with the measured value under different substrate concentrations. A value of 0.224 ± 0.002 was obtained for k with the R^2 estimate of 0.99. The model presented in Equation (8.1) has been found to describe the methane production in sewer sediment under different flow velocity (shear stress) conditions (Liu *et al.*, 2016).

The proposed half-order kinetic model can be easily used in the determination of the contribution of sewer sediments to the overall sewer network emissions. The model is very simple as it involves only one parameter to be calibrated that is, k. However, more field data is required to examine the accuracy of the proposed model and understand the dependency of k on key sewer conditions including the sediment properties and wastewater characteristics.

8.2.4 Empirical models predicting methane production in sewers

The mechanistic model for CH_4 production described in Section 8.2.1 requires a large amount of data and is not suitable for quick CH_4 estimation for a sewer pipe. Alternatively, an empirical model could be a useful tool in such a case.

Foley *et al.* (2009) proposed a simple empirical model for estimating CH_4 production in a rising main sewer using data collected from Australian sewers. This simple empirical model represented a correlation of measured CH_4 data from a limited number of rising main sewers with the pipe properties and hydraulic conditions (Equation (8.2)). It was intended for application to similar rising mains with 'similar operational characteristics', which included temperature and organic matter content of the wastewater. The model was a best fit of measured dissolved CH_4 production to hydraulic residence time (HRT) and the ratio of biofilm area to water volume in the sewer.

$$C_{CH4} = 5.24 \times 10^{-5} \times \left(\frac{A}{V} \times \text{HRT} \right) + 0.0015 \tag{8.2}$$

where C_{CH4} is the concentration of dissolved methane (kg/m^3); 5.24×10^{-5} kg/m^2/h represents the rate of methanogenic activity of the pipeline biofilm; and 0.0015 kg/m^3 is the average residual concentration of dissolved methane. This empirical model is based on field observations and considers that the CH_4 production is a function of the wastewater HRT and the biofilm area to water volume (A/V) ratio of the pipe. This simple equation offers a valuable tool for water authorities to predict methane emissions from a rising main sewer. It should be noted that the methane production rate (5.24×10^{-5}) is expected to be affected by many other factors such as the wastewater composition (specifically the COD concentration) and temperature, and it likely varies from system to system. Therefore, more field data is required to further calibrate and validate this empirical model for its generalized application.

With regard to gravity sewers, Chaosakul *et al.* (2014) developed an empirical model to predict methane formation in gravity sewers based on the A/V ratio, HRT and wastewater temperature (Equation (8.3)). The model parameters were estimated using the field data collected in central Thailand.

$$C_{CH4} = 6.0 \times 10^{-5} \times \left(\frac{A}{V} \times \text{HRT} \right) \times 1.05^{(T-20)} + 0.0015 \tag{8.3}$$

where C_{CH4} is the concentration of dissolved methane (kg/m^3); 6.0×10^{-5} kg/m^2/h is the rate of methanogenic activity of the pipeline biofilm; 0.0015 kg/m^3 is the average residual concentration of dissolved methane; and $1.05^{(T-20)}$ is a function of temperature (in °C). This model has been calibrated with measured methane data from the field and partially validated using rising main sewer data. However, the fit of the model predictions with the measured data was poor as R^2 was found to be only 0.06, which is very low. A number of different possible reasons, including limited range of A/V and HRT used in the study, and variation in weather conditions, have been postulated for this observation. By comparing the two equations presented here for rising main and gravity sewer, respectively, it appears that the gravity sewers in Thailand would produce more CH_4 than the rising main sewers in Australia for the same HRT and A/V ratio, which itself is quite surprising.

Xu *et al.* (2018) attempted to further improve the model proposed by Chaosakul *et al.* (2014) by introducing a biomass term and removing the A/V ratio term as shown in Equation (8.4).

$$Q_{CH4} = Y_{CH4/x} \cdot X \cdot HRT \cdot 1.05^{(T-20)} \quad (8.4)$$

where Q_{CH4} is the methane production in mg/L·day; $Y_{CH4/x}$ is the yield coefficient (mg methane/kg biomass); and X is the amount of biomass (kg). The biomass amount is estimated by considering the wall shear-stress, which depends upon sewer slope, degree of fullness of sewer flow, and velocity of flow. The details of the equations used for estimating the biomass amount can be found in Xu *et al.* (2018).

None of the three empirical equations described above consider the impacts of substrate concentration (COD), and this could lead to some errors in methane prediction.

Recently, Water Research Foundation (WRF) has published a methodology for sewer methane estimation in the form of a technical report (Willis *et al.*, 2020). In an attempt to develop the tools for the quantification of methane emissions from gravity as well as the rising main sewers, separate empirical equations taking into account the key field variables such as wastewater flow, pipe diameter, slope and temperature have been proposed for the gravity and rising main sewers. These equations have been developed using the data generated from a large number of simulations with the SeweX model for a range of the variables representing a wide variety of sewer design conditions. The parameters of the model were estimated by carrying out regression and fitting the parameter values to minimize the sum of the square of errors among the two data sets.

The proposed equation for the prediction of methane production in a gravity sewer is:

$$r_{CH_4} = 0.419 \cdot 1.06^{(T-20)} \cdot Q^{0.26} \cdot D^{0.28} \cdot S^{-0.138} \quad (8.5)$$

where, r_{CH_4} is the methane production rate (kg/km·day); Q is the average flow over a day (m^3/s); D is the pipe diameter (m); and S is the pipe slope (m/m).

The equation for the estimation of methane production in a rising main sewer is:

$$r_{CH_4} = 3.45 \cdot 1.06^{(T-20)} \cdot D \cdot N_P{}^{0.202} \cdot 0.396^{(1-N_P \times P_I/1440)} \quad (8.6)$$

where, r_{CH_4} is the methane production rate (kg/km·day); T is the temperature (°C); D is the pipe diameter (m); N_P is the number of pumping events per day; and P_I is the average pumping interval (min). This equation could be used for intermittently running and continuously running rising main sewers as well as the surcharged sewer pipes.

Once the characteristics of a sewer network are known, the above equations could be used to estimate the overall CH_4 emission from the entire sewer network, with an assumption that all the CH_4 produced in the sewer network ultimately gets emitted to the atmosphere. Although, there have been some efforts made towards the validation of these models, more work is needed.

8.2.5 Methane emission in sewers

The mass transfer of CH_4 from the liquid phase to the sewer headspace is the key process for CH_4 emission. Like oxygen, the mechanism of CH_4 liquid-gas mass transfer in assumed to be controlled by the transfer in the liquid film as, similarly to oxygen, methane is poorly soluble in water. The following relationship is commonly used for modelling the liquid-gas transfer of methane.

$$\frac{dC_{CH_4,L}}{dt} = -k_L a \cdot \left(C_{CH_4,L} - \frac{C_{CH_4,g}}{H} \right) \quad (8.7)$$

where $k_L a$ is the mass transfer coefficient (1/day); $C_{CH_4,L}$ is the liquid phase methane concentration (mg/L); $C_{CH_4,g}$ is the methane concentration in the gas phase (mg/L); H is the Henry's law constant; and $dC_{CH_4,L} / dt$ is the volumetric mass flux of methane (mg/L·day).

The mass transfer coefficient depends upon several factors including temperature, water quality and the thickness of the interfacial liquid layer (Liss & Slater, 1974). A number of different relationships are available for the estimation of the $k_L a$ value for oxygen transfer as a function of physical and hydraulic properties of sewer pipes and streams (Jensen, 1995; Lahav *et al.*, 2004; Owens *et al.*, 1964; Parkhurst & Pomeroy, 1972). Once the mass transfer coefficient for oxygen is known, the same for methane could be estimated based on the ratio of the coefficient of molecular diffusion of CH_4 to that of O_2 (Liss & Slater, 1974) as follows.

$$\frac{k_L a, CH_4}{k_L a, O_2} = \left(\frac{D_{CH_4}}{D_{O_2}} \right)^n \tag{8.8}$$

where $k_L a, CH_4$ is the mass transfer coefficient for CH_4; $k_L a, O_2$ is the mass transfer coefficient for O_2; D_{CH_4} is the molecular diffusion coefficient for CH_4; D_{O_2} is the molecular diffusion coefficient for O_2; and n is the constant, which could be taken as 0.5 under turbulent flow conditions (Liss & Slater, 1974; Carrera *et al.*, 2016).

8.2.6 Model calibration and validation

There has been some work done in relation to calibrating and validating the methane models for the collection system (Chaosakul *et al.*, 2014; Foley *et al.*, 2009; Liu *et al.*, 2015a). However, only a limited data set (either from a single system or data over a limited period) has been used in the calibration, and hence the validity of the model parameters is questionable. It is therefore warranted that the models are calibrated with the data collected from the field and such a calibrated model be applied to estimate the methane generation in and emission from a sewer network.

Different models presented in previous sections would require different data sets for their calibration. Generally, sewer data (pipe size, slope, length etc.), hydraulic data (flow, velocity, water depth, pump operation information etc.), environmental data (temperature etc.), and wastewater characteristics are required as inputs for the calibration. The empirical models require quantification of the parameters and variables involved in the model and generally use average values for the variables, whereas a dynamic model would require the information on dynamic variation of flow and the wastewater characteristics. For comparison, dissolved CH_4 concentration needs to be monitored at selected locations along the sewer network. This data can be used for both calibration and validation of the model. Normally data collected from one system is used for calibration of model parameters and the data from a separate system is used for validation.

8.2.7 Further model development

Liu *et al.* (2015a) have highlighted the limitations of the current CH_4 models for sewer CH_4 production. For instance, the potential for biological CH_4 oxidation has not been factored in in the current models mainly because of the lack of understanding of those processes. In addition, other processes which serve as a sink for methane in sewers should be included in the models once such processes are identified and a proper understanding is established.

Another potential development is related to the integrated management of urban water-wastewater systems as there is an increasing interest in understanding the effect of the interactions among urban water system components. This can be enabled through integrating the WWTP model and the sewer models, such as SeweX, resulting in better prediction of methane emission over the entire wastewater system (Guo *et al.*, 2012). With further development of the sewer models, the integrated modelling approach will provide more reliable information in relation to GHG emissions from the entire urban water system.

8.3 METHANE MODELLING FOR ACTIVATED SLUDGE PROCESS

8.3.1 Incorporating aerobic methane oxidation in activated sludge models

Because of the aerobic environment prevailing in the activated sludge process, the only process that is relevant to methane emission is the aerobic methane oxidation due to the presence of methanotrophs. The well-established Activated Sludge Model n°1 (ASM1, Henze *et al.*, 1987) has been extended by Daelman *et al.* (2014) to include the biological methane oxidation. The resulting model, named ASM1m, adds two processes to ASM1: aerobic growth and decay of methanotrophs. The two additional state variables in the model are methane as a substrate (S_{CH4}) and methane oxidizing bacteria (X_{MOB}) as the biomass component. Methanotrophic bacteria are singled out from the other heterotrophic organisms (X_{BH}) and are therefore described by a separate state variable, X_{MOB}, as in Arcangeli and Arvin (1999). The details of the reaction stoichiometry and kinetic parameters used in ASM1m are available in Arcangeli and Arvin (1999) and Daelman *et al.* (2014). The original parameter values of ASM1 are preserved and the list has been extended with additional parameters to be used in the equations that describe methanotrophic growth and decay, taken from Arcangeli and Arvin (1999). The details of the reaction stoichiometry and the process rates used in ASM1m model are presented in Tables 8.2 and 8.3, respectively.

In ASM1m, the growth of methanotrophs is modelled using Monod kinetics for methane and oxygen similar to those used in a number of publications (Alvarez-Cohen & McCarty, 1991; Arcangeli & Arvin, 1999; Broholm *et al.*, 1992; Oldenhuis *et al.*, 1991; Yoon *et al.*, 2009). Unlike in Yoon *et al.* (2009), oxygen is also considered as a limiting substrate. Ammonia inhibition, as considered by Arcangeli and Arvin (1999), is not included in the model.

The effect of the ammonium concentration on the methane oxidation rate by methanotrophs is ambiguous. A number of studies have reported an inhibitory effect of ammonium (Begonja & Hrsak, 2001; Hanson & Hanson, 1996; Nyerges & Stein, 2009), while others have reported no such effect (van der Ha *et al.*, 2010, 2011). In contrast, Noll *et al.* (2008) observed selective stimulation of methanotrophs by ammonium. These observations have been made under ammonium concentrations at least one order of magnitude higher than the concentration commonly encountered in an activated sludge system and in systems described in models such as BSM1. Ammonium inhibition is therefore omitted in the model. Decay of methanotrophic biomass is described in the same manner as the other biomass groups, using the concept of death-regeneration. First-order reaction kinetics has been used for the biomass decay.

8.3.2 Modelling methane gas-liquid mass transfer

The modelling of gas-liquid transfer of methane is illustrated considering a completely mixed reactor, with the reactor influent as the sole source of methane, dissolved methane leaving with the effluent, methane stripping (transfer from the liquid to the gas phase) and biological methane conversion (Figure 8.5).

A typical mass balance for dissolved methane, m_{CH4} (g COD) then reads as Equation (8.9).

$$\frac{dm_{CH4}(t)}{dt} = Q_{in}(t) \cdot S_{CH4,in}(t) - Q_{out}(t) \cdot S_{CH4}(t) - \dot{m}_{CH4}^{L\text{-}G(t)} - \dot{R}_{CH4(t)} \quad (8.9)$$

Q_{in} and Q_{out} (m³/d) are the imposed liquid flows into and out of the reactor, respectively, $S_{CH4,in}$ and S_{CH4} (g COD/m³) are the respective incoming and outgoing methane concentrations, $\dot{m}_{CH4}^{L\text{-}G}$ (g COD/d) is the stripping rate and $\dot{R}_{CH4}$ (g COD/d) is the conversion rate. The concentration of methane in the liquid volume, V (m³), relates to its total mass via Equation (8.10).

$$S_{CH4}(t) = \frac{m_{CH4}(t)}{V(t)} \quad (8.10)$$

It is important to realize that the liquid-gas transfer rate, $\dot{m}_{CH4}^{L\text{-}G}$, is affected by gradients in the gas phase composition and pressure, which can be taken into account through comprehensive expressions

Table 8.2 Matrix representation of ASM1m (Daelman *et al.*, 2014). The processes and reactions added to the original ASM1 are shaded.

A_{ij}	S_I gCOD/m³	S_S gCOD/m³	S_{CH4} gCOD/m³	X_I gCOD/m³	X_S gCOD/m³	X_{BH} gCOD/m³	X_{BA} gCOD/m³	X_{MOB} gCOD/m³	X_P gCOD/m³	S_O gCOD/m³	S_{NO} gN/m³	S_{NH} gN/m³	S_{ND} gN/m³	X_{ND} gN/m³	S_{ALK} mole HCO_3^-/m³
PROCESS								**STOICHIOMETRIC MATRIX**							
1. Aerobic growth of heterotrophs		$-1/Y_H$				1				$-(1\text{-}Y_H)/Y_H$		$-i_{XB}$			$-i_{XB}/14$
2. Anoxic growth of heterotrophs		$-1/Y_H$				1					$-(1\text{-}Y_H)/(2.86Y_H)$	$-i_{XB}$			$(1\text{-}Y_H)/(14\times 2.86\times Y_H) -i_{XB}/14$
3. Aerobic growth of autotrophs							1			$-(4.57\text{-}Y_A)/Y_A$	$1/Y_A$	$-i_{XB} -1/Y_A$			$-i_{XB}/14-1/(7\times Y_A)$
4. Aerobic growth of methanotrophs			$-1/Y_{MOB}$					1		$-(1\text{-}Y_{MOB})/Y_{MOB}$		$-i_{XB}$			$-i_{XB}/14$
5. Decay of heterotrophs					$1\text{-}f_P$	–1			f_P					$i_{XB}\text{-}f_P\times i_{XP}$	
6. Decay of autotrophs					$1\text{-}f_P$		–1		f_P					$i_{XB}\text{-}f_P\times i_{XP}$	
7. Decay of methanotrophs					$1\text{-}f_P$			–1	f_P					$i_{XB}\text{-}f_P\times i_{XP}$	
8. Ammonification of soluble organic nitrogen												1	–1		1/14
9. Hydrolysis of entrapped organics		1			–1										
10. Hydrolysis of entrapped organic nitrogen													1	–1	
CONSERVATIVES								**COMPOSITION MATRIX**							
COD	1	1	1	1	1	1	1	1	1	–1	–4.57	0			
N						i_{XB}	i_{XB}	i_{XB}	i_{XP}	0	1	1	1	1	
Charge											–1/14	1/14			–1

Table 8.3 Process rates for ASM1m (Daelman *et al.*, 2014). The rates for the processes added to the original ASM1 are shaded.

j	Process	Process Rate (ρ_j)
1.	Aerobic growth of heterotrophs	$\mu_H^{\max} \cdot \frac{S_S}{K_S + S_S} \cdot \frac{S_O}{K_O^H + S_O} \cdot X_{BH}$
2.	Anoxic growth of heterotrophs	$\mu_H^{\max} \cdot \frac{S_S}{K_S + S_S} \cdot \frac{K_O^H}{K_O^H + S_O} \cdot \frac{S_{NO}}{K_{NO} + S_{NO}} \cdot \eta_{y.g} \cdot X_{BH}$
3.	Aerobic growth of autotrophs	$\mu_A^{\max} \cdot \frac{S_{NH}}{K_{NH}^A + S_{NH}} \cdot \frac{S_O}{K_O^A + S_O} \cdot X_{BA}$
4.	Aerobic growth of methanotrophs	$\mu_{MOB}^{\max} \cdot \frac{S_{CH_4}}{K_{NH_4} + S_{CH_4}} \cdot \frac{S_O}{K_O^{MOB} + S_O} \cdot X_{MOB}$
5.	Decay of heterotrophs	$b_H \cdot X_{BH}$
6.	Decay of autotrophs	$b_A \cdot X_{BA}$
7.	Decay of methanotrophs	$b_{MOB} \cdot X_{MOB}$
8.	Ammonification of soluble organic nitrogen	$k_a \cdot S_{ND} \cdot X_{BH}$
9.	Hydrolysis of entrapped organics	$k_H \cdot \frac{X_S / X_{BH}}{K_X + X_S / X_{BH}} \cdot \left[\frac{S_O}{K_O^H + S_O} + \eta_{y.h} \cdot \frac{K_O^H}{K_O^H + S_O} \cdot \frac{S_{NO}}{K_{NO} + S_{NO}}\right] \cdot X_{BH}$
10.	Hydrolysis of entrapped organic nitrogen	$k_H \cdot \frac{X_S / X_{BH}}{K_X + X_S / X_{BH}} \cdot \left[\frac{S_O}{K_O^H + S_O} + \eta_{y.h} \cdot \frac{K_O^H}{K_O^H + S_O} \cdot \frac{S_{NO}}{K_{NO} + S_{NO}}\right] \cdot X_{BH} \cdot \frac{X_{ND}}{X_S}$

(Baeten *et al.*, 2020). However, in the case of methane, the stripping rate can be very well approximated with a liquid-gas transfer model (Equation (8.11)) that considers the mean gas phase mole fraction and mean pressure along the reactor height (Baeten *et al.*, 2020).

$$\dot{m}_{CH4}^{L\text{-}G(t)} = K_{La_{O2(t)}} \cdot V(t) \cdot \frac{S_{CH4}(t) - i_{COD,CH4} \cdot h_{CH4} \cdot \left(\left(\left(p_{atm}^{G+pg}(H/2)\right) M_{CH4}\right) / RT\right) \cdot x_{in,CH4}^{G}}{\sqrt{(D_{O2} / D_{CH4}) + 0.6} \cdot h_{CH4} \cdot (H/2)} \quad (8.11)$$

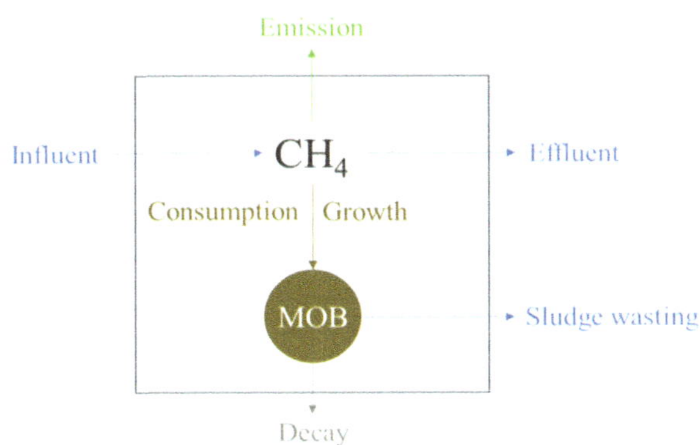

Figure 8.5 Sinks and sources of methane (CH_4) and methane oxidizing bacteria (MOB) considered in a simple completely stirred tank reactor (CSTR) model (Baeten *et al.*, 2021).

$K_L a_{O2}$ (1/d) denotes the volumetric overall transfer coefficient of oxygen, $i_{COD,CH4}$ (g COD/g) is the COD content of methane, h_{CH4} (g/m^3 in the liquid phase per g/m^3 in the gas phase) is the Henry coefficient of methane, p^G_{atm} (Pa) is the atmospheric pressure, ρ (kg/m^3) is the density of water, g (m/s^2) is the gravitational acceleration, H (m) is the water column height during aeration, M_{CH4} (g/mol) is the molecular mass of methane, R (J/mol·K) is the universal gas constant, T is the reactor temperature (K), $x^G_{in,CH4}$ (mole/mole) is the mole fraction of methane in the atmosphere and D_{O2} and D_{CH4} (m^2/d) are the respective diffusion coefficients of oxygen and methane.

8.4 METHANE MODELLING FOR ANAEROBIC DIGESTION

Anaerobic Digestion Model No. 1 (ADM1), developed and published by IWA Anaerobic Digestion Modelling Task Group (Batstone *et al.*, 2002), is widely used as the model for methane production and emission during anaerobic digestion. The model considers disintegration and hydrolysis, acidogenesis, acetogenesis and methanogenesis steps.

ADM1 comprises a large number of simultaneous and sequential processes with a complex reaction kinetics. The processes are primarily classified as either biochemical or physicochemical. The biochemical reactions are considered to be catalysed by extra-cellular enzymes involving organic substrates. Empirical based first-order reaction kinetics is used for all the extra-cellular biochemical reactions, while all the intra-cellular biochemical reactions follow the Monod-type kinetics. Typical to any biological reaction, substrate uptake reaction rates are considered to be a function of the biomass growth rate and biomass concentration. The model considers pH inhibition for acetogenic and acetolactic methanogenic bacterial groups through H_2 and free ammonia inhibition, respectively.

The details of the processes, kinetic expressions, and stoichiometric and kinetic parameters used in the model are available in IWA (2002). Since ADM1 has been widely reported in the literature, no further description has been provided in this chapter.

REFERENCES

Alvarez-Cohen L. and McCarty P. L. (1991). Product toxicity and cometabolic competitive-inhibition modeling of chloroform and trichloroethylene transformation by methanotrophic resting cells. *Applied and Environmental Microbiology*, **57**(4), 1031–1037. https://doi.org/10.1128/aem.57.4.1031-1037.1991

Arcangeli J. P. and Arvin E. (1999). Modelling the growth of a methanotrophic biofilm: estimation of parameters and variability. *Biodegradation*, **10**(3), 177–191. https://doi.org/10.1023/A:1008317906069

Baeten J. E., van Loosdrecht M. C. M. and Volcke E. I. P. (2020). When and why do gradients of the gas phase composition and pressure affect liquid-gas transfer? *Water Research*, **178**, 115844. https://doi.org/10.1016/j.watres.2020.115844

Baeten J. E., Walgraeve C., Granja R., van Loosdrecht M. C. M. and Volcke E. I. P. (2021). Unaerated feeding alters the fate of dissolved methane during aerobic wastewater treatment. *Water Research*, **204**, 117619. https://doi.org/10.1016/j.watres.2021.117619

Batstone D. J., Keller J., Angelidaki I., Kalyuzhnyi S. V., Pavlostathis S. G., Rozzi A., Sanders W. T. M., Siegrist H. and Vavilin V. A. (2002). The IWA anaerobic digestion model no 1 (ADM1). *Water Science and Technology*, **45**(10), 65–73. https://doi.org/10.2166/wst.2002.0292

Begonja A. and Hrsak D. (2001). Effect of growth conditions on the expression of soluble methane monooxygenase. *Food Technology and Biotechnology*, **39**(1), 29–35.

Broholm K., Christensen T. H. and Jensen B. K. (1992). Modeling TCE degradation by a mixed culture of methane-oxidizing bacteria. *Water Research*, **26**(9), 1177–1185. https://doi.org/10.1016/0043-1354(92)90178-7

Carrera L., Springer F., Lipeme-Kouyi G. and Buffiere P. (2016). A review of sulfide emissions in sewer networks: overall approach and systemic modelling. *Water Science and Technology*, **73**(6), 1231–1242.

Cesca J., Sharma K., Vuong L., Yuan Z., Hamer G. and McDonald A. (2015). South Australia Water Corporation's pro-active corrosion and odour management strategy development. *Proceedings of the Water Environment Federation*, **2015**(9), 919–935. https://doi.org/10.2175/193864715819555463

Chaosakul T., Koottatep T. and Polprasert C. (2014). A model for methane production in sewers. *Journal of Environmental Science and Health, Part A*, **49**, 1316–1321. https://doi.org/10.1080/10934529.2014.910071

Daelman M. R. J., Van Eynde T., van M. C. M. and Volcke E. I. P. (2014). Effect of process design and operating parameters on aerobic methane oxidation in municipal WWTPs. *Water Research*, **66**, 308–319. https://doi.org/10.1016 j.watres.2014.07.034

Foley J., Yuan Z. and Lant P. (2009). Dissolved methane in rising main sewer systems: field measurements and simple model development for estimating greenhouse gas emissions. *Water Science & Technology*, **60**, 2963–2971. https://doi.org/10.2166/wst.2009.718

Guisasola A., Sharma K. R., Keller J., Yuan Z. and Jiang G. (2009). Development of a model for assessing methane formation in rising main sewers. *Water Research*, **43**, 2874–2884. https://doi.org/10.1016/j.watres.2009.03.040

Guo L., Porro J., Sharma K. R., Amerlinck Y., Benedetti L., Nopens I., Shaw A., Van Hulle S. W. H., Yuan Z. and Vanrolleghem P. A. (2012). Towards a benchmarking tool for minimizing wastewater utility greenhouse gas footprints. *Water Science & Technology*, **66**, 2483–2495. https://doi.org/10.2166/wst.2012.495

Hanson R. S. and Hanson T. E. (1996). Methanotrophic bacteria. *Microbiological Reviews*, **60**(2), 439–471. https://doi.org/10.1128/mr.60.2.439-471.1996

Henze M., Grady C. P. L. J., Gujer W., Marais G. v. R. and Matsuo T. (1987). Activated Sludge Model No. 1, IAWQ Scientific and Technical Report No. 1, IAWQ, London.

IWA Task Group for Mathematical Modelling of Anaerobic Digestion Processes (2002). Anaerobic Digestion Model No.1 (ADM1), IWA Scientific and Technical Report, IWA Publishing, London

Jensen N. A. (1995). Empirical modeling of air-to-water oxygen transfer in gravity sewers. *Water Environment Research*, **67**(6), 979–991. https://doi.org/10.2175/106143095X133211

Lahav O., Lu Y., Shavit U. and Loewenthal R. (2004). Modeling hydrogen sulphide emission rates in gravity sewage collection systems. *Journal of Environmental Engineering*, **130**(11), 1382–1389. https://doi.org/10.1061/(ASCE)0733-9372(2004)130:11(1382)

Liss P. S. and Slater P. G. (1974). Flux of gases across the air-sea interface. *Nature*, **247**(5438), 181–184. https://doi.org/10.1038/247181a0

Liu Y., Ni B. J., Sharma K. R. and Yuan Z. (2015a). Methane emission from sewers. *Science of the Total Environment*, **524–525**, 40–51. https://doi.org/10.1016/j.scitotenv.2015.04.029

Liu Y., Sharma K. R., Fluggen M., O'Halloran K., Murthy S. and Yuan Z. (2015b). Online dissolved methane and total dissolved sulfide measurement in sewers. *Water Research*, **68**, 109–118. https://doi.org/10.1016/j.watres.2014.09.047

Liu Y., Tugtas A. E., Sharma K. R., Ni B. J. and Yuan Z. (2016). Sulfide and methane production in sewer sediments: field survey and model evaluation. *Water Research*, **89**, 142–150. https://doi.org/10.1016/j.watres.2015.11.050

Mohanakrishnan J., Gutierrez O., Sharma K. R., Guisasola A., Werner U., Meyer R. L., Keller J. and Yuan Z. (2009). Impact of nitrate addition on biofilm properties and activities in rising main sewers. *Water Research*, **43**, 4225–4237. https://doi.org/10.1016/j.watres.2009.06.021

Nguyen T., Sharma K., Jiang G., Ganigue R., Cesca J., Vuong L. and Yuan Z. (2015). SeweX modelling to support corrosion and odour management in sewers. *Water (Journal of Australian Water Association)*, **42**(7), 71–78.

Noll M., Frenzel P. and Conrad R. (2008). Selective stimulation of type I methanotrophs in a rice paddy soil by urea fertilization revealed by RNA-based stable isotope probing. *Fems Microbiology Ecology*, **65**(1), 125–132. https://doi.org/10.1111/j.1574-6941.2008.00497.x

Nyerges G. and Stein L. Y. (2009). Ammonia cometabolism and product inhibition vary considerably among species of methanotrophic bacteria. *FEMS Microbiology Letters*, **297**(1), 131–136. https://doi.org/10.1111/j.1574-6968.2009.01674.x

Oldenhuis R., Oedzes J. Y., Vanderwaarde J. J. and Janssen D. B. (1991). Kinetics of chlorinated hydrocarbon degradation by Methylosinus trichosporium OB3B and toxicity of trichloroethylene. *Applied and Environmental Microbiology*, **57**(1), 7–14. https://doi.org/10.1128/aem.57.1.7-14.1991

Owens M., Edwards E. W. and Gibbs J. W. (1964). Some reaeration studies in streams. *International Journal of Air Pollution*, **8**, 469–486.

Parkhurst J. D. and Pomeroy R. D. (1972). Oxygen absorption in streams. *Journal of the Sanitary Engineering Division, ASCE*, **98**(SA1), 101–124. https://doi.org/10.1061/JSEDAI.0001366

Schierbaum K. D., Weimar U. and Göpel W. (1992). Comparison of ceramic, thick-film and thin-film chemical sensors based upon SnO_2. *Sensors & Actuators B Chemical*, **7**, 709–716. https://doi.org/10.1016/0925-4005(92)80390-J

Sharma K., de Haas D., Corrie S., O'Halloran K., Keller J. and Yuan Z. (2008). Predicting hydrogen sulfide formation in sewers: a new model. *Water*, **35**, 132–137.

van der Ha D., Hoefman S., Boeckx P., Verstraete W. and Boon N. (2010). Copper enhances the activity and salt resistance of mixed methane-oxidizing communities. *Applied Microbiology and Biotechnology*, **87**(6), 2355–2363. https://doi.org/10.1007/s00253-010-2702-4

van der Ha D., Bundervoet B., Verstraete W. and Boon N. (2011). A sustainable, carbon neutral methane oxidation by a partnership of methane oxidizing communities and microalgae. *Water Research*, **45**(9), 2845–2854. https://doi.org/10.1016/j.watres.2011.03.005

Willis J., Yuan Z., Sharma K., Jiang G., Murthy S., DeClippelleir H., Kumar A. and Satyadev A. (2020). Conveyance Asset Prediction System: Sewer Methane Estimation Methodology and Significance Determination, The Water Research Foundation, USA.

Xu J., He Q., Li H., Yang C., Wang Y. and Ai H. (2018). Modelling of methane formation in gravity sewer system: the impact of microorganism and hydraulic condition. *AMB Express*, **8**(1), 1–10.

Yoon S., Carey J. N. and Semrau J. D. (2009). Feasibility of atmospheric methane removal using methanotrophic biotrickling filters. *Applied Microbiology and Biotechnology*, **83**(5), 949–956. https://doi.org/10.1007/s00253-009-1977-9

NOMENCLATURE

α	Temperature coefficient
ADM1	Anaerobic Digestion Model No. 1
A/V	Sewer biofilm area to water volume ratio
ASM1	Activated Sludge Model 1
$C_2H_4O_2$	Acetate
$C_3H_6O_2$	Propionate
$C_6H_{12}O_6$	Glucose
C_{CH4}	Dissolved methane concentration
$C_{CH_4,L}$	Liquid phase methane concentration
$C_{CH_4,g}$	Gas phase methane concentration
CH_4	Methane
CO_2	Carbon di-oxide
CSTR	Completely stirred tank reactor
D	Pipe diameter
D_{CH_4}	Molecular diffusion coefficient for CH_4
D_{O_2}	Molecular diffusion coefficient for O_2
GHG	Greenhouse gases
H	Henry's law constant
H_2	Hydrogen gas
H_2O	Water
H_2S	Hydrogen sulfide
H_2SO_4	Sulfuric acid
HRT	Hydraulic residence time
IWA	International Water Association

k	Rate constant for methane production
k_{CH_4,H_2}	Rate constant for methane production with hydrogen
$k_{CH_4,SAC}$	Rate constant for methane production with acetate
k_{H_2S,H_2}	Rate constant for sulfide production with hydrogen
$k_{H_2S,PROP}$	Rate constant for sulfide production with propionate
$k_{H_2S,SAC}$	Rate constant for sulfide production with acetate
$k_La_{,CH_4}$	Mass transfer coefficient for CH_4
$k_La_{,O_2}$	Mass transfer coefficient for O_2
$K_{AC,SRB}$	Half saturation constant for acetate (sulfidogenesis)
K_F	Half saturation constant for fermentable substrate
$K_{H_2,MA}$	Half saturation constant for hydrogen (methanogenesis)
$K_{H_2,SRB}$	Half saturation constant for hydrogen (sulfidogenesis)
$K_{PROP,SRB}$	Half saturation constant for propionate (sulfidogenesis)
$K_{SAC,MA}$	Half saturation constant for acetate (methanogenesis)
K_{SO_4}	Half saturation constant for sulfate (sulfidogenesis)
N_P	Number of pumping events per day
O_2	Oxygen
P_I	Average pumping interval
q_{ACETOG}	Rate constant for acetogenesis
q_{ACIDOG}	Rate constant for acidogenesis
Q	Average daily flow rate
Q_{CH4}	Methane production
r_{CH_4}	Methane production rate
S	Pipe slope
S_{AC}	Acetate concentration
S_{CH_4}	Dissolved methane concentration
S_F	Fermentable substrate concentration
S_{H_2}	Hydrogen concentration
S_{PROP}	Propionate concentration
SRB	Sulfate reducing bacteria
S_{SO_4}	Sulfate concentration
T	Wastewater temperature (°C)
X	Amount of biomass
X_{BA}	Autotrophic biomass concentration
X_{BH}	Heterotrophic biomass concentration
X_{MOB}	Concentration of methane oxidizing bacteria
$Y_{CH4/X}$	Yield coefficient (mg CH_4/kg biomass)

doi: 10.2166/9781789060461_213

Chapter 9

Benchmarking strategies to control GHG production and emissions

Xavier Flores-Alsina[1], Magnus Arnell[2,3], Lluís Corominas[4], Chris Sweetapple[5], Guangtao Fu[5], David Butler[5], Peter A. Vanrolleghem[6], Krist V. Gernaey[1] and Ulf Jeppsson[3]

[1]Process and Systems Engineering (PROSYS) Centre, Department of Chemical and Biochemical Engineering, Technical University of Denmark, Søltofts Plads, Building 228 A, Kgs. Lyngby 2800, Denmark
[2]Unit of Urban Water Management, RISE Research Institutes of Sweden, Gjuterigatan 1D, Linköping SE-58273, Sweden
[3]Division of Industrial Electrical Engineering and Automation (IEA), Department of Biomedical Engineering, Lund University, PO Box 118, Lund SE-22100, Sweden
[4]ICRA, Catalan Institute for Water Research, Scientific and Technological Park, H_2O Building, Emili Grahit 101, Girona 17003, Spain
[5]Centre for Water Systems, College of Engineering, Mathematics and Physical Sciences, University of Exeter, North Park Road, Exeter, Devon EX4 4QF, UK
[6]modelEAU, Département de génie civil et de génie des eaux, Université Laval, 1065 Avenue de la Médecine, Québec, Canada G1V 0A6

SUMMARY

Benchmarking has been a useful tool for unbiased comparison of control strategies in wastewater treatment plants (WWTPs) in terms of effluent quality, operational cost and risk of suffering microbiology-related total suspended solids (TSS) separation problems. This chapter presents the status of extending the original Benchmark Simulation Model No 2 (BSM2) towards including greenhouse gas (GHG) emissions. A mathematical approach based on a set of comprehensive models that estimate all potential on-site and off-site sources of CO_2, CH_4 and N_2O is presented and discussed in detail. Based upon the assumptions built into the model structures, simulation results highlight the potential undesirable effects on increased GHG emissions when carrying out local energy optimization in the activated sludge section and/or energy recovery in the anaerobic digester. Although off-site CO_2 emissions may decrease in such scenarios due to either lower aeration energy requirement or higher heat and electricity production, these effects may be counterbalanced by increased N_2O emissions, especially since N_2O has a 300-fold stronger greenhouse effect than CO_2. The reported results emphasize the importance of using integrated approaches when comparing and evaluating (plant-wide) control strategies in WWTPs for more informed operational decision-making.

Keywords: Carbon footprint, control strategies, GHG, modelling, multi-criteria evaluation, plant-wide, sustainability

TERMINOLOGY

Term	Definition
Greenhouse gas	Gas that absorbs and emits radiant energy within the thermal infrared range.
Benchmarking	Objective comparison of two items.

9.1 INTRODUCTION

The main focus in assessing the operation of wastewater treatment plants (WWTPs) has historically been the effluent water quality under constraints of technical feasibility and cost. This certainly still holds, but discussions on sustainability in general, and the impact on climate change due to greenhouse gas (GHG) emissions in particular, have widened the scope for utilities and regulators (Gustavsson & Tumlin, 2013). An increasing interest in GHG emissions calls for novel approaches to evaluate the performance of control and operational strategies in order to include additional performance indicators related to GHG emissions (Mannina *et al.*, 2016; Nguyen *et al.*, 2020).

Aside from evaluating control and operational strategies (Gernaey *et al.*, 2014) before full-scale implementation (Ayesa *et al.*, 2006), dynamic activated sludge models (ASMs) (Henze *et al.*, 2000) have been widely used for multiple purposes in wastewater engineering, such as control and monitoring (Olsson 2012), benchmarking (Jeppsson *et al.*, 2007; Solon *et al.*, 2017), diagnosis (Rodriguez-Roda *et al.*, 2002), design (Flores-Alsina *et al.*, 2012), teaching (Hug *et al.*, 2009), optimization (Feldman *et al.*, 2018; Rivas *et al.*, 2008), and regulatory policy development of wastewater treatment plants (Meng *et al.*, 2016, 2020). Based on new knowledge on the chemical and biochemical mechanisms of GHG production, several efforts have been made to capture emissions of CO_2, CH_4 and N_2O, and to integrate these processes in the traditional ASMs (Domingo-Felez *et al.*, 2017; Ni & Yuan, 2015; Poquet *et al.*, 2016).

In recent years, an increasing number of studies have discussed the need for adding a new dimension related to GHG production and emission to the traditional effluent quality and operational cost indices within the performance evaluation procedures of activated sludge control strategies (Flores-Alsina *et al.*, 2011, 2014; Guo et al., 2012; Sweetapple *et al.*, 2015). In this chapter, an extended version of the International Water Association (IWA) Benchmark Simulation Model No 2 (BSM2), that is BSM2G, is used to show how decision making about the most suitable control/operational strategies may change when a GHG emission dimension is added. The model based methodology includes all major contributions to assess the carbon footprint of the plant under study. Two case studies (case study#1 and #2) are presented involving changes in the following operational variables: (i) the dissolved oxygen (DO) set-point of the aeration system in the activated sludge section; (ii) the removal efficiency of the total suspended solids (TSS) in the primary clarifier; (iii) the temperature in the anaerobic digester (AD); and (iv) the control of the flow of anaerobic digester supernatants from the sludge treatment section of the plant. Furthermore, we consider the main interactions between the water and sludge line. Finally, changes in effluent quality index (*EQI*), operational cost index (*OCI*) and CO_2, CH_4 and N_2O emissions are analysed by means of different graphical representations. As a side effect, synergies and trade-offs between local energy optimization and the overall GHG production are studied in detail.

9.2 BENCHMARK PLANT DESCRIPTION

The WWTP under study has the same layout as the IWA Benchmark Simulation Model No 2 platform proposed by Gernaey *et al.* (2014) (see Figure 9.1). The BSM2 was initially conceived for unbiased

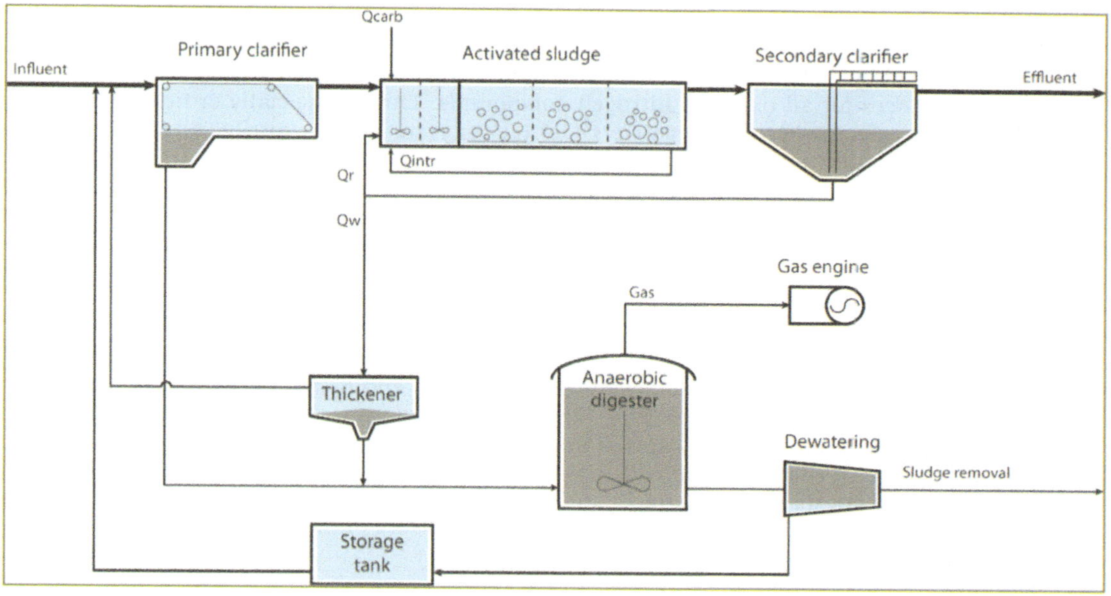

Figure 9.1 Schematic representation of the BSM2 plant layout.

comparison of control strategies based on predefined process and sensor models, influent disturbances and evaluation criteria. More specifically, the plant treats an influent flow rate of 20 648 $m^3 \cdot day^{-1}$ and total COD and N loads of 12 240 and 1 140 $kg \cdot day^{-1}$, respectively. Influent characteristics are generated following the principles stated in Gernaey *et al.* (2011). The activated sludge (AS) unit is a modified Ludzack-Ettinger configuration consisting of five tanks in series. Tanks 1 (ANOX1) and 2 (ANOX2) are anoxic (total volume= 3 000 m^3), while tanks 3 (AER1), 4 (AER2) and 5 (AER3) are aerobic (total volume=9 000 m^3). AER3 and ANOX1 are linked by means of an internal recycle with the purpose of nitrate recycle for pre-denitrification. The BSM2 plant further contains a primary (PRIM, 900 m^3) and a secondary (SEC, 6 000 m^3) clarifier, a sludge thickener (THK), an anaerobic digester (AD, 3 400 m^3), a storage tank (160 m^3) and a dewatering unit (DW). Additional information about the plant design and operational conditions can be found in Gernaey *et al.* (2014).

9.3 BENCHMARK MODEL UPGRADES AND MODIFICATIONS

9.3.1 Activated sludge model (ASM)

The Activated Sludge Model No. 1 (ASM1) (Henze *et al.*, 2000) has been expanded based on the principles proposed by Hiatt and Grady (2008) and Mampaey *et al.* (2013). The Hiatt and Grady model incorporates two nitrifying populations: ammonia oxidizing bacteria (AOB) and nitrite oxidizing bacteria (NOB) using free ammonia (FA) and free nitrous acid (FNA) as nitrogen substrate, respectively. The model also considers sequential reduction of nitrate (NO_3^-) to nitrogen gas (N_2) via nitrite (NO_2^-), nitric oxide (NO) and nitrous oxide (N_2O) using individual reaction-specific parameters. Additionally, the ideas summarized in Mampaey *et al.* (2013) are used to consider NO and N_2O formation from the nitrification pathway assuming ammonia (NH_3) as the electron donor. Parameter values were adjusted according to Guo and Vanrolleghem (2014).

9.3.2 ASM/ADM interface

The interfaces presented in Nopens *et al.* (2010) have been modified to link the upgraded ASM and the default Anaerobic Digestion Model No. 1 (ADM1), by considering chemical oxygen demand (COD) and N balances for all oxidized nitrogen compounds. This is especially critical in Step 1 of the ASM–ADM interface where all negative COD (i.e. oxygen, nitrate, nitrite, nitrous oxide and nitrogen monoxide) is subtracted from the COD pool with an associated loss of substrate (S_s, X_s, X_{BH}, X_{BA} in that order). The last step, where inorganic carbon (S_{IC}) is calculated as part of the assumption of charge conservation at both sides of the interface is upgraded with the new (oxidized) nitrogen species, that is S_{NO3} and S_{NO2}. There are no modifications to the original formulation of the ADM–ASM interface.

9.3.3 Mass transfer

Mass transfer between the liquid and the gas phase in the ASM is modelled for selected compounds (S_{N2}, S_{NO} and S_{N2O}). Specific transfer coefficients (K_La_{N2}, K_La_{NO} and K_La_{N2O}) are estimated using the ratio of the squared roots of diffusivities (Foley *et al.*, 2011). The transport rates are formulated as a function of the difference between the saturation concentration and the actual concentration of the gas dissolved in the liquid (Batstone *et al.* 2012). The saturation concentration of the gas in the liquid is given by Henry's law of dissolution, which states that the saturation concentration is equal to the product of Henry's constant (K_H) multiplied by the partial pressure of the gas (P_i).

9.3.4 Temperature correction

To account for seasonal variability, liquid-gas saturation constants, kinetic parameters, transfer coefficients and equilibrium reactions are temperature dependent. More specifically, growth and decay rates are modelled according to the Ratkowsky equations. Equilibrium constants to calculate FA and FNA are adjusted using Van't Hoff corrections. Finally, K_La is kinetically adjusted with temperature changes (Gernaey *et al.*, 2014).

9.3.5 Other ancillary models

The other models have not been modified from their original description in Gernaey *et al.* (2014). The primary clarifier is modelled in accordance with Otterpohl and Freund (1992). The double exponential settling velocity function of Takács *et al.* (1991) is used to model the secondary settling process through a one-dimensional model consisting of 10 layers. Regarding the thickener and dewatering units, these are modelled as ideal, continuous processes with no biological activity, and with a constant percentage of TSS in the concentrated sludge flows leaving the thickening and dewatering units. The widely recognized ADM1 (Batstone *et al.*, 2002) is the dynamic anaerobic digestion model implemented.

9.4 EVALUATION CRITERIA

9.4.1 Effluent quality (*EQI*) and operational cost (*OCI*) indices

The overall pollution removal efficiency is obtained using the effluent quality index (*EQI*) from the standard BSM2 (Nopens *et al.*, 2010). *EQI* is an aggregated weighted index of all pollution loads: TSS, COD, 5-day biochemical oxygen demand (BOD_5), total Kjeldahl nitrogen (TKN) and the oxidized forms of nitrogen (NO_X), leaving the plant. The economic objectives are evaluated using the operational cost index (*OCI*) (Gernaey *et al.*, 2014). It consists of the sum of all major operating costs in the plant: aeration energy (AE), pumping energy (PE), mixing energy (ME), sludge production (SP), external carbon addition (EC), methane production (MP) and the net heating energy (HE^{net}). *EQI* and *OCI* are based on simulation results with the 609 days of dynamic influent data generated following the principles outlined in Gernaey *et al.* (2011). Only the last 364 days are used for the evaluation itself.

9.4.2 On-site/off-site GHG emissions

The comprehensive approach suggested by Flores-Alsina *et al.* (2011, 2014) is used to estimate all potential GHG emissions from the studied WWTP that cannot be obtained from the explicit results of the modified BSM2. The overall GHG evaluation comprises the estimation of GHG emissions from the following sources: (i) direct secondary treatment, (ii) sludge processing, (iii) net power and chemical use, (iv) sludge disposal and reuse, and (v) receiving waters. It is important to highlight that the GHGs are converted into units of CO_2 equivalent (CO_2e) to properly deal with the different natures of the generated GHGs (CO_2, CH_4 and N_2O). Further information can be found in Flores-Alsina *et al.* (2011, 2014) and Corominas *et al.* (2012).

9.4.2.1 Direct secondary treatment emissions

Direct emissions from the activated sludge process are calculated in the bioprocess model, including CO_2 generation from microbial respiration, and production and emission of N_2O. The CO_2 is credited for growth of autotrophic nitrifying organisms with a factor of 0.31 kg CO_2/kg $N_{\text{nitrified}}$ (Tchobanoglous *et al.*, 2003). Most of the produced N_2O will be stripped to the surrounding gas phase in reactors with forced aeration and only a small fraction remains in solution. In the BSM2G, the dissolved N_2O is assumed to follow the plant effluent to the receiving waters.

9.4.2.2 Sludge processing emissions

The GHG emissions from sludge treatment are mainly generated in the anaerobic digester. Direct biogas CO_2 and CH_4 emissions are quantified using the ADM1. In this case it is assumed that the biogas is fed directly into a gas-fired combustion turbine converting the CH_4 into CO_2 and generating electricity and heat (in turn used to heat the anaerobic digester). The CO_2 generated during anaerobic digestion and the CO_2 produced in the combustion are assumed to be released to the atmosphere.

In addition, direct emissions of CO_2, CH_4 and N_2O from the sludge train of the BSM2G plant (i.e. thickener, digester, dewatering and sludge storage) are accounted for. From the AD and the gas system a leakage of 1% of the raw biogas is assumed (Avfall Sverige Utveckling, 2009). The gas flow available for utilization is therefore reduced by the corresponding amount. The CH_4 that remains dissolved in the sludge after digestion is assumed to be stripped at the dewatering unit and adds to the emissions.

Based on common practice in countries applying dewatered sludge as fertilizer on productive land (e.g. crops or forest), it is assumed that the sludge is stored uncovered at the plant for 12 months (for hygienic purposes) before use. During storage, post digestion occurs, leading to GHG emissions. The emissions of CH_4 and N_2O are set to 8.7 g CH_4/kg volatile solids (VS) and 0.36% of total nitrogen (TN) as N_2O-N (Jönsson *et al.*, 2015). Corresponding amounts of carbon and nitrogen are subtracted from the sludge (S_S, X_S and S_{NH}).

In the gas engine, the CH_4 in the biogas is combusted and converted to CO_2 which is emitted with the fumes along with a fraction of the gas that passes through the engine un-combusted. The emission factor (EF) for un-combusted biogas is set to 1.7% of the gas fed to the gas engine (Liebetrau *et al.*, 2010). Consequently, the energy production is reduced by the same factor.

9.4.2.3 Net power and chemical use emissions

Net energy is calculated as the difference between energy consumption (aeration, pumping, mixing and heating) and energy production. The electricity generated by the turbine is calculated by using a factor for the energy content of the methane gas (50 014 MJ/kg CH_4) and assuming a 43% efficiency for electricity generation. For external electricity required, a value of 0.359 kg CO_2e/kWh is assumed (Arnell *et al.*, 2017).

It is assumed that methanol is used as external carbon source for denitrification. A common type of methanol, sourced from fossil resources, is assumed with an emission factor of 1.54 kg CO_{2e}/kg MeOH. Methanol is the only chemical included in the BSM2G (Dong & Steinberg, 1997).

9.4.2.4 Sludge disposal and reuse emissions

After 12 months of storage, the sludge is transported for final disposal or reuse. While reuse options, transport distance and specific emissions vary widely with location, the following mix of disposal options is chosen (Arnell *et al.*, 2017):

(i) Crop land – 38% of the sludge; 150 km transport; N_2O Emission Factor (EF): 0.01 kg N_2O-N/kg TN.
(ii) Composting – 45% of the sludge; 20 km transport; N_2O EF: 0.01 kg N_2O-N/kg TN CH_4 EF: 0.0075 kg CH_4/kg total organic carbon (TOC)
(iii) Forest – 17% of the sludge; 144 km transport; N_2O EF: 0.01 kg N_2O-N/kg TN

GHG emissions resulting from transport of sludge are calculated with an emission factor for Euro 4 class trucks running on diesel for all disposal options, where unloaded return trips are assumed.

9.4.2.5 Receiving water emissions

Indirect emissions of N_2O from the recipient due to residual nitrogen in the effluent are calculated and presented (Arnell *et al.*, 2017). Studies of N_2O emissions from natural waters show a large variability depending on climate and type of water system (lake, river, sea, etc.). For BSM2G, an emission factor corresponding to an inland lake or river was included, 0.0003 kg N_2O-N/kg $TN_{effluent}$ (IPCC, 2013).

9.4.3 Sustainability indicators

Additional indicators can be calculated for economic and environmental aspects of sustainability – these are inspired by the work of Molinos-Senante *et al.* (2014) and detailed fully by Sweetapple *et al.* (2014a, 2015). Societal aspects are not considered since typical indicators (such as noise, odour and visual impact) cannot be determined from the model, and are not expected to be subject to any perceivable change as a result of adjusting control strategies.

Operational costs, represented by the *OCI*, are considered within economic sustainability. Capital/investment costs associated with implementation of a new control strategy cannot be quantified from the model, but are expected to be small relative to the long-term operational costs.

Environmental sustainability is assessed based on (i) treatment efficiency, (ii) net energy consumption, (iii) sludge production, and (iv) GHG emissions. The percentage of influent COD, TSS and total nitrogen removed (or not removed), based on the modelled influent and effluent concentrations, provide three indicators for treatment efficiency. Net energy consumption is quantified as in the calculation of GHG emissions, based on energy uses included in the *OCI* and credit from the energy content of recovered methane (Section 9.4.2). Sludge production is calculated as in the *OCI* (Gernaey *et al.*, 2014), and emissions from secondary treatment, sludge processing, net power, chemical use and sludge disposal and reuse are included in the calculation of total GHG emissions, as detailed in Section 9.4.2.

9.5 EXAMPLES/CASE STUDIES

9.5.1 Case study #1: evaluation of plant-wide control strategies

In the first case study, the BSM2G is simulated in a closed loop regime, which includes two defined proportional integral (PI) control loops. The first loop controls the dissolved oxygen (DO) concentration in AER2 by manipulating the air supply rate, here implemented as the oxygen transfer coefficient K_La4 (set-point = 2 g O_2·m^{-3}). K_La3 is set equal to K_La4 and K_La5 is set to half its value. The second loop controls the nitrate concentration in ANOX2 by manipulating the internal recycle flow rate (Q_{intr}). Two different waste sludge flow rates (Q_{W_winter} = 300 m^3·day^{-1} // Q_{W_summer} = 450 m^3·day^{-1}) are imposed in SEC depending on temperature (above or below 15°C) in order to sustain the nitrifying biomass in the system during the winter period. Noise and delays are applied to sensor and actuator models to give the simulations more realism. The external recirculation flow rate (Q_r) and carbon source addition (Q_{carb}) remain constant throughout the simulations. Additional details about the default operational strategy can be found in Flores-Alsina *et al.* (2011). The selection of the different scenarios is intended to demonstrate the relative effects of logical control strategies that may be implemented by operators to increase energy efficiency and/or improve overall plant performance. The following four control scenarios are simulated in the previously predefined case study:

(i) *Impact of DO control* (commonly used to reduce aeration costs) by varying the set-point value between 1 and 3 g O_2·m^{-3} (default value 2 g O_2·m^{-3}).
(ii) *Impact of primary clarifier efficiency* by varying the TSS removal efficiency in PRIM from 33% to 66% (default value 50%). Although in reality this does not happen without chemical addition, the effect of improving TSS removal, such as through chemical addition, is the change of interest.
(iii) *Impact of the anaerobic digester operating mode* by changing the temperature in the anaerobic digester from mesophilic (35°C) to thermophilic (55°C) (default value 35°C).
(iv) *Impact of anaerobic digester supernatants* by controlling the return flow rate originating from the DW unit. This timer-based control strategy stores the dewatering liquor during daytime (when the plant is experiencing high load) and returns it at night (when the plant is at low load). Note that the default BSM2 strategy does not use this control approach and liquors from dewatering are simply returned as they are generated.

EQI, *OCI* and GHG values for the different simulated scenarios are shown in Figure 9.2. Hence, it is possible to see how the overall picture changes when: (i) *EQI* and *OCI* are considered only (Figure 9.2a, b); or (ii) when adding the total quantity of CO_2, N_2O and CH_4 emissions (quantified in kg CO_2e. m^{-3} of treated wastewater) (Figure 9.2c, d). From the generated results one can see that: (i) the DO set-point in the activated sludge section is of paramount importance to the plant's total GHG emissions (*z*-axis) next to the well-known impacts on effluent quality and operating costs; (ii) better TSS removal efficiency in PRIM mainly improves effluent quality and operational cost (*x*- and *y*-axis), but the total GHG emissions remain almost equal; (iii) thermophilic conditions in the anaerobic digester reveal that a higher operating temperature appears to be a more expensive way to operate the plant (with higher operational cost, *y*-axis) without having substantial benefits in terms of increased gas production; and (iv) control of the anaerobic digester supernatants return flow rate improves effluent quality, slightly increases cost but does not have an effect on the GHG emissions unless DO is very low (see dotted lines in Figure 9.2b).

The study presents an important result to the wastewater community by demonstrating the potential impacts of *energy optimization*, particularly in the aeration/anaerobic digester system, and by showing the importance of *plant-wide evaluation*. For example, based on the reported results, operating a plant at low DO concentrations cannot be recommended due to the decrease in effluent quality despite the substantial savings in *OCI* (see Figure 9.2a, b). The situation becomes even worse when GHG emissions are included in the analysis (Figure 9.2c, d) and the substantial contribution of N_2O in the total plant's global warming potential would rank that alternative even lower. Another example in Figure 9.2 illustrates the potential of improving the % TSS removal in the PRIM. Firstly, the load to the activated sludge section is substantially reduced (and thus the off-site CO_2 emissions due to aeration) when the % TSS removal in PRIM increases. Secondly, there is an increase in energy recovery from the anaerobic digestion (higher CO_2 credit). However, in terms of total quantity of GHG emissions the strategy does not seem to pay off since there is a substantial increase in N_2O emissions due to the inadequate C/N ratios that result (poor denitrification). These two examples demonstrate the usefulness of a third GHG dimension for deciding on the optimum operational setting to meet a specific plant's objectives.

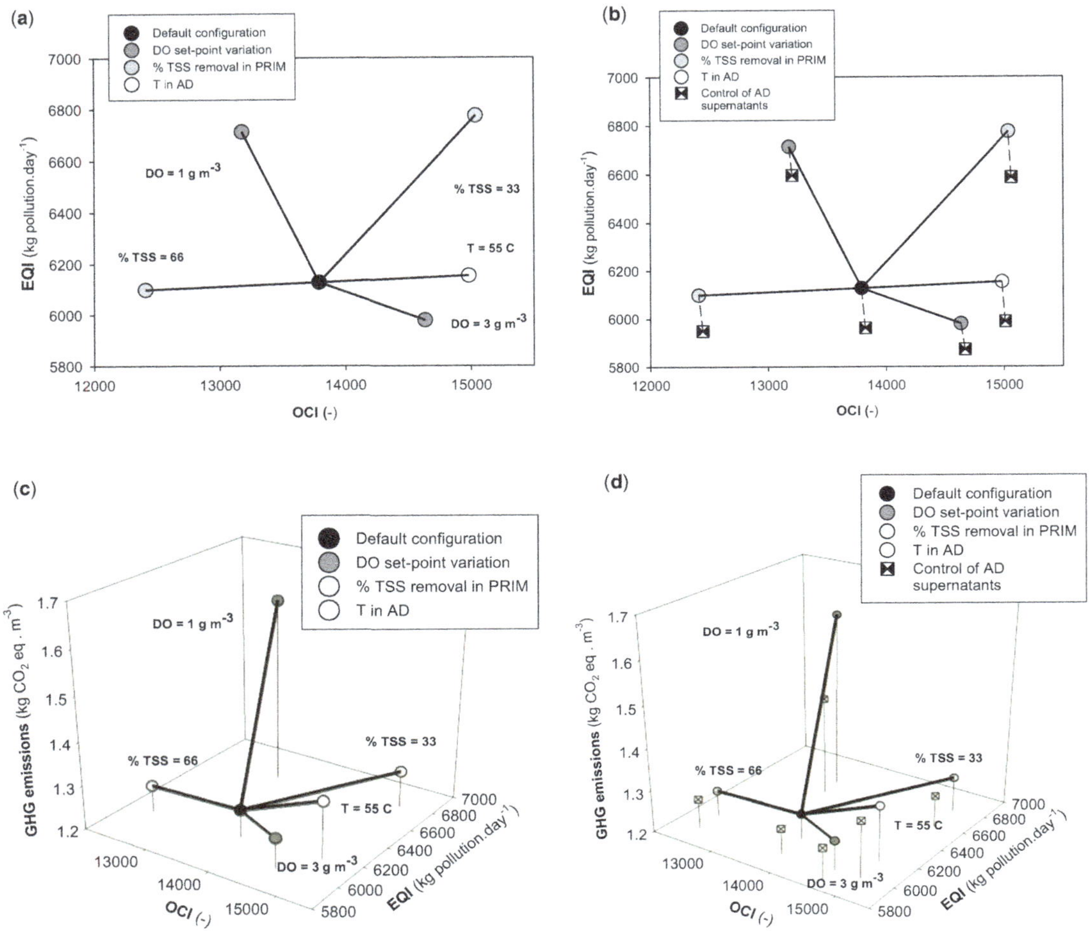

Figure 9.2 Effluent, cost and emission criteria for all the evaluated strategies. (a), (b) *EQI* & *OCI*, (c), (d) *EQI*, *OCI* & *GHG*.

9.5.2 Case study #2: investigating the impact of net energy reduction on sustainability

This case study aims to more broadly investigate the effect on sustainability of modifying the plant control system to reduce net energy use. It considers both economic and environmental aspects of sustainability (including GHG emissions), using the indicators presented in Section 9.4.3.

Two different control strategies are considered, each with multiple variants. In the first control strategy (CL1), DO is controlled using a single PI control loop, as described for Case Study #1. In the second (CL2), three independent PI control loops are used to control the DO spatial distribution, with oxygen transfer coefficients K_La3, K_La4 and K_La5 manipulated to control the DO concentration in AER1, AER2 and AER3 respectively (based on previous findings by Jeppsson *et al.* (2007) that this uses less energy for aeration than a range of other alternatives). In both control strategies, an additional PI control loop controls the nitrate concentration in ANOX2 by manipulating the internal recycle flow rate and different waste sludge flow rates are imposed depending on the time of year. This controller was not included in case study #1.

Factorial sampling is used to generate sets of waste sludge flow rates and DO set-points to implement in these control strategies (within the range 240–360 $m^3 \cdot day^{-1}$ for winter waste sludge flow rate, 360–540 $m^3 \cdot day^{-1}$ for summer waste sludge flow rate, 1–3 g $O_2 \cdot m^{-3}$ for the DO set-point in CL1, and 0.5–2.0 g $O_2 \cdot m^{-3}$ for the DO set-point in CL2). This provides a total of 315 control options for evaluation. CL1 with waste sludge flow rates and DO set-point as in Case Study #1 represents the base case option. Further details on the control options can be found in Sweetapple *et al.* (2015).

A pair-wise comparison of sustainability indicators for control options which provide a reduction in net energy (with respect to the base case) and a compliant effluent is shown in Figure 9.3. Whilst

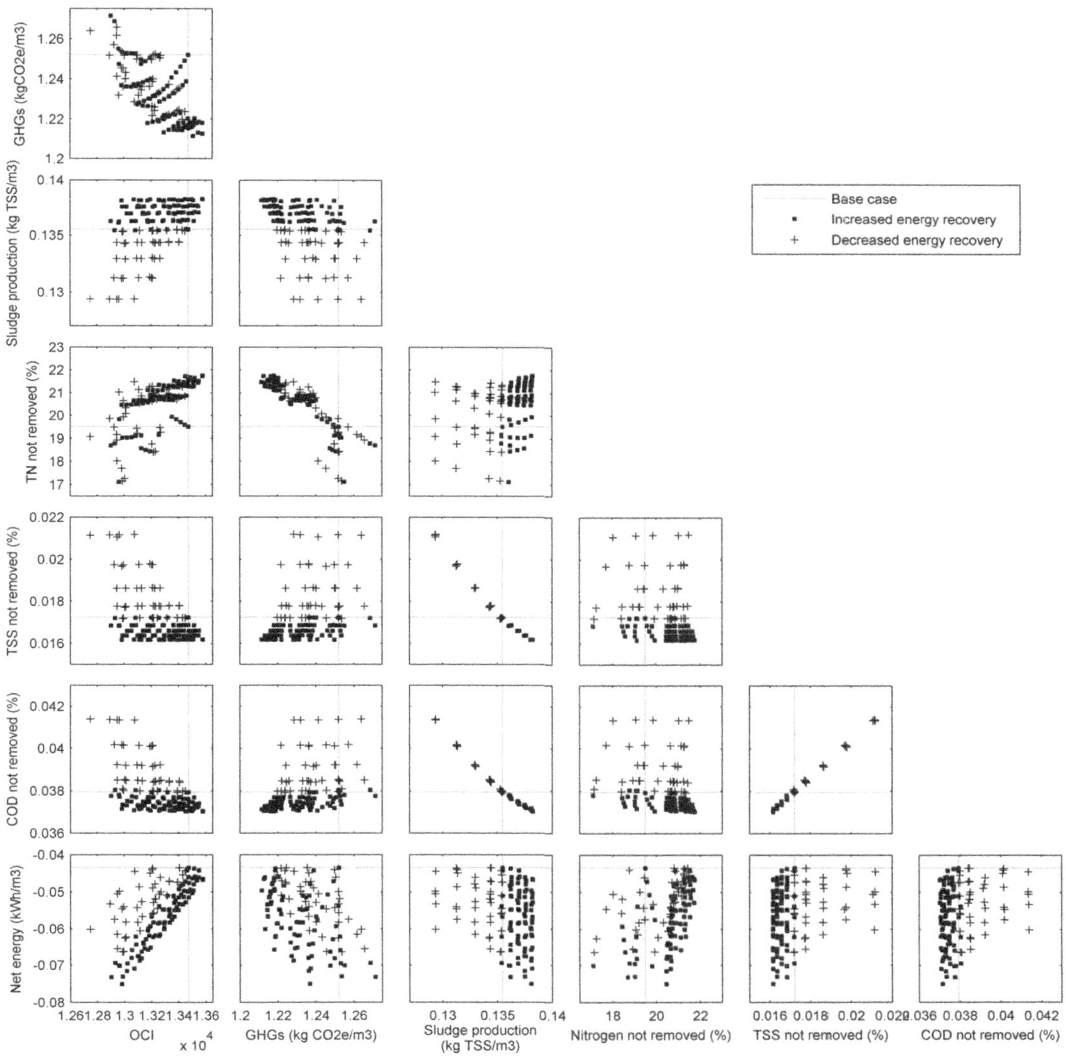

Figure 9.3 Sustainability indicator values for control options providing a reduction in net energy and a compliant effluent.

this shows that control options selected to reduce net energy use also reduce GHG emissions in a high proportion of cases, it also highlights that increased GHG emissions are a potential adverse side effect (as observed in 10% of cases). This may be attributed to lower DO set-points providing lower energy consumption but increased N_2O formation (Flores-Alsina *et al.*, 2014), and raises the important issue that reducing energy use cannot be seen as a reliable approach for reducing total GHG emissions. These results also show that few of the considered solutions (11%) both maintain/improve nitrogen removal and reduce GHG emissions. This highlights a particular challenge since N_2O is emitted during nitrification and denitrification and, whilst these emissions can be curbed to some extent by ensuring sufficient DO (Kampschreuer *et al.*, 2009), this will have implications on energy use.

The best solution with respect to energy reduction is the CL2 control strategy with a 20% increase in the waste flow rates and DO set-points of 1, 1, and 0.5 g $O_2 \cdot m^{-3}$ in AER1, AER2 and AER3 respectively, which uses 73% less energy than the base case. However, this does not provide a universal improvement in sustainability, with both nitrogen removal and sludge production worsened. None of the control options evaluated in fact provides an improvement in all of the sustainability indicators, although seven provide an improvement in all but one. The performance of these control options, alongside the solution that provides the minimum net energy use are shown in Figure 9.4.

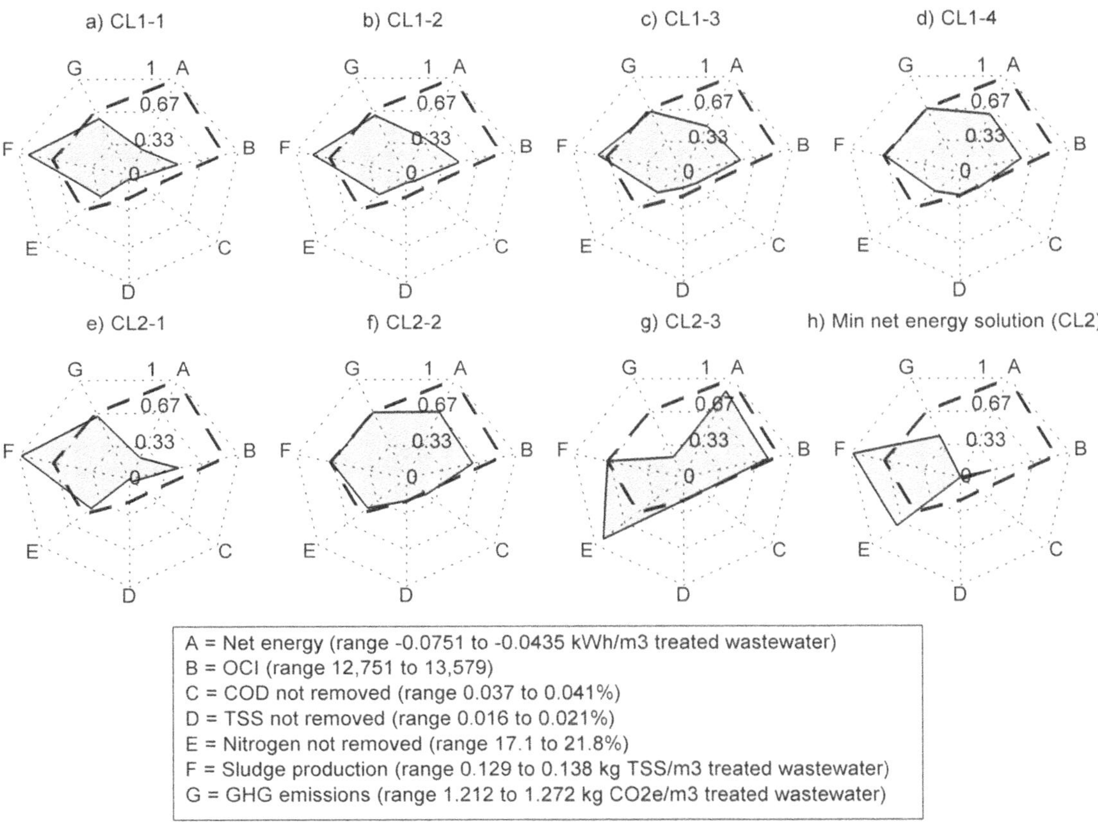

Figure 9.4 Sustainability indicator values for selected solutions, with the dashed lines representing the base case and values closer to the centre of the plot being preferable. (a) CL1-1, (b) CL1-2, (c) CL1-3, (d) CL1-4, (e) CL2-1, (f) CL2-2, (g) CL2-3, (h) Min net energy solution (CL2).

This illustrates that, although each solution reduces net energy use, the type and magnitude of their sustainability impacts vary considerably and trade-offs must be considered.

The results of this case study are of importance because they highlight that, whilst improving both DO control and waste flow rate selection can provide substantial energy savings and increase economic sustainability, the impacts on environmental sustainability are not universally beneficial. Trade-offs are identified and it is shown that nitrogen removal and sludge production in particular are likely to be detrimentally affected in the lowest energy solutions. The ability to model and include both on-site and off-site GHG emissions is particularly valuable since solutions are identified in which a significant reduction in net energy corresponds with an increase in total GHG emissions; in the absence of emissions modelling, such solutions might be assumed to have a more beneficial environmental impact than is the case in reality.

9.5.3 Other relevant case studies

Besides case study #1 and #2 presented in this chapter, there have been many other investigations where the BSM2G has been applied to evaluate control strategies with different purposes, for example, the evaluation of control strategies in further studies using multi-objective optimization (Sweetapple *et al.*, 2015). Control strategies have also been developed which are particularly focused on reducing N_2O emissions (Boiocchi *et al.*, 2017a, 2017b; Santin *et al.*, 2017). It is important to mention the different types of sensitivity analysis that have been implemented in the platform in order to gain understanding of the main interactions amongst model parameters (Boiocchi *et al.*, 2017b; Sweetapple *et al.*, 2014b). Finally, the original WWTP layout has been modified by including sewer and catchment models to also account for CH_4 emissions within the sewer network (Guo *et al.*, 2012). Additional modifications consist of adding two extra anaerobic tanks to allow for biological phosphorus removal (Solis *et al.*, 2019).

9.6 LIMITATIONS

It is important to highlight that the N_2O models used in the study are still under development and are in the process of being validated with full-scale data. Results thus far have been promising (Lindblom *et al.*, 2016). In this paper, the N_2O production by AOB is based on denitrification with NH_4^+ as the electron donor. Other possible mechanisms, such as the formation of N_2O as a by-product of incomplete oxidation of hydroxylamine (NH_2OH) to NO_2^-, are not considered (Wunderlin *et al.*, 2013). Recent investigations demonstrate that both the autotrophic denitrification and the NH_2OH oxidation are involved in N_2O production, although the latter to only a minor degree (Domingo-Felez *et al.*, 2017; Ni & Yuan, 2015; Poquet *et al.*, 2016). Therefore, the results reflect the assumptions built into the N_2O model structure of Mampaey *et al.* (2013), and Guo and Vanrolleghem (2014).

The recent advances when modelling physicochemical processes (Batstone *et al.*, 2012; Flores-Alsina *et al.*, 2015) would allow: (i) pH calculation according to influent conditions and process conditions; and (ii) FA and FNA calculation accounting for ion strength and ion pairing. Indeed, several investigations showed the substantial impact that weak acid chemistry might have on N_2O emissions (Su *et al.*, 2019). Another important aspect would be the quantification of biogenic CO_2 emissions and their inclusion in the overall carbon balance within the plant. Preliminary results can be found in Solis *et al.* (2022).

9.7 CONCLUSIONS AND PERSPECTIVES

The key observations of the presented study can be summarized in the following points:

- The inclusion of GHG emissions provides an additional criterion when evaluating control/operational strategies in a WWTP, offering a better idea about the overall ‘sustainability’ of plant control/operational strategies.

- Simulation results show the risk of energy-related (aeration energy in AS/energy recovery from AD) optimization procedures, and the opposite effect that N_2O and its 300-fold stronger GHG effect (compared to CO_2) might have on the overall global warming potential (GWP) of the WWTP.
- The importance of considering the water and sludge lines together and their impact on the total quantity of GHG emissions are shown when the temperature regime is modified and the anaerobic digester supernatants return flows are controlled.
- While these observations are WWTP specific, the use of the developed tools is demonstrated and can be applied to other systems.

REFERENCES

Arnell M., Rahmberg M., Oliveira F. and Jeppsson U. (2017). Multi-objective performance assessment of wastewater treatment plants combining plant-wide process models and life cycle assessment. *Journal of Water and Climate Change*, **8**(4), 715–729, https://doi.org/10.2166/wcc.2017.179

Avfall Sverige Utveckling (2009). Voluntary commitment – survey of methane losses from biogas plants 2007–2008 (In Swedish: frivilligt åtagande – kartläggning av metanförluster från biogasanläggningar 2007–2008). Avfall Sverige AB, Stockholm, Sweden. ISSN 1103-4092.

Ayesa E., De la Sota A., Grau P., Sagarna J. M., Salterain A. and Suescun J. (2006). Supervisory control strategies for the new WWTP of Galindo-Bilbao: the long run from the conceptual design to the full-scale experimental validation. *Water Science & Technology*, **53**(4–5), 193–201, https://doi.org/10.2166/wst.2006.124

Batstone D. J., Keller J., Angelidaki I., Kayuznyi S. V., Pavlostathis S. G., Rozzi A., Sanders W. T. M., Siegrist H. and Vavilin V. A. (2002). Anaerobic Digestion Model No 1, IWA Scientific and Technical Report No. 13, IWA Publishing, London, UK.

Batstone D. J., Amerlinck Y., Ekama G., Goel R., Grau P., Johnson B., Kaya I., Steyer J.-P., Tait S., Takács I. and Vanrolleghem P. A. (2012). Towards a generalized physicochemical framework. *Water Science & Technology*, **66**(6), 1147–1161, https://doi.org/10.2166/wst.2012.300

Boiocchi R., Gernaey K. V. and Sin G. (2017a). A novel fuzzy-logic control strategy minimizing N_2O emissions. *Water Research*, **123**, 479–494, https://doi.org/10.1016/j.watres.2017.06.074

Boiocchi R., Gernaey K. V. and Sin G. (2017b). Understanding N_2O formation mechanisms through sensitivity analyses using a plant-wide benchmark simulation model. *Chemical Engineering Journal*, **317**, 935–951, https://doi.org/10.1016/j.cej.2017.02.091

Corominas L., Flores-Alsina X., Snip L. and Vanrolleghem P. A. (2012). Comparison of different modelling approaches to better understand greenhouse gas emissions from wastewater treatment plants. *Biotechnology and Bioengineering*, **109**(11), 2855–2863, https://doi.org/10.1002/bit.24544

Domingo-Felez C., Calderó-Pascual M., Sin G., Plósz B. G. and Smets B. F. (2017). Calibration of the comprehensive NDHA-N_2O dynamics model for nitrifier-enriched biomass using targeted respirometric assays. *Water Research*, **126**, 29–39, https://doi.org/10.1016/j.watres.2017.09.013

Dong Y. and Steinberg M. (1997). Hynol: an economical process for methanol production from biomass and natural gas with reduced CO_2 emission. *International Journal of Hydrogen Energy*, **22**(10–11), 971–977, https://doi.org/10.1016/S0360-3199(96)00198-X

Feldman H., Flores-Alsina X., Ramin P., Kjellberg K., Batstone D. J., Jeppsson U. and Gernaey K. V. (2018). Optimization of full-scale anaerobic granular sludge reactor using a model based approach. *Biotechnology Bioengineering*, **115**(11), 2726–2739, https://doi.org/10.1002/bit.26807

Flores-Alsina X., Corominas L., Snip L. and Vanrolleghem P. A. (2011). Including greenhouse gas emissions during benchmarking of wastewater treatment plant control strategies. *Water Research*, **45**(16), 4700–4710, https://doi.org/10.1016/j.watres.2011.04.040

Flores-Alsina X., Corominas L., Neumann M. B. and Vanrolleghem P. A. (2012). Assessing the use of activated sludge processs design guidelines in wastewater treatment plant projects. *Environmental Modelling & Software*, **38**, 50–58, https://doi.org/10.1016/j.envsoft.2012.04.005

Flores-Alsina X., Arnell M., Amerlinck Y., Corominas Ll., Gernaey K. V., Guo L., Lindblom E., Nopens I., Porro J., Shaw A., Snip L., Vanrolleghem P. A. and Jeppsson U. (2014). Balancing effluent quality, economical cost and greenhouse gas emissions during the evaluation of plant-wide wastewater treatment plant control strategies. *Science of the Total Environment*, **466–467**, 616–624, https://doi.org/10.1016/j.scitotenv.2013.07.046

Flores-Alsina X., Kazadi-Mbama C., Solon K., Vrecko D., Tait S., Batstone D., Jeppsson U. and Gernaey K. V. (2015). A plant wide aqueous phase chemistry module describing pH variations and ion speciation/pairing in wastewater treatment models. *Water Research*, **85**, 255–265, https://doi.org/10.1016/j.watres.2015.07.014

Foley J., Yuan Z., Keller J., Senante E., Chandran K., Willis J., van Loosdrecht M. and van Voorthuizen E. (2011). N_2O and CH_4 Emission From Wastewater Collection and Treatment Systems, Technical report, Global Water Research Coalition, London, UK.

Gernaey K. V., Flores-Alsina X., Rosen C., Benedetti L. and Jeppsson U. (2011). Dynamic influent pollutant disturbance scenario generation using a phenomenological modelling approach. *Environmental Modelling & Software*, **26**(11), 1255–1267, https://doi.org/10.1016/j.envsoft.2011.06.001

Gernaey K. V., Jeppsson U., Vanrolleghem P. A. and Copp J. B. (2014). Benchmarking of Control Strategies for Wastewater Treatment Plants, IWA Scientific and Technical Report No. 23, IWA Publishing, London, UK.

Guo L. and Vanrolleghem P. A. (2014). Calibration and validation of an Activated Sludge Model for Greenhouse Gases no. 1 (ASMG1): prediction of temperature-dependent N_2O emission dynamics. *Bioprocess and Biosystems Engineering*, **37**, 151–163, https://doi.org/10.1007/s00449-013-0978-3

Guo L., Porro J., Sharma K. R., Amerlinck Y., Benedetti L., Nopens I., Shaw A., Van Hulle S. W. H., Yuan Z. and Vanrolleghem P. A. (2012). Towards a benchmarking tool for minimizing wastewater utility greenhouse gas footprints. *Water Science & Technology*, **66**(11), 2483–2495, https://doi.org/10.2166/wst.2012.495

Gustavsson D. J. I. and Tumlin S. (2013). Carbon footprints of Scandinavian wastewater treatment plants. *Water Science & Technology*, **68**(4) 887–893, https://doi.org/10.2166/wst.2013.318

Henze M., Gujer W., Mino T. and van Loosdrecht M. C. M. (2000). Activated Sludge Models ASM1, ASM2, ASM2d and ASM3, IWA Scientific and Technical Report No. 9, IWA Publishing, London, UK.

Hiatt W. C. and Grady C. P. L., Jr. (2008). An updated process model for carbon oxidation, nitrification, and denitrification. *Water Environment Research*, **80**(11), 2145–2156, https://doi.org/10.2175/106143008X304776

Hug T., Benedetti L., Hall E. R., Johnson B. R., Morgenroth E., Nopens I., Rieger L., Shaw A. and Vanrolleghem P. A. (2009). Wastewater treatment models in teaching and training: The mismatch between education and requirements for jobs. *Water Science & Technology*, **59**(4), 745–753, https://doi.org/10.2166/wst.2009.595

IPCC (2013). Climate change 2013: The physical science basis. In: Contribution of Working Group I to the Fifth Assessment Report of the Intergovernmental Panel on Climate Change, T. F. Stocker, D. Qin, G.-K. Plattner, M. Tignor, S. K. Allen, J. Boschung, A. Nauels, Y. Xia, V. Bex and P. M. Midgley (eds.), Cambridge University Press, Cambridge, England and New York, NY, USA.

Jeppsson U., Pons M. N., Nopens I., Alex J., Copp J. B., Gernaey K. V., Rosen C., Steyer J. P. and Vanrolleghem P. A. (2007). Benchmark simulation model no. 2: general protocol and exploratory case studies. *Water Science & Technology*, **56**(8), 67–78, https://doi.org/10.2166/wst.2007.604

Jönsson H., Junestedt C., Willén A., Yang J., Tjus K., Baresel C., Rodhe L., Trela J., Pell M. and Andersson S. (2015). Reduce greenhouse gas emissions from wastewater treatment and sewage sludge (In Swedish: Minska utsläpp av växthusgaser från rening av avlopp och hantering av avloppsslam), Report SVU 2015-02, The Swedish Water and Wastewater Association, Stockholm, Sweden.

Kampschreur M. J., Temmink H., Kleerebezem R., Jetten M. S. M. and van Loosdrecht M. C. M. (2009). Nitrous oxide emission during wastewater treatment. *Water Research*, **43**(17), 4093–4103, https://doi.org/10.1016/j.watres.2009.03.001

Liebetrau J., Clemens J., Cuhls C., Hafermann C., Friehe J., Weiland P. and Daniel-Gromke J. (2010). Methane emissions from biogas-producing facilities within the agricultural sector. *Engineering in Life Sciences* **10**(6), 595–599, https://doi.org/10.1002/elsc.201000070

Lindblom E., Arnell M., Flores-Alsina X., Stenström F., Gustavsson D. J. I. and Jeppsson U. (2016). Dynamic modelling of nitrous oxide emissions from three Swedish full-scale reject water treatment systems. *Water Science & Technology*, **73**(4), 798–806, https://doi.org/10.2166/wst.2015.534

Mampaey K. E., Beuckels B., Kampschreur M. J., Kleerebezem R., van Loosdrecht M. C. M. and Volcke E. I. P. (2013). Modelling nitrous and nitric oxide emissions by autotrophic ammonia-oxidizing bacteria. *Environmental Technology*, **34**(12), 1555–1566, https://doi.org/10.1080/09593330.2012.758666

Mannina G., Ekama G., Caniani D., Cosenza A., Esposito G., Gori R., Garrido-Baserba M., Rosso D. and Olsson G. (2016). Greenhouse gases from wastewater treatment – a review of modelling tools. *Science of the Total Environment*, **551–552**, 254–270, https://doi.org/10.1016/j.scitotenv.2016.01.163

Meng F., Fu G. and Butler D. (2016). Water quality permitting: from end-of-pipe to operational strategies. *Water Research*, **101**, 114–126, https://doi.org/10.1016/j.watres.2016.05.078

Meng F., Fu G. and Butler D. (2020). Regulatory implications of integrated real-time control technology under environmental uncertainty. *Environmental Science & Technology*, **54**(3), 1314–1325, https://doi.org/10.1021/acs.est.9b05106

Molinos-Senante M., Gomez T., Garrido-Baserba M., Caballero R. and Sala-Garrido R. (2014). Assessing the sustainability of small wastewater treatment systems: a composite indicator approach. *Science of the Total Environment*, **497–498**, 607–617, https://doi.org/10.1016/j.scitotenv.2014.08.026

Nopens I., Benedetti L., Jeppsson U., Pons M. N., Alex J., Copp J. B., Gernaey K. V., Rosen C., Steyer J. P. and Vanrolleghem P. A. (2010). Benchmark Simulation Model No 2: Finalisation of plant layout and default control strategy. *Water Science & Technology*, **62**, 1967–1974. https://doi.org/10.2166/wst.2010.044

Ni B. J. and Yuan Z. (2015). Recent advances in mathematical modeling of nitrous oxides emissions from wastewater treatment processes. *Water Research*, **87**, 336–346, https://doi.org/10.1016/j.watres.2015.09.049

Nopens I., Batstone D. J., Copp J. B., Jeppsson U., Volcke E., Alex J. and Vanrolleghem P. A. (2009). An ASM/ADM model interface for dynamic plant-wide simulation. *Water Research*, **43**(7), 1913–1923, https://doi.org/10.1016/j.watres.2009.01.012

Olsson G. (2012). ICA And me: a subjective review. *Water Research*, **46**(6), 1586–1624, https://doi.org/10.1016/j.watres.2011.12.054

Otterpohl R. and Freund M. (1992). Dynamic models for clarifiers of activated sludge plants with dry and wet weather flows. *Water Science and Technology*, **26**, 1391–1400, https://doi.org/10.2166/wst.1992.0582

Pocquet M., Wu Z., Queinnec I. and Spérandio M. (2016). A two pathway model for N_2O emissions by ammonium oxidizing bacteria supported by the NO/N_2O variation. *Water Research*, **88**(1), 948–959, https://doi.org/10.1016/j.watres.2015.11.029

Rivas A., Irizar I. and Ayesa E. (2008). Model-based optimisation of wastewater treatment plants design. *Environmental Modelling & Software*, **23**(4), 435–450, https://doi.org/10.1016/j.envsoft.2007.06.009

Rodriguez-Roda I., Sànchez-Marrè M., Comas J., Baeza J., Colprim J., Lafuente J., Cortes U. and Poch M. (2002). A hybrid supervisory system to support WWTP operation: implementation and validation. *Water Science & Technology*, **45**(4–5), 289–297, https://doi.org/10.2166/wst.2002.0608

Santin I., Barbu M., Pedret C. and Vilanova R. (2017). Control strategies for nitrous oxide emissions reduction on wastewater treatment plants operation. *Water Research*, **125**, 466–477, https://doi.org/10.1016/j.watres.2017.08.056

Solis B., Guisasola A., Flores-Alsina X., Jeppsson U. and Baeza A. (2019). A Plant-Wide Model Describing Energy/Nutrient Recovery Options and GHG Emissions for Water Resource Recovery Facilities. In: Proceedings of 10th IWA Symposium on Modelling and Integrate Assessment. IWA Publishing.

Solis B., Guissasola A., Flores-Alsina X., Jeppsson U. and Baeza J. A. (2022). A plant wide model describing GHG emissions and nutrient recovery options for water resource recovery facilities. *Water Resarch*. Accepted. https://doi.org/10.1016/j.watres.2022.118223

Solon K., Flores-Alsina X., Kazadi-Mbamba C., Ikumi D., Volcke E. I. P., Vaneeckhaute C., Ekama G., Vanrolleghem P. A., Batstone D. J., Gernaey K. V. and Jeppsson U. (2017). Plant-wide modelling of phosphorus transformations in biological nutrient removal wastewater treatment systems: impacts of control and operational strategies. *Water Research*, **113**, 97–110, https://doi.org/10.1016/j.watres.2017.02.007

Su Q., Domingo Felez C., Zhang Z., Blum J. M., Jensen M. M. and Smets B. F. (2019). The effect of pH on N_2O production in intermittently-fed nitritation reactors. *Water Research*, **156**, 223–231, https://doi.org/10.1016/j.watres.2019.03.015

Sweetapple C., Fu G. and Butler D. (2014a). Multi-objective optimisation of wastewater treatment plant control to reduce greenhouse gas emissions. *Water Research*, **55**, 52–62, https://doi.org/10.1016/j.watres.2014.02.018

Sweetapple C., Fu G. and Butler D. (2014b). Identifying key sources of uncertainty in the modelling of greenhouse gas emissions from wastewater treatment. *Water Research*, **47**(13), 4652–4665, https://doi.org/10.1016/j.watres.2013.05.021

Sweetapple C., Fu G. and Butler D. (2015). Does carbon reduction increase sustainability? A study in wastewater treatment. *Water Research*, **87**, 522–530, https://doi.org/10.1016/j.watres.2015.06.047

Takács I., Patry G.G. and Nolasco D. (1991). A dynamic model of the clarification-thickening process. *Water Research*, **25**, 1263–1271, https://doi.org/10.1016/0043-1354(91)90066-Y

Tchobanoglous G., Burton F. L. and Stensel H. D. (2003). Wastewater Engineering: Treatment, Disposal and Reuse. McGraw-Hill, New York, USA.

Wunderlin P., Lehmann M. F., Siegrist H., Tuzson B., Joss A., Emmenegger L. and Mohn J. (2013). Isotope signatures of N_2O in a mixed microbial population system: constraints on N_2O producing pathways in wastewater treatment. *Environmental Science and Technology*, **47**(3), 1339–1348.

NOMENCLATURE

AD	Anaerobic digester
ADM	Anaerobic digestion model
AE	Aeration energy ($kWh\cdot day^{-1}$)
AER	Aerobic section
AOB	Ammonium oxidizing bacteria
ANOX	Anoxic section
ASM	Activated sludge model
BOD	Biochemical oxygen demand ($g\cdot m^{-3}$)
BSM2	Benchmark Simulation Model No 2
CH_4	Methane (kg $CH_4\cdot day^{-1}$)
CO_2	Carbon dioxide (kg $CO_2\cdot day^{-1}$)
CO_2e	Equivalent carbon dioxide (kg $CO_{2e}\cdot day^{-1}$)
COD	Chemical oxygen demand ($g\cdot m^{-3}$)
DO	Dissolved oxygen concentration ($g\cdot m^{-3}$)
DW	Dewatering unit
EC	Consumption of external carbon source (kg $COD\cdot day^{-1}$)
EQI	Effluent quality index (kg $pollution\cdot day^{-1}$)
GHG	Greenhouse gas
GWP	Global warming potential
HE	Heating energy ($kWh\cdot day^{-1}$)
K_La	Volumetric oxygen transfer coefficient (day^{-1})
ME	Mixing energy ($kWh\cdot day^{-1}$)
MP	Methane production ($kgCH_4\cdot day^{-1}$)
N	Nitrogen
NH_4^+	Ammonium nitrogen (g $N\cdot m^{-3}$)
NO	Nitric oxide nitrogen (g $N\cdot m^{-3}$)
N_2O	Nitrous oxide nitrogen (kg $N\cdot day^{-1}$)
NOB	Nitrite oxidizing bacteria
NO_2^-	Nitrite nitrogen (g $N\cdot m^{-3}$)
NO_3^-	Nitrate nitrogen (g $N\cdot m^{-3}$)
NO	Oxidized forms of nitrogen (g N m^{-3})
OCI	Operational cost index (-)
PE	Pumping energy ($kWh\cdot day^{-1}$)

PRIM	Primary clarifier
PI	Proportional integral controller
Q_e	Effluent flow rate ($m^3 \cdot day^{-1}$)
Q_{carb}	External carbon source flow rate ($m^3 \cdot day^{-1}$)
Q_{intr}	Internal recycle flow rate ($m^3 \cdot day^{-1}$)
Q_r	External recirculation flow rate ($m^3 \cdot day^{-1}$)
Q_W	Waste sludge flow rate ($m^3 \cdot day^{-1}$)
SEC	Secondary clarifier
SP	Sludge production (kg TSS$\cdot day^{-1}$)
ST	Storage tank
SRT	Sludge retention time (days)
TKN	Total Kjeldahl nitrogen ($g \cdot m^{-3}$)
TN	Total nitrogen ($g \cdot m^{-3}$)
THK	Thickener
TOC	Total organic carbon ($g \cdot m^{-3}$)
TSS	Total suspended solids ($g \cdot m^{-3}$)
VS	Volatile solids ($g \cdot m^{-3}$)
WWTP	Wastewater treatment plant

doi: 10.2166/9781789060461_229

Chapter 10

Knowledge-based and data-driven approaches for assessing greenhouse gas emissions from wastewater systems

Jose Porro[1], Vasileia Vasilaki[2], Giacomo Bellandi[3] and Evina Katsou[2]

[1]Cobalt Water Global, 77 Sands Street, Brooklyn, NY 11201, USA. E-mail: jose.porro@cobaltwater-global.com
[2]Department of Civil and Environmental Engineering, Brunel University London, Uxbridge UB8 3PH, UK. E-mail: Vasileia.Vasilaki@brunel.ac.uk; Evina.Katsou@brunel.ac.uk
[3]AM-TEAM, Oktrooiplein 1 – Box 601, 9000 Gent, Belgium. E-mail: giacomo.bellandi@am-team.com

SUMMARY

This chapter provides an overview of modelling approaches other than the mechanistic activated sludge model (ASM) framework for assessing greenhouse gas (GHG) emissions from urban wastewater systems. Examples include knowledge-based artificial intelligence, integrating mechanistic modelling and computational fluid dynamics (CFD) with artificial intelligence (AI), and data-driven and machine learning (ML) methods for assessing and mitigating nitrous oxide (N_2O) emissions from wastewater treatment.

Keywords: Artificial intelligence, knowledge-based systems, machine learning, nitrous oxide, principal component analysis, support vector machines

TERMINOLOGY

Term	Definition
Artificial intelligence	Study of the human mechanisms, which provide the behaviour that can be considered intelligent, and emulation of these mechanisms, called cognitive tasks (reason, problem-solving, remember or learn), in a computer software.
Data mining	The process of uncovering patterns and other valuable information from large data sets.
Fuzzy logic	Approach to variable processing that allows for multiple possible truth values, or degrees of membership to a specific classification (e.g., high, medium, and low risk), to be processed through the same variable.
Greenhouse gas	Gas that absorbs and emits radiant energy within the thermal infrared range.
Knowledge-based system	A computer program that reasons and uses a knowledge base (normally representing expert knowledge) to solve complex problems.

Machine learning	The use of algorithms and statistical models to analyse and draw inferences from patterns in data and make predictions on the value of specific parameters included in the dataset.
Mechanistic model	A mathematical model, usually comprised of differential equations, describing physical, chemical, biochemical or biological processes based on the knowledge of these processes.
Principal component analysis	PCA is a multivariate statistical method for data mining that simplifies the complexity in high-dimensional data, while retaining trends and patterns, by transforming the data into fewer dimensions, which act as summaries of features.
Support vector machines	A type of machine learning, for supervised non-parametric classification and regression algorithms.

10.1 INTRODUCTION

The preceding chapters have covered mechanistic modelling of methane (CH_4) and nitrous oxide (N_2O) production/emissions from sewers and water resource recovery facilities, mainly using the activated sludge model (ASM) framework. However, there are other modelling approaches that can provide additional insight into what might be happening from mechanistic, hydrodynamic, and process points of views that the ASM models alone cannot readily provide. This chapter provides an overview of knowledge-based and data-driven modelling approaches that can be used in gaining additional valuable insights for the purpose of greenhouse gas (GHG) emissions assessment and mitigation from urban wastewater systems. In today's Digital Water age, these approaches also illustrate how data that are currently being recorded, but not being fully valorized, can be leveraged for reducing GHG emissions and the benefit of the planet.

Other modelling approaches that have not been covered in the preceding chapters include knowledge-based artificial intelligence (AI), the coupling of AI with computational fluid dynamics (CFD), and data-driven approaches such as data mining and machine learning. Knowledge-based AI can shed light on the pathways and influencing factors of GHG production in wastewater systems and how they can be mitigated to reduce GHG emissions. In a hybrid modelling approach, integrating knowledge-based AI with both process modelling and CFD can uncover the effect that different process control strategies and the mixing conditions in reactors can have on GHG production, respectively. Data-driven approaches such as principal component analysis (PCA) and machine learning (ML) can identify correlations and patterns in process data to categorize conditions leading to higher or lower GHG emissions, and corresponding potential mitigation strategies. Although this chapter summarizes how each of these approaches have been implemented for assessing N_2O emissions from wastewater treatment specifically, the approaches can also be followed for assessing and mitigating CH_4 emissions.

10.2 KNOWLEDGE-BASED ARTIFICIAL INTELLIGENCE

Knowledge-based systems are used to represent knowledge of a particular process or system within a software tool, and then use this knowledge, with a particular set of conditions as inputs, to make decisions. Knowledge-based systems can also be considered expert systems because they can mimic the decision-making or reasoning ability of on expert when confronted with data for a particular condition or set of conditions. Combined with fuzzy logic (Zadeh, 1965), another artificial intelligence technique, an expert system can look at numerical data, and using fuzzy logic as part of its inference engine along with the knowledge represented in a knowledge base, express a qualitative diagnosis (e.g., high risk, medium risk, or low risk) of the system with a numerical score between 0 and 1. Because the expert system can be coded with a fuzzy logic rule base, it enables application of the knowledge of a particular process, phenomena, or condition to large data sets (historical or real-time) to dynamically diagnose a system, automating the same reasoning process an expert would follow.

Knowledge-based systems have been applied to wastewater treatment in several cases (Comas *et al.*, 2003; Rodríguez-Roda *et al.*, 2002; Sànchez-Marrè *et al.*, 1996). Comas *et al.* (2003) and Rodríguez-Roda *et al.* (2002) applied the expert knowledge of the conditions relating to the risk of problems of a microbiological nature, because the conditions leading to problems such as filamentous sludge bulking are well-known from expert knowledge and published literature. Similarly, the conditions (or influencing factors) for the various N_2O pathways in wastewater treatment are well known and have been reported in the literature as being useful for qualitatively determining the risk of N_2O production and emissions (Ahn *et al.*, 2010; Colliver and Stephenson, 2000; Foley *et al.*, 2010; GWRC, 2011; Kampschreur *et al.*, 2008, 2009). Therefore, Porro *et al.* (2014) proposed a knowledge-based approach to assess the risk of producing and emitting N_2O from a wastewater treatment process (from nitrification and denitrification) through a fuzzy logic, expert system, which is now known as the N_2O Risk Model, to use values of the various influencing factors, such as dissolved oxygen (DO), nitrite, pH, and chemical oxygen demand (COD):N, as inputs to then dynamically give a risk score between 0 and 1. As each influencing factor is linked to a specific N_2O pathway, and a risk score is generated for each influencing factor, the N_2O Risk Model gives insights into which pathways are occurring and how they can be mitigated.

This approach was implemented for the Eindhoven WWTP (Netherlands). Figure 10.1 shows the risk versus the measured aqueous N_2O concentration in the aeration (nitrification) reactor. As seen in Figure 10.1, N_2O is being produced due to risk of both low-DO conditions (ammonia oxidizing bacteria (AOB) denitrification pathway), and high-DO conditions (hydroxylamine pathway), as evidenced from the peak N_2O concentrations corresponding with both peaks in risk due to low-DO conditions and high-DO conditions.

The clear control action to mitigate the peaks of risk and the peaks of N_2O based on Figure 10.1 was to adjust the DO to avoid high risk due to both low-DO and high-DO conditions. Therefore, this mitigation strategy was implemented in full-scale, which resulted in a 40% reduction in the overall GHG emissions for the facility, and over a 90% reduction in N_2O emissions (Porro *et al.*, 2017). Figure 10.2 illustrates the aqueous N_2O concentrations inside the aeration reactor at the Eindhoven WWTP before, during, and after a mitigation test in which the dissolved oxygen control was changed based upon the outputs of the N_2O Risk Model. As can be seen in Figure 10.2, the N_2O concentration drops significantly during the N_2O risk-based control test.

As can also be seen in Figure 10.2, not only was N_2O reduced, but also the ammonia peaks were effectively eliminated, and the DO no longer increased to 6 mg O_2/L as was previously occurring. Therefore, examining the process from an N_2O risk point of view inherently also provides clear process benefits such as improved nitrification, lower ammonia effluent concentrations, and more efficient aeration.

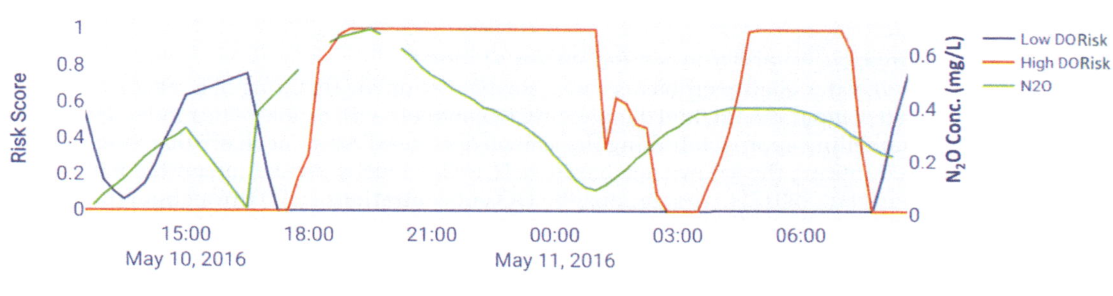

Figure 10.1 N_2O risk and aqueous N_2O measurement results for Eindhoven WWTP (NL).

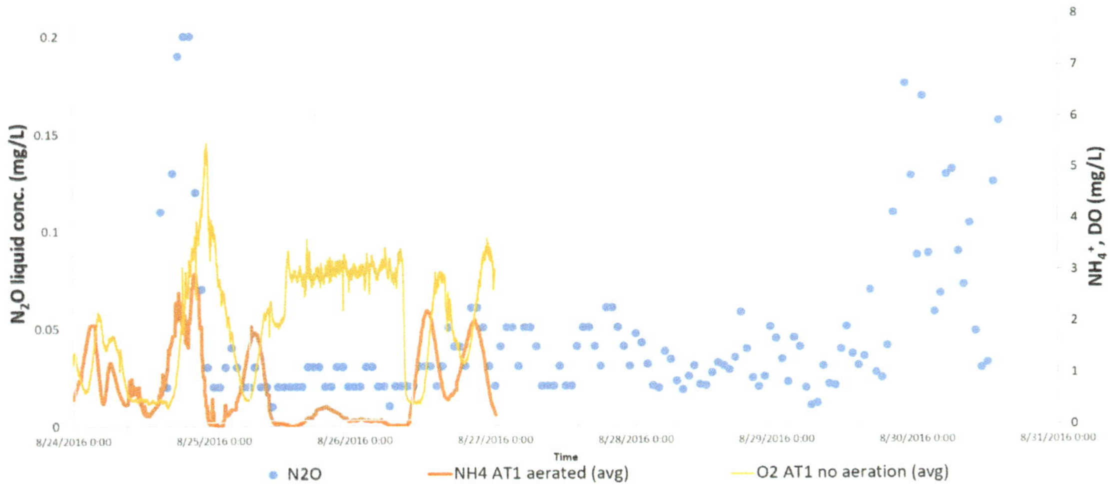

Figure 10.2 Aqueous N_2O (blue), ammonium (red), and dissolved oxygen (yellow) concentrations in Eindhoven WWTP (NL) aeration reactor before (grey shaded area), during (blue shaded area), and after (grey shaded area) mitigation control test.

10.2.1 Integrating knowledge-based AI with mechanistic process models

In a hybrid modelling approach, Porro *et al.* (2014) proposed linking the knowledge-based N_2O Risk Model with the BSM2 platform detailed in Chapter 9. This allows comparing of the resulting N_2O risk for different control strategies on a whole-plant level, as well as on an individual reactor level, which is helpful since the air can be distributed or controlled differently among the different reactors at a plant. Furthermore, you can have different processes and environmental conditions occurring in the different reactors, which can influence N_2O emissions differently. Although the BSM2 platform is used in the case of Porro *et al.* (2014), the example highlights how any mechanistic model, of the ASM type that does not include N_2O mechanistic models, but does have the state variables (DO, COD/N, pH, NO_2^-) for computing N_2O risk, can be used to first identify when, where, and why there is N_2O risk for different scenarios; to identify control actions to mitigate the risk; and then to test that the identified risk-based control actions do not negatively impact the effluent quality with the mechanistic model before implementing them in the real plant. However, using the ASM outputs for the N_2O Risk Model inputs can also complement mechanistic models that do include N_2O processes by providing knowledge-based insight into why the predicted N_2O emissions might be higher or lower, as well as to provide insights into how they can be mitigated, so that the mitigation actions can be tested with the mechanistic model to confirm how much N_2O can be reduced with the new (mitigated) set of N_2O emissions.

10.2.2 Hybrid biokinetic/CFD and knowledge-based AI model

To show how localized water quality problems can arise due to non-ideal mixing in a reactor, Rehman *et al.* (2017) integrated computational fluid dynamics (CFD) models with a biokinetic model. Therefore, in a similar hybrid modelling approach linking the knowledge-based AI to mechanistic models, Porro *et al.* (2019) proposed linking the knowledge-based N_2O Risk Model with the integrated biokinetic/CFD model of Rehman *et al.* (2017) to first predict the DO concentrations in 3D throughout the aerobic and anoxic volumes of an aeration reactor, specifically for the Eindhoven WWTP (Netherlands). The DO in each cell of the CFD geometry mesh was then used as the input for the N_2O Risk Model of Porro *et al.* (2014) to generate a risk score for each cell of the aerobic zone to determine risk of N_2O emissions due to low-DO conditions (low-DO risk) and high-DO conditions (high-DO risk), which

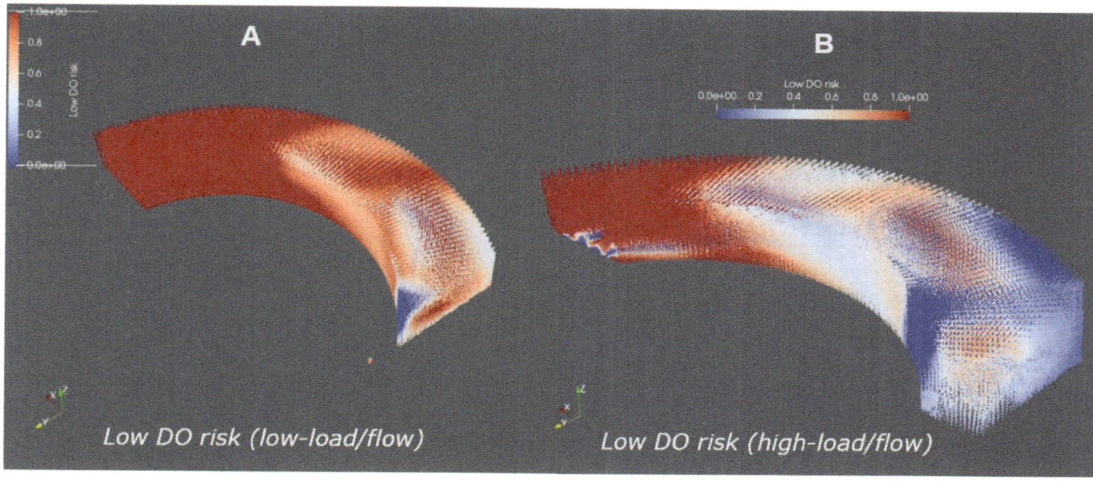

Figure 10.3 N_2O risk in 3D for aerobic zone in Eindhoven WWTP (NL) aeration reactor for A (low-load, low-flow, left) and B (high-load/high-flow, right).

implicate nitrification N_2O production due to the nitrifier (AOB) denitrification N_2O pathway and the hydroxylamine N_2O pathway, respectively. Similarly for the anoxic zone, an anoxic high-DO risk score, which signals N_2O production from incomplete heterotrophic denitrification, was generated for each cell of the CFD geometry mesh. N_2O risk was assessed in 3D for the reactor based upon the hydrodynamic conditions for a low-flow, low-load condition, as well as a high-flow, high-load condition.

Figure 10.3a shows how the N_2O risk propagates through the aerobic zone for the low-flow/low-load condition. As seen in Figure 10.3a, a significant portion of the aerobic zone has high risk due to low-DO conditions and diminishes in the direction of flow, from higher to lower risk, with the lowest risk near the end of the aerobic zone. This supports what has been seen during previous N_2O measurement campaigns (Bellandi *et al.*, 2017), when N_2O was measured at the beginning and end of the aerobic zone and higher emissions were seen at the beginning of the aerobic zone. Risk had also been calculated based upon the plant DO sensor at the end of the aerobic zone and a portable DO probe placed at the beginning of the aerobic zone, which revealed higher low-DO risk (higher risk due to low-DO conditions) at the beginning of the aerobic zone. This may be due to multiple factors: higher oxygen demand at the beginning where ammonia is fed in and, based on recirculation rate/plug flow conditions, because a concentration gradient for DO and ammonia has been observed between the beginning and end of the aerobic zone.

For the high-load/high-flow condition (Figure 10.3b), there is slightly less low-DO risk in general, but there is still the gradient in risk and a high low-DO risk at the beginning of the aerobic zone. This is most likely because oxygen demand is now higher due to higher concentrations, but because there is more aeration, higher DO to begin with, and better mixing from higher flows, the high low-DO risk does not propagate through the aerobic zone as much. This indicates that under this condition, it is possible that N_2O at the beginning of the aerobic zone can be produced by low-DO and high-DO conditions, and at the end of the aerobic zone from most likely high-DO conditions since there is mostly blue (low or zero low-DO risk) seen, which was confirmed when Porro *et al.* (2019) quantified the high-DO risk for the same conditions.

Figure 10.4 shows the low-DO risk for the same conditions as in Figure 10.3, but now taking a vertical slice from the beginning of the aerobic zone to see how the risk varies due to mixing conditions

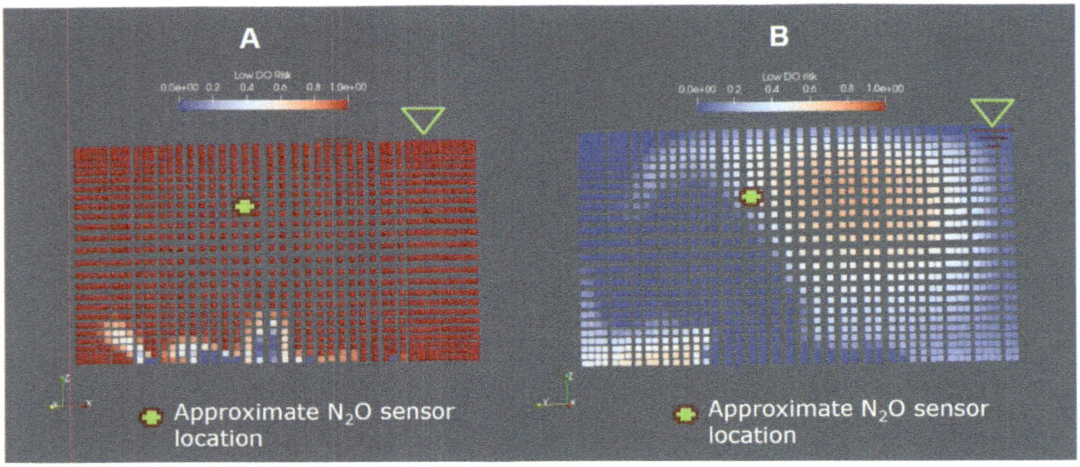

Figure 10.4 N_2O risk (low DO risk) in vertical slice at beginning of aerobic zone in Eindhoven WWTP (NL) aeration reactor for A (low-load, low-flow, left) and B (high-load/high-flow, right).

across a cross section of the reactor. For the low-load/low-flow condition (Figure 10.4a), almost the entire cross section has high risk due to low-DO conditions. This supports what has previously been observed, that when ammonia starts to increase, it is under low-DO risk conditions, and we see N_2O production/emissions start to increase as seen previously in Figure 10.1. Now it is clear that it is because there is low-DO everywhere in this cross section that the N_2O production/emissions start to increase.

For the high-load/high-flow condition (Figure 10.4b), approximately half of the cross section has low risk (blue area); however, there is still a considerable area (upper right) with relatively higher low-DO risk. The high-load/high-flow condition is when we have seen the highest emissions and was previously thought to be only due to high-DO risk and not low-DO risk. Based on the approximate location of the N_2O sensor, this makes sense because a big pocket of low (blue) low-DO risk can be seen, right near where the sensor is. This pocket was seen to be occupied by high high-DO risk when generating similar plots as in Figure 10.4 but for high-DO risk (Porro *et al.*, 2019), which supports what had already been seen and measured (Porro *et al.*, 2017). This cross section indicates that the increased N_2O measured in both the liquid and the off-gas may have been from N_2O production due to both high-DO and low-DO conditions; hence, due to both AOB pathways.

It is not suggested that this level of hybrid modelling is always required, especially when N_2O risk, measurements, and mitigation all seem to make sense based on the available knowledge; however, it is clear from Figures 10.3 and 10.4 that integrating mechanistic, CFD, and knowledge-based AI can give greater insights into how N_2O production and risk can vary in a reactor due to hydrodynamic conditions when it is suspected that localized effects from uneven mixing can potentially be impacting N_2O dynamics, but not possible to confirm this based on sensor locations.

Qiu *et al.* (2019) followed a similar approach, integrating a mechanistic model that included the biokinetics of N_2O production as described by Domingo-Félez and Smets (2016) and Domingo-Félez *et al.* (2017) with CFD, to give predicted N_2O production/emissions as opposed to the N_2O risk in 3D for the reactor. These two approaches can complement each other because you gain insight not only into why there are N_2O emissions from the risk, but also into the expected N_2O emissions, which can then be used to model scenarios to mitigate the risk and emissions.

10.3 DATA-DRIVEN APPROACHES

Reliable estimation of N_2O emissions from wastewater treatment processes can provide guidance on N_2O mitigation measures and support WWTPs towards achieving carbon neutrality goals. It provides an alternative to theoretical methodologies for quantifying and reporting N_2O emission factors (EFs) that have been widely criticized (Cadwallader & VanBriesen, 2017). This is very important for wastewater treatment processes and specifically for sidestream technologies, given that they can contribute significantly to the operational carbon footprint of WWTPs (Schaubroeck *et al.*, 2015). Long-term monitoring campaigns of over one year are required though for the development of process-based reliable N_2O EFs (Gruber *et al.*, 2019; Vasilaki *et al.*, 2019). However, there are high costs and complexities related to long-term, online monitoring of N_2O in wastewater treatment processes.

Two additional pitfalls of the conventional methods of managing and analysing N_2O data were identified (Vasilaki *et al.*, 2018). First, simple descriptive statistics, and univariate and bi-variate graphical representations are incapable of succinctly explaining the combined role of operational conditions of N_2O emissions. The second weakness of long-term multivariate timeseries is overcrowding and clutter of information that could lead to obscuring important events and short-term dependencies. The clutter limits knowledge extraction from the wastewater sensor signals. One practical approach to overcome such pitfalls is to deploy structured approaches that use readily available wastewater data (i.e., from sensors and actuators) that collect information about the pattern of N_2O emissions. Combined with advanced visualization and dimensionality reduction techniques, the interpretation of the long-term N_2O emissions could become more accurate and achievable.

Extraction of the information hidden in the raw sensor signals can facilitate the identification of patterns and hidden structures and reveal significant information on the behaviour of N_2O emissions. However, past reviews on knowledge discovery and data-mining techniques in the wastewater sector have shown that WWTP data are conventionally underutilized without being translated into actional information to provide feedback to decision making (Corominas *et al.*, 2018; Newhart *et al.*, 2019; Olsson, 2012).

Specifically related to GHG, Vasilaki *et al.* (2020a) applied a knowledge discovery framework to valorize information hidden in sensor and laboratory data from a wastewater treatment process and explain the long-term N_2O emissions dynamics and triggering operational conditions. The latter is important for the wastewater sector to showcase the evolution of WWTPs towards Industry 4.0, where 'smartification' of processes is being demonstrated through smart sensing, data analytics and responsive monitoring and control. The framework couples wastewater treatment domain knowledge with data mining techniques, to extract useful information from sensor data and laboratory analyses, with the goal of maximizing insights into the long-term carbon footprint dynamics and support carbon footprint minimization. Abnormal events detection, structural changepoint detection, clustering, classification and regression algorithms have been used to translate data into actionable information, link N_2O emissions ranges with specific operational conditions, predict the range of emissions based on operational and environmental conditions and provide feedback to monitoring campaigns for the minimization of sampling requirements.

Figure 10.5 shows methodological steps that can be implemented for knowledge discovery based on Vasilaki *et al.* (2020b).

Vasilaki *et al.* (2018) used multivariate statistical techniques to extract information from the long-term N_2O monitoring campaign of a full-scale Carrousel reactor (duration over one year) that exhibited strong temporal variability of N_2O emissions. The analysis showed that data mining techniques, including principal component analysis (PCA), can be used to assist operators to detect, understand and visualize the temporal behaviour and characteristics of the operational and environmental variables monitored online and their impact on measured N_2O formation. Additionally, the segmentation of the system in different time periods, based on differences in the behaviour of N_2O emissions, enabled the detection of strong and varying local dependencies with the operational variables that were not visible when

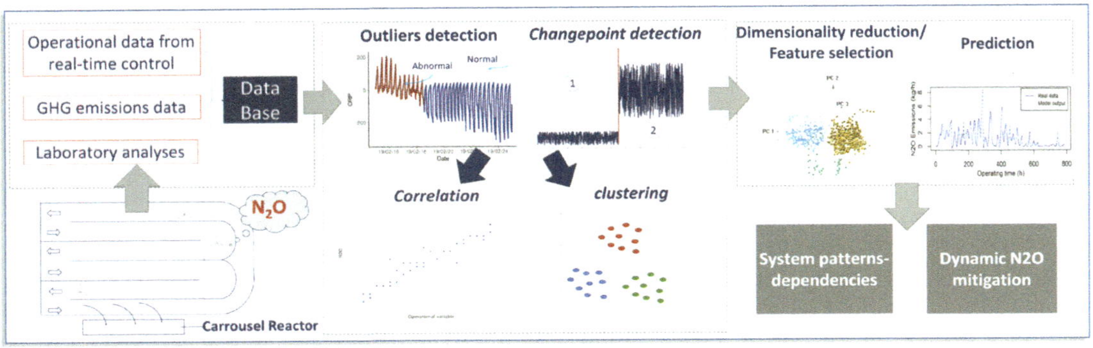

Figure 10.5 Knowledge discovery framework.

the complete time series were considered. The investigation of dependencies between N_2O emissions and operational variables in biological processes needs to account for (i) the temporal variability of operational and environmental conditions that results in changes of the N_2O triggering mechanisms, (ii) system disturbances (e.g., extreme precipitation events) that can influence, in the short-term (i.e., 1 day) or longer periods (i.e., one week), both the system performance and the N_2O generation, and (iii) the combined effect of the operational variables on N_2O emissions (Vasilaki *et al.*, 2018).

Bellandi *et al.* (2020) used a similar approach to Vasilaki *et al.* (2018) and also found that PCA could isolate the main known relations between operational variables and N_2O formation.

A methodology coupling changepoint detection of operational variables and data-driven modelling, to minimize N_2O sampling requirements for the reliable quantification of annual N_2O EFs, and ultimately to predict the range of N_2O emissions, has been also demonstrated in the same system, as shown in Figure 10.6 (Vasilaki *et al.*, 2020b). The authors have provided a practical approach that can

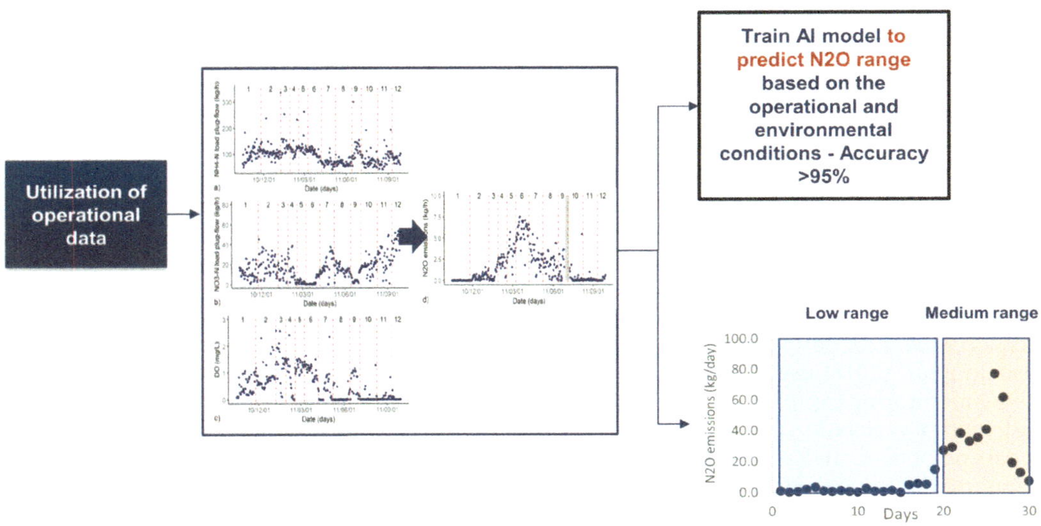

Figure 10.6 Changepoint detection of operational variables to identify changes in the behaviour of N_2O and prediction of N_2O emissions.

help utilities to quantify, understand and report the N_2O EF and decide when the sampling campaigns need to take place. The detection of changepoints for operational variables conventionally monitored in WWTPs (i.e., mean, standard deviation of hydraulic retention time (HRT), solids retention time (SRT)…) can guide N_2O sampling campaigns in terms of sampling requirements. N_2O emission ranges are expected to be similar under specific operational and environmental conditions. This means, that monitoring N_2O for approximately three days between the different segments in conditions, is sufficient to get a representative EF for the whole segment. The AI-guided sampling strategy resulted in the most reliable annual N_2O EF compared to the other method not only in terms of emission accuracy, but also in terms of the probability of underestimating emissions.

Support vector machine (SVM) classification models can predict the state of the system (based on the segments of time periods identified by the changepoint detection method). Since the different segments were characterized by relatively stationary N_2O fluxes, the SVM predicted classes provided a good approximation of the expected range of N_2O emission loads. Because operational states have been identified and linked with the specific ranges of N_2O emissions, SVM can be used to predict the N_2O emission ranges (low, medium, high) of different facilities with similar operational states. The analysis of historical data and investigation of seasonal effects can be of paramount importance in the planning of monitoring campaigns sampling frequency and duration. The approach demonstrated in the work of Vasilaki *et al.* (2020b) can be applied when long-term online sampling is not feasible (due to budget or equipment limitations) to identify N_2O emissions 'hotspot' periods and guide towards the identification of the operational periods requiring extensive investigation of N_2O pathways and mitigation measures.

The majority of the studies on N_2O generation in full-scale sidestream technologies have focused on short-term monitoring campaigns and investigation of the effect of different control strategies on N_2O emissions (i.e., Castro-Barros *et al.*, 2015; Kampschreur *et al.*, 2008, 2009; Mampaey *et al.*, 2016). Similarly, lab-scale studies have mainly focused on investigating the effect of specific parameters (i.e., NO^{2-} accumulation, pH, DO, COD/N etc.) on N_2O generation in a controlled laboratory environment (i.e., Alinsafi *et al.*, 2008; Itokawa *et al.*, 2001; Su *et al.*, 2019; Tallec *et al.*, 2006; Zhou *et al.*, 2008). A different approach is presented by Vasilaki *et al.* (2020a), investigating the fluctuations of dissolved N_2O concentration from the monitoring of a full-scale sidestream SBR (sequencing batch reactor), while the control was kept constant. As shown in Figure 10.7, SVM classification and support vector regression (SVR) models were trained to predict the real dissolved N_2O behaviour and concentration during the different phases of SBR operation. During aerobic phases, elevated average dissolved N_2O concentration was linked with DO less than 1 mg/L and increased conductivity decrease rates (conductivity values represent NH_4-N concentration values in the reactor). Therefore, cycles with increased conductivity decrease rate indicate higher NH_4-N removal efficiency and NO_2^--N accumulation.

Based on the findings of Vasilaki *et al.* (2020a), increasing the reactor DO concentration to values higher than 1.3 mg/L can result in decreased aerobic N_2O generation. However, with the current anaerobic supernatant feeding strategy, blowers operate at maximum flowrate, so it is not possible to increase the aeration in the system. On the other hand, the implementation of a step-feeding strategy could foster the reduction of N_2O emissions thanks to the lower NH_4-N and free ammonia (FA) concentration at the beginning of the cycle, which has been recognized as a triggering factor for N_2O production (Desloover *et al.*, 2012). Conductivity at the end of the cycle can act as surrogate to estimate the effluent NH_4-N concentration of the reactor and optimize the anaerobic supernatant feeding load. Consequently, the aerobic initial NH_4-N concentration could be controlled to avoid either FA accumulation or high ammonia oxidation rate (AOR) with subsequent N_2O generation. Additionally, frequent alternation of aerobic/anoxic phases can be introduced in order to avoid high nitrite accumulation. Vasilaki *et al.* (2020a) demonstrated that low-cost sensors, conventionally used to monitor SBR systems (i.e., pH, DO, oxidation reduction potential (ORP)) can be useful for predicting the dissolved N_2O concentration; therefore, the models developed in the study can be used to rapidly

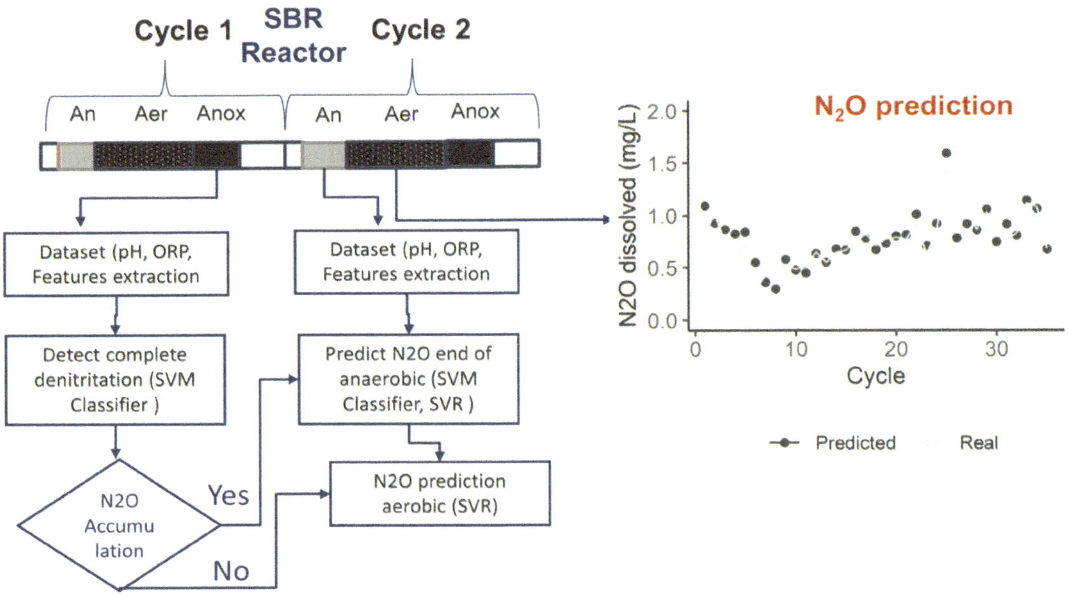

Figure 10.7 SVM classification and regression models to predict N_2O in a sidestream sequencing batch reactor (SBR).

and precisely estimate the hard-to-measure N_2O concentrations. These findings, together with the models developed, can be the basis for the development of intelligent control algorithms to integrate emissions control in sidestream SBR reactors performing nitritation/partial nitritation.

Historical data contain great value for understanding specific plant performances and detecting unwanted behaviours. Data mining can help understand and isolate plant-specific behaviours while clustering simplifies the data mining output into a useful input for a plant control logic. In the studies mentioned, the application of data-driven methods, SVM, PCA, and clustering, have demonstrated the opportunity for incorporating these techniques into an advanced plant-wide control or warning system to avoid implementing operational settings leading to N_2O production and emissions. Vasilaki *et al.* (2020a), for example, showed how a data-driven model can be trained to link the commonly online-measured state variables of a wastewater treatment plant to the state of N_2O as an alternative to describing the mechanistic pathways, which is not straightforward and is complex to understand. The model can be used as a soft sensor and embedded in control strategies for multi-criteria optimization of energy expenditure and carbon footprint.

Table 10.1 summarizes studies that have applied data-driven methods for linking operational conditions to N_2O emissions and for identifying opportunities for monitoring and control of N_2O emissions in full-scale systems. The specific methods implemented in the studies are also provided in Table 10.1.

10.4 CONCLUSIONS AND PERSPECTIVES

Data-driven models can reliably estimate emissions behaviour in wastewater processes under given operational conditions. The introduction of knowledge-based AI and data-driven models allows water professionals to understand and improve the performance and environmental profile of their operations. It creates knowledge and understanding that can facilitate the introduction of better regulations and enables regulators to better set targets and requirements.

Table 10.1 Studies that have applied data-driven models for monitoring and control of N_2O emissions and target mitigation measures in full-scale systems.

Study – system	Methods	Main findings	Mitigation measures – actionable insights
Survey of 12 WWTPs (Ahn *et al.*, 2010)	• Multivariate regression modelling	• Positive correlation with N_2O emissions from aerobic zones, were $NH4^+$-N, NO_2^--N, and DO concentrations (isolated effect), and $NH4^+$-N, NO_2^--N concentrations (interactive effect)	• Correlations between N_2O emissions and $NH4^+$-N, NO_2^--N, and DO consistent with known pathways/mechanistic behaviour • $NH4^+$-N, NO_2^--N, and DO concentrations are inherently linked with process parameters such as TKN loadings, SRT and wastewater composition; therefore, the potential for N_2O emissions from any given WWTP site can only be evaluated within the joint framework of its process operation and performance characteristics.
Carrousel reactor (Vasilaki *et al.* 2018)	• Spearman's rank correlation (Spearman, 1904) • Binary segmentation – changepoint detection (Scott and Knott, 1974) • Hierarchical k-means clustering (Arai and Barakbah, 2007) • PCA (Jolliffe, 2002)	• Low correlation coefficients can indicate non-monotonic interrelationships that Spearman's rank correlation cannot identify • Fluctuating dependencies of the N_2O emissions with the operational variables • N_2O emission peaks were linked with the diurnal behaviour of the nutrients' concentrations and with rain weather events, whereas low nitrate concentrations in the preceding plug flow reactor (<1 mg/l) resulted in increased loadings in the subsequent Carrousel and high N_2O emissions.	• Insights into the combined operational conditions linked with specific ranges of N_2O emissions. • Clusters can guide towards the identification of possible N_2O triggering mechanisms. For instance, clusters characterized with high dissolved oxygen (DO) and peaks in nitrite and nitrate concentrations indicated insufficient denitrification zones in the reactor. • The methodology enables the identification ranges of operating variables that have historically resulted in low or high ranges of N_2O emissions.
Carrousel reactor (Vasilaki *et al.* 2020a)	• E-divisive – changepoint detection (James and Matteson, 2013) • SVM (Cortes and Vapnik, 1995)	• Changepoints of the operational variables coincide with the changes of the N_2O emissions range and behaviour. Abrupt decreases of the N_2O emission profile were linked with drops in the ammonium load, increase in the nitrate-nitrogen load of the plug-flow reactor and increase in the DO concentration of the Carrousel reactor • Limited 24-h N_2O samples between the changepoint intervals are sufficient to estimate the average N_2O emission factor (EF) for the whole year, while conventional strategies resulted in lower accuracy of the N_2O EF.	• A data-driven N_2O sampling approach is proposed to reduce sampling requirements for the quantification of annual EFs at WWTPs • A classification model based on SVMs has been developed, tested and validated using data from variables monitored online to predict the range of N_2O emission loads (i.e. low, medium, high). The SVM classifier can be used to detect periods with operational behaviour that has been historically linked with elevated emissions. The development of mitigation measures in the predicted high-risk N_2O emission periods, can be supported with the integration of mechanistic models or practical, simplified theoretical approaches.

(*Continued*)

Table 10.1 Studies that have applied data-driven models for monitoring and control of N_2O emissions and target mitigation measures in full-scale systems (*Continued*).

Study – system	Methods	Main findings	Mitigation measures – actionable insights
Carrousel reactor (Bellandi *et al.* 2020)	• PCAs (Johnson and Wichern, 1992)Clustering: k-means (Han and Kamber, 2001); agglomerative (Murtagh and Legendre, 2014); HDBSCAN (Melvin *et al.* 2016)	• PCA could isolate the key known relations between N_2O production and plant operation. • AOB denitrification and incomplete NH_2OH oxidation were clearly identified by PCA.	• If system shifts to the cluster responsible for N_2O production due to incomplete NH_2OH oxidation, this can be used to suggest the option of reducing the DO. • If system is shifting towards the cluster responsible for N_2O production due to AOB denitrification, operators can be prompted to evaluate the possibility of increasing the DO concentration.
SCENA (Vasilaki *et al.* 2020b)	• Outlier detection – DBSCAN, (Ester *et al.* 1996) • SVM (Cortes and Vapnik, 1995) • SVR (Cortes and Vapnik, 1995)	• Range of emissions: 7.6% of the NH_4-N load (1.3–19% of NH_4-N load) • Abnormal cycles linked with limitations in the anaerobic supernatant or fermentation liquid. • The accumulation of N_2O at the end of the SBR anoxic phase was stripped in the subsequent aerobic phase and had a significant impact on the amount of N_2O emitted. • DO and conductivity linked with emissions • After the integration of SCENA, the EF of the WWTP slightly increased; mitigation measures have been proposed to minimize the risk of elevated emissions. • Energy consumption variation mainly due to the duration of the aerobic phase; high installed power of the blower and the actual electricity consumption of the process is expected to be lower	• Data mining and knowledge discovery framework applied; a model developed, tested and validated that can help operators control and mitigate emissions • Increase DO >1.3 mg/L • Step feeding and use of conductivity as a surrogate to estimate the effluent NH_4-N concentration of the reactor and optimize the anaerobic supernatant feeding load (avoid either FA accumulation or high AOR that trigger N_2O) • Frequent alternation of aerobic/anoxic phases to avoid nitrite accumulation • The developed model can be used to estimate rapidly and precisely the hard-to-measure N_2O concentrations during aeration and detect N_2O accumulation in non-aerated phase (help the control of anoxic duration and avoid N_2O accumulation).

HDBSCAN: Hierarchical Density-Based Spatial Clustering of Applications with Noise; DBSCAN: Density-Based Spatial Clustering of Applications with Noise; SCENA: Short-Cut Enhanced Nutrient Abatement.

Further research is required for the investigation of the optimal sensor location and the optimal combination of monitored variables for N_2O emissions control for different wastewater configurations. Wastewater processes are characterized by non-stationarity, high dynamics and variations at different scales in time. Therefore, the development of novel methods and standardized frameworks that inherently consider these features while still being practical is necessary. Future research can also explore the possibility of coupling sophisticated statistical tools (e.g., multivariate statistics) and ML algorithms with multiple-pathway mechanistic models for full-scale applications, to facilitate the fast and adaptable online implementation of model-predictive control and forecasting decision support tools.

For instance, ML models trained with outputs of mechanistic models can enhance the generalization capabilities of these models. On the other hand, computationally universal mechanistic models can be used to simulate variables not conventionally monitored in WWTPs (i.e., NO_2^-, NO). Simulations of key variables coupled with the raw sensor signals can be used in the knowledge discovery process to enhance the reliability of the findings and improve the generalization capabilities of the data-driven models. Machine learning could be a powerful instrument for better calibrating the parameters of N_2O emissions models. In the case of hybrid biokinetic/CFD modelling, a trained ML model predicting N_2O concentrations/emissions can be used in lieu of the N_2O biokinetic model, by using the state variables of the non-N_2O biokinetic/CFD model as inputs to the ML model to predict N_2O concentrations/emissions for each cell in the CFD geometry mesh.

Multivariate statistics and pattern recognition algorithms can be applied to the online monitored variables in WWTPs. They differentiate operational conditions and guide towards different calibration tactics for the same process. Finally, multivariate statistics can be applied to identify and isolate complex relationships between system variables allowing better representative models. Such integrated, practical tools can help plant operators to design effective mitigation strategies. Integrating the knowledge-based AI into these tools can also help validate the strategies identified by the data-driven techniques.

Additionally, several aspects need to be considered before the integration of AI and data mining techniques into the data management practices of water utilities. Water utilities have been dominated by traditional operations focusing on long-term investments and continuity. Historically, water utilities have separate departments doing separate jobs. Data analysis and algorithmic calculations on all data of all departments are not performed; standardized approaches are missing. The techniques presented in Table 10.1 need to be advanced to practical tools and interfaces that can provide the desired information in a simple and intuitive way. For instance, segmentation, clustering, classification and regression techniques integrated into supervisory control and data acquisition (SCADA) systems can be used to benchmark and predict key performance indicators (KPIs) (performance, cost, environmental aspects) under different operational conditions (i.e., based on seasonal influent composition variations, different process rates affected by environmental conditions, system shocks etc.). User-friendly dashboards, combining methods (both knowledge-based and data-driven) and communicating the results in a simple and informative way will help operators to detect operational modes in which the system is underperforming, analyse and visualize risks and prioritize optimization needs providing a platform for continuous internal multivariate benchmarking of WWTP performance. In practice, we are already seeing ML and the N_2O Risk Model being coupled for better accounting of N_2O, where N_2O has not been measured, and prioritizing sites for monitoring and mitigation (Porro *et al.*, 2021).

Mechanistical models as we have seen in the previous chapters can be powerful tools. However, there are equally powerful tools applying knowledge-based and data-driven techniques that can valorize data that are already available and can provide additional insights that mechanistic models alone cannot provide. There are clear benefits in applying the techniques separately, but even greater benefits can be gained by combining knowledge-based and data-driven techniques, as well as combining knowledge-based and/or data-driven techniques with mechanistic models, and, if needed, even greater insight like hydrodynamic impacts on N_2O emissions, with CFD. Although the methods detailed in this chapter were applied to assessment of N_2O emissions, the same approaches can be applied when looking at methane emissions from sewers and WWTPs.

ACKNOWLEDGEMENTS

The authors (J. Porro and G. Bellandi) would like to acknowledge Waterboard De Dommel for their support through multiple measurements campaigns. J. Porro would also like to acknowledge Waterboard De Dommel's support throughout the development of the N_2O Risk Model. The research work of J. Porro on the N_2O Risk Model was financed by People Program (Marie Curie Actions) of the European Union's Seventh Framework Programme FP7/2007–2013, 579 under REA agreement 289193 (SANITAS). This research of E. Katsou and V. Vasilaki was supported by the Horizon 2020 research and innovation program SMART-Plant (grant agreement No 690323).

REFERENCES

Ahn J. H., Kim S., Park H., Rahm B., Pagilla K. and Chandran K. (2010). N_2O emissions from activated sludge processes, 2008–2009: results of a national monitoring survey in the United States. *Environmental Science & Technology*, **44**, 4505–4511, https://doi.org/10.1021/es903845y

Alinsafi A., Adouani N., Béline F., Lendormi T., Limousy L. and Sire O. (2008). Nitrite effect on nitrous oxide emission from denitrifying activated sludge. *Process Biochemistry*, **43**, 683–689, https://doi.org/10.1016/j.procbio.2008.02.008

Arai K. and Barakbah A. R. (2007). Hierarchical K-means: an algorithm for centroids initialization for K-means. *Rep. Fac. Sci. Eng.*, **36**, 25–31.

Bellandi G., Porro J., Senesi E., Caretti C., Caffaz S., Weijers S., Nopens I. and Riccardo G. (2017). Multi-point monitoring of nitrous oxide emissions in three full-scale conventional activated sludge tanks in Europe. *Water Science and Technology*, **77**, 880–890, https://doi.org/10.2166/wst.2017.560

Bellandi G., Weijers S., Gori R. and Nopens I. (2020). Towards an online mitigation strategy for N_2O emissions through principal components analysis and clustering techniques. *Journal of Environmental Management*, **261**(110219), 0301–4797, https://doi.org/10.1016/j.jenvman.2020.110219

Cadwallader A. and VanBriesen J. M. (2017). Incorporating uncertainty into future estimates of nitrous oxide emissions from wastewater treatment. *Journal of Environmental Engineering*, **143**, 04017029, https://doi.org/10.1061/(ASCE)EE.1943-7870.0001231

Castro-Barros C. M., Daelman M. R. J., Mampaey K. E., van Loosdrecht M. C. M. and Volcke E. I. P. (2015). Effect of aeration regime on N2O emission from partial nitritation-anammox in a full-scale granular sludge reactor. *Water Research*, **68**, 793–803, https://doi.org/10.1016/j.watres.2014.10.056

Colliver B. B. and Stephenson T. (2000). Production of nitrogen oxide and dinitrogen oxide by autotrophic nitrifiers. *Biotechnology Advances*, **18**, 219–232, https://doi.org/10.1016/S0734-9750(00)00035-5

Comas J., Rodríguez-Roda I., Sànchez-Marrè M., Cortés U., Freixó A., Arráez J. and Poch M. (2003). A knowledge-based approach to the deflocculation problem: integrating on-line, off-line, and heuristic information. *Water Research*, **37**, 2377–2387, https://doi.org/10.1016/S0043-1354(03)00018-6

Corominas L., Garrido-Baserba M., Villez K., Olsson G., Cortés U. and Poch M. 2018. Transforming data into knowledge for improved wastewater treatment operation: A critical review of techniques. *Environmental Modelling & Software*, **106**, 89–103, https://doi.org/10.1016/j.envsoft.2017.11.023

Cortes C. and Vapnik V. 1995. Support-vector networks. *Machine Learning*, **20**, 273–297, https://doi.org/10.1007/BF00994018

Desloover J., Vlaeminck S. E., Clauwaert P., Verstraete W. and Boon N. (2012). Strategies to mitigate N_2O emissions from biological nitrogen removal systems. *Current Opinion in Biotechnology*, **23**, 474–482, https://doi.org/10.1016/j.copbio.2011.12.030

Domingo-Félez C. and Smets B. F. (2016). A consilience model to describe N_2O production during biological N removal. *Environmental Science: Water Research and Technology*, **2**, 923–930, https://doi.org/10.1039/C6EW00179C

Domingo-Félez C., Calderó-Pascual M., Plósz B. G., Smets B. F. and Sin G. (2017). Calibration of the comprehensive NDHA-N2O dynamics model for nitrifier-enriched biomass using targeted respirometric assays. *Water Research*, **126**, 29–39, https://doi.org/10.1016/j.watres.2017.09.013

Ester M., Kriegel H.-P., Sander J. and Xu X. (1996). A density-based algorithm for discovering clusters in large spatial databases with noise. Kdd. pp. 226–231.

Foley J., de Haas D., Yuan Z., Lant P. (2010). Nitrous Oxide Generation in Full-Scale Biological Nutrient Removal Wastewater treatment Plants. *Water Res.*, **44**, 831– 844.

Global Water Research Coalition (2011). N_2O and CH_4 Emission from Wastewater Collection and Treatment Systems State of the Science Report: Report of the GWRC Research Strategy Workshop. Montreal, Canada, September 2010, Report 29.

Gruber W., Villez K., Kipf M., Wunderlin P., Siegrist H., Vogt L. and Joss A. (2019). N_2O emission in full-scale wastewater treatment: proposing a refined monitoring strategy. *Science of the Total Environment*, **699**, 134157, https://doi.org/10.1016/j.scitotenv.2019.134157

Han J., Kamber M. and Pei J. (2011). Data Mining: Concepts and Techniques. Elsevier Science, Netherlands.

Itokawa H., Hanaki K. and Matsuo T. (2001). Nitrous oxide production in high-loading biological nitrogen removal process under low cod/n ratio condition. *Water Research*, **35**, 657–664, https://doi.org/10.1016/S0043-1354(00)00309-2

James N. A. and Matteson D. S. (2013). ecp: An R Package for Nonparametric Multiple Change Point Analysis of Multivariate Data. ArXiv13093295 Stat.

Johnson R. A. and Wichern D. W. (1992). Applied Multivariate Statistical Analysis. 5th Edition. Prentice Hall, Inc., New York.

Jolliffe I. T. (ed.) (2002). Principal Component Analysis. Springer Series in Statistics. Springer New York, New York, NY, pp. 167–198, https://doi.org/10.1007/0-387-22440-8_8

Kampschreur M. J., van der Star W. R. L., Wielders H. A., Mulder J. W., Jetten M. S. M. and van Loosdrecht M. C. M. (2008). Dynamics of nitric oxide and nitrous oxide emission during full-scale reject water treatment. *Water Research*, **42**, 812–826, https://doi.org/10.1016/j.watres.2007.08.022

Kampschreur M. J., Poldermans R., Kleerebezem R., van der Star W. R. L., Haarhuis R., Abma W. R., Jetten M. S. M., Jetten M. S. M. and van Loosdrecht M. C. M. (2009). Emission of nitrous oxide and nitric oxide from a full-scale single-stage nitritation-anammox reactor. *Water Science and Technology*, **60**, 3211–3217, https://doi.org/10.2166/wst.2009.608

Mampaey K. E., De Kreuk M. K., van Dongen U. G. J. M., van Loosdrecht M. C. M. and Volcke E. I. P. (2016). Identifying N_2O formation and emissions from a full-scale partial nitritation reactor. *Water Research*, **88**, 575–585, https://doi.org/10.1016/j.watres.2015.10.047

Melvin R. L., Godwin R. C., Xiao J., Thompson W. G., Berenhaut K. S. and Salsbury F. R. (2016). Uncovering large-scale conformational change in molecular dynamics without prior knowledge. *Journal of Chemical Theory and Computation*, **12**, 6130–6146, https://doi.org/10.1021/acs.jctc.6b00757

Murtagh F. and Legendre P. (2014). Ward's hierarchical agglomerative clustering method: which algorithms implement Ward's criterion? *Journal of Classification*, **31**, 274–295, https://doi.org/10.1007/s00357-014-9161-z

Newhart K. B., Holloway R. W., Hering A. S. and Cath T. Y. (2019). Data-driven performance analyses of wastewater treatment plants: a review. *Water Research*, **157**, 498–513, https://doi.org/10.1016/j.watres.2019.03.030

Olsson G. (2012). ICA and me – a subjective review. *Water Research*, **46**, 1585–1624, https://doi.org/10.1016/j.watres.2011.12.054

Porro J., Milleri C., Comas J., Rodriguez-Roda I., Pijuan M., Corominas L., Guo L. S., Daelman M. R. J., Volcke E. I. P., van Loosdrecht M. C. M., Vanrolleghem P. A. and Nopens I. (2014). Risk assessment modelling of nitrous oxide in activated sludge systems: quality not quantity. 4th IWA/WEF Wastewater Treatment Modelling Seminar, (WWTmod 2014), Spa, Belgium.

Porro J., Bellandi G., Rodriguez-Roda I., Deeke A., Weijers S., Vanrolleghem P., Comas J. and Nopens I. (2017). Developing an artificial intelligence-based WRRF nitrous oxide mitigation road map: The Eindhoven N_2O mitigation case study. Proceedings, 90th Annual Conference and Exposition of the Water Environment Federation (WEFTEC 2017), 1–4 October 2017, Chicago IL.

Porro J., Rehman U., Flameling T., Bellandi G., Deeke A., Audenaert W., Weijers S. and Nopens I. (2019). An integrated CFD/Biokinetic/N_2O risk modelling approach for mitigating N_2O emissions and optimal WWTP design. 10th IWA Symposium on Modelling and Integrated Assessment, (Watermatex 2019), Copenhagen, Denmark.

Porro J., Tessier M., Deeke A. and Weijers S. (2021). Using artificial intelligence for water utility-level nitrous oxide reduction and net zero emissions targets. 5th IWA Specialized International Conference of Ecotechnologies for Wastewater Treatment (EcoSTP 2021), Milan, Italy.

Qiu Y., Griffin C. T., Ekström S. E. M., Smets B. F., Valverde Pérez B., Climent J. and Plósz B. G. (2019). Numerical modelling of N_2O emission from surface aerated oxidation ditch activated sludge reactors. 10th IWA Symposium on Modelling and Integrated Assessment, (Watermatex 2019), Copenhagen, Denmark.

Rehman U., Audenaert W., Amerlinck Y., Maere T., Arnaldos Orts M. and Nopens I. (2017). How well-mixed is well mixed? : Hydrodynamic-biokinetic model integration in an aerated tank of a full-scale water resource recovery facility. *Water Science and Technology*, **76**(8), 1950–1965, https://doi.org/10.2166/wst.2017.330

Rodríguez-Roda I., Comas J., Colprim J., Poch M., Sànchez-Marrè M., Cortés U., Baesa J. and Lafuente J. (2002). A hybrid supervisory system to support wastewater treatment plant operation: implementation and validation. *Water Science and Technology*, **45**, 289–297, https://doi.org/10.2166/wst.2002.0608

Sànchez-Marrè M., Cortés U., Lafuente J., Rodriguez-Roda I. and Poch M. (1996). DAI-DEPUR: an integrated and distributed architecture for wastewater treatment plants supervision. *Artificial Intelligence in Engineering*, **10**, 275–285, https://doi.org/10.1016/0954-1810(96)00004-0

Schaubroeck T., De Clippeleir H., Weissenbacher N., Dewulf J., Boeckx P., Vlaeminck S. E. and Wett B. (2015). Environmental sustainability of an energy self-sufficient sewage treatment plant: improvements through DEMON and co-digestion. *Water Research*, **74**, 166–179, https://doi.org/10.1016/j.watres.2015.02.013

Scott A. J. and Knott M. (1974). A cluster analysis method for grouping means in the analysis of variance. *Biometrics*, **30**, 507–512, https://doi.org/10.2307/2529204

Spearman C. (1904). 'General intelligence', objectively determined and measured. *American Journal of Psychology*, **15**, 201–292, https://doi.org/10.2307/1412107

Su Q., Domingo-Félez C., Zhang Z., Blum J.-M., Jensen M. M. and Smets B. F. (2019). The effect of pH on N_2O production in intermittently-fed nitritation reactors. *Water Research*, **156**, 223–231, https://doi.org/10.1016/j.watres.2019.03.015

Tallec G., Garnier J., Billen G. and Gousailles M. (2006). Nitrous oxide emissions from secondary activated sludge in nitrifying conditions of urban wastewater treatment plants: effect of oxygenation level. *Water Research*, **40**, 2972–2980, https://doi.org/10.1016/j.watres.2006.05.037

Vasilaki V., Volcke E. I. P., Nandi A. K., Loosdrecht M. C. M. and Katsou E. (2018). Relating N_2O emissions during biological nitrogen removal with operating conditions using multivariate statistical techniques. *Water Research*, **140**, 387–402, https://doi.org/10.1016/j.watres.2018.04.052

Vasilaki V., Massara T. M., Stanchev P., Fatone F. and Katsou E. (2019). A decade of nitrous oxide (N_2O) monitoring in full-scale wastewater treatment processes: a critical review. *Water Research*, **161**, 392–412, https://doi.org/10.1016/j.watres.2019.04.022

Vasilaki V., Conca V., Frison N., Eusebi A. L., Fatone F. and Katsou E. (2020a). A knowledge discovery framework to predict the N2O emissions in the wastewater sector. *Water Research*, **178**, 115799, https://doi.org/10.1016/j.watres.2020.115799

Vasilaki V., Danishvar S., Mousavi A. and Katsou E. (2020b). Data-driven versus conventional N_2O EF quantification methods in wastewater; how can we quantify reliable annual EFs? *Computers & Chemical Engineering*, **141**, 106997, https://doi.org/10.1016/j.compchemeng.2020.106997

Zadeh L. A. (1965). Fuzzy Sets. *Information and Control*, **8**, 338–353. http://dx.doi.org/10.1016/S0019-9958(65)90241-X

Zhou Y., Pijuan M., Zeng R. J. and Yuan Z. (2008). Free nitrous acid inhibition on nitrous oxide reduction by a denitrifying-enhanced biological phosphorus removal sludge. *Environmental Science & Technology*, **42**, 8260–8265, https://doi.org/10.1021/es800650j

NOMENCLATURE

AI	Artificial intelligence
AOB	Ammonia oxidizing bacteria
DO	Dissolved oxygen
EF	Emission factor
ML	Machine learning
PCA	Principal component analysis
SVM	Support vector machines

doi: 10.2166/9781789060461_245

Chapter 11

Perspectives on fugitive GHGs reduction from urban wastewater systems

Ingmar Nopens[1,2], *Jose Porro*[3] *and Liu Ye*[4]

[1]*Biomath, Ghent University, Coupure Links 653, 9000 Gent, Belgium, ingmar.nopens@ugent.be*
[2]*Capture, Centre for Advanced Process Technology for Urban REsource Recovery, Frieda Saeysstraat 1, 9052 Gent, Belgium, ingmar.nopens@ugent.be*
[3]*Cobalt Water Global, Inc. 77 Sands Street, Brooklyn, NY 11209, USA, jose.porro@cobaltwater-global.com*
[4]*School of Chemical Engineering, The University of Queensland, St Lucia, Brisbane, QLD 4072, Australia, l.ye@uq.edu.au*

SUMMARY

The *Perspectives* chapter provides a summary of previous chapters in terms of the state-of-the-art knowledge on urban water system greenhouse gas (GHG) emissions. The chapter also explains some issues in the GHG quantification, modelling and reporting guidelines. The knowledge gaps between the fundamental and practical implementation, as well as potential mitigation strategies are discussed in this chapter. Finally, this chapter provides some perspectives on the future direction for GHG reduction from urban wastewater systems.

Keywords: Mitigation; modelling; perspectives; state-of-the-art knowledge

TERMINOLOGY

Term	Definition
Artificial intelligence	Study of the human mechanisms, which provide the behaviour that can be considered intelligent, and emulation of these mechanisms, called cognitive tasks (reason, problem-solving, remember or learn), in a computer software.
CH_4	Methane, a potent GHG, with a global warming potential 25-fold stronger than that of carbon dioxide (CO_2).
Greenhouse gas	Gas that absorbs and emits radiant energy within the thermal infrared range.
N_2O	Nitrous oxide, a potent GHG, with a global warming potential 265-fold stronger than that of carbon dioxide (CO_2).
Soft sensor	Getting access to a cumbersome or even unmeasurable variable through a model that can predict it based on the measurement of another variable.

11.1 A SUMMARY ON THE STATE-OF-THE-ART KNOWLEDGE IN QUANTIFICATION AND MODELLING OF FUGITIVE GHG EMISSIONS FROM URBAN WASTEWATER SYSTEMS

This book intends to provide an overview with regard to the sources, production mechanisms and operational factors influencing urban wastewater system (UWS) emissions of both methane and nitrous oxide, the most important greenhouse gases being released from UWSs.

Despite extensive investigations over the last decade, both experimental and model-based, and the vast progress made, the detailed mechanisms of N_2O production from wastewater treatment systems are still not easily identified due to the system complexity, dynamics, and the several possible pathways based on operational and environmental factors. Direct emissions of methane occur through anaerobic breakdown of organics contained in sewage (either naturally in sewer systems depleted of oxygen, or artificially induced in wastewater treatment plants (WWTPs) through anaerobic digestion (AD) to reduce solids volume and capture/recover energy). Methane formation pathways and identification of emission spots in engineered UWSs are well identified.

In terms of quantification and reporting of N_2O, generic emission factors set out in the internationally accepted Intergovernmental Panel on Climate Change (IPCC) methodology, are still being widely used. However, as the consensus among the scientific community is that applying a single emission factor (e.g., IPCC Tier 1 as a global factor or Tier 2 as a country-level factor) is challenging based on the science, efforts to derive alternative and more reliable country-specific emission factors have been made through numerous national bottom-up monitoring initiatives. For CH_4, quantification and reporting using IPCC guidelines is more straightforward and reliable as there is only one pathway, and it only occurs when above a certain temperature. However, guidelines for accounting and reporting CH4 emissions from sewers are lacking entirely.

The mathematical modelling of N_2O production has reached a maturity that facilitates the estimation of site-specific N_2O emissions and the development of mitigation strategies for a WWTP taking into account the specific design and operational conditions of the plant. Given the sound understanding of pathways of CH4 production, predictive models for both WWTPs and sewers are well established.

Next to the efforts in mechanistic modelling of greenhouse gas (GHG) emissions, data-driven models have been vastly underused, not in the least due to lack of data. Apart from an effort to use principal component analysis (PCA) and clustering techniques that have been proven to be able to highlight N_2O production mechanisms among the variables that are normally measured on a full-scale water resource recovery facility (WRRF), very few efforts have been reported. Results, however, confirm the potential for defining a new monitoring system for N_2O emissions based on historical or online plant data.

Finally, benchmark modelling has long been proven a useful tool for unbiased comparison of control strategies in WWTPs in terms of effluent quality, operational cost and risk of suffering microbiology-related total suspended solids (TSS) separation problems. An extended version of the original Benchmark Simulation Model No. 2 (BSM2) aiming towards including GHG emissions is available as a tool to evaluate the impact of control actions on the above-mentioned criteria. Often, this leads to a trade-off between these. Reported results emphasize the importance of using integrated approaches when comparing and evaluating (plant-wide) control strategies in WWTPs for more informed operational decision-making.

11.2 ISSUES, KNOWLEDGE GAPS AND PERSPECTIVES ON GHG QUANTIFICATION METHODS AND THE REPORTING GUIDELINES

11.2.1 GHG quantification

Although many full-scale quantification methods have been developed and reported in the past decade, a widely implemented protocol for measuring GHG emissions from different WWTPs is still lacking (Mannina *et al.*, 2018). Current quantification methods typically only partially depict the high spatial-temporal variability of GHG emissions, which are strongly influenced by environmental and process conditions. This holds for both N_2O and CH_4. As a result, different measurement strategies could affect the outcome of GHG quantification (Daelman *et al.*, 2013; Lim & Kim, 2014). The duration of the measurement campaign (long-term, short-term), the sampling strategies (online sampling, grab samples) (Daelman *et al.*, 2013), the monitoring locations (Gruber *et al.*, 2019), and the scale of the measurement (partly covered, fully covered) (Kosonen *et al.*, 2016; Marques *et al.*, 2016) could all contribute to the GHG quantification gaps on different levels. Long-term monitoring campaigns have repeatedly shown strong seasonal variations of N_2O emissions (Daelman *et al.*, 2015; Gruber *et al.*, 2019). Consequently, short-term GHG quantification could hardly represent the N_2O emissions patterns identified from a long-term campaign (Daelman *et al.*, 2013, 2015; Massara *et al.*, 2017), at least for sites where there is a significant variation in seasonal temperatures. For quantification methods with small footprints like flux chamber or liquid-gas mass transfer estimation methods (Chandran, 2011; Foley *et al.* 2010; Ye *et al.* 2014), determining the representative monitoring locations is critical to quantifying the overall emissions. Coupling computational fluid dynamics (CFD) with knowledge-based artificial intelligence (AI) to visualize risk across a reactor, as demonstrated by Porro *et al.* (2019), can also be helpful for determining where, in a reactor, emissions can vary due to localized mixing conditions and where it is critical to sample. To counteract the variability caused by operational and environmental disturbances (non-localized) and increase the representativeness of the monitoring results, more studies based on long-term monitoring campaigns should be conducted and a refined monitoring protocol should be proposed (Gruber *et al.*, 2019; Kosonen *et al.*, 2016). Furthermore, the AI-guided monitoring approach proposed by Vasilaki *et al.* (2020) can be useful for minimizing the amount of time needed to represent the full range of emissions for a particular site.

Extensive sampling and long-term monitoring of WWTPs and sewer networks are necessary to capture the complexity of the targeted systems, thus requiring a significant input of resources on site. To make maximal use of resources, future measurement campaigns at full-scale WWTPs and in sewer systems should be well designed with regard to their objective. A guidance document on what, where and how frequently to measure based on the specific objectives of the monitoring campaign would be useful. There are various methods, each with advantages and disadvantages; however, it is the objectives that will dictate what methods should be used. The guidance can also include how to fill gaps in whatever method is being used. For example if a flux chamber or microsensor is used, guidance on which locations should be measured simultaneously to get a representative emissions characterization can be provided. Moreover, it would be good for utilities to better share their experiences, such as through participation in the International Water Association Climate Smart Utilities forum. One idea could be to populate a database system with data and meta-data of the design and results of measurement campaigns. As more GHG emissions monitoring campaigns will be launched in efforts to reach net zero, it will be critical to have consensus-based guidance to maintain consistency and maximize confidence that what is being measured is representative for each site across the water sector. It will also be critical that experience and knowledge in other off-gas measurement techniques, such as for oxygen transfer efficiency (ASCE, 1991; Rosso *et al.*, 2005) are fully leveraged.

But should utilities then keep monitoring indefinitely? Likely not, as with the transition to the digital age, soft sensors and digital twins are likely to come to the rescue. Indeed, once insights have been obtained through vast monitoring campaigns, emissions can be linked to process parameters or performance indicators that can be monitored with less effort or are already monitored in a standard way. That is the principle of a soft sensor, that is getting access to a cumbersome or even unmeasurable variable through a model that can predict it based on the measurement of another variable. Soft sensors can be mechanistic or data-driven (e.g., AI). Efforts towards the development of soft sensors should be increased as these are really the low hanging fruit.

Another gap is data collection for systems other than suspended sludge systems. Full-scale data for other types of systems are rare and deserve more attention. A final remark concerns industrial treatment plants that biologically remove nitrogen loads. It is unclear whether these are on the radar

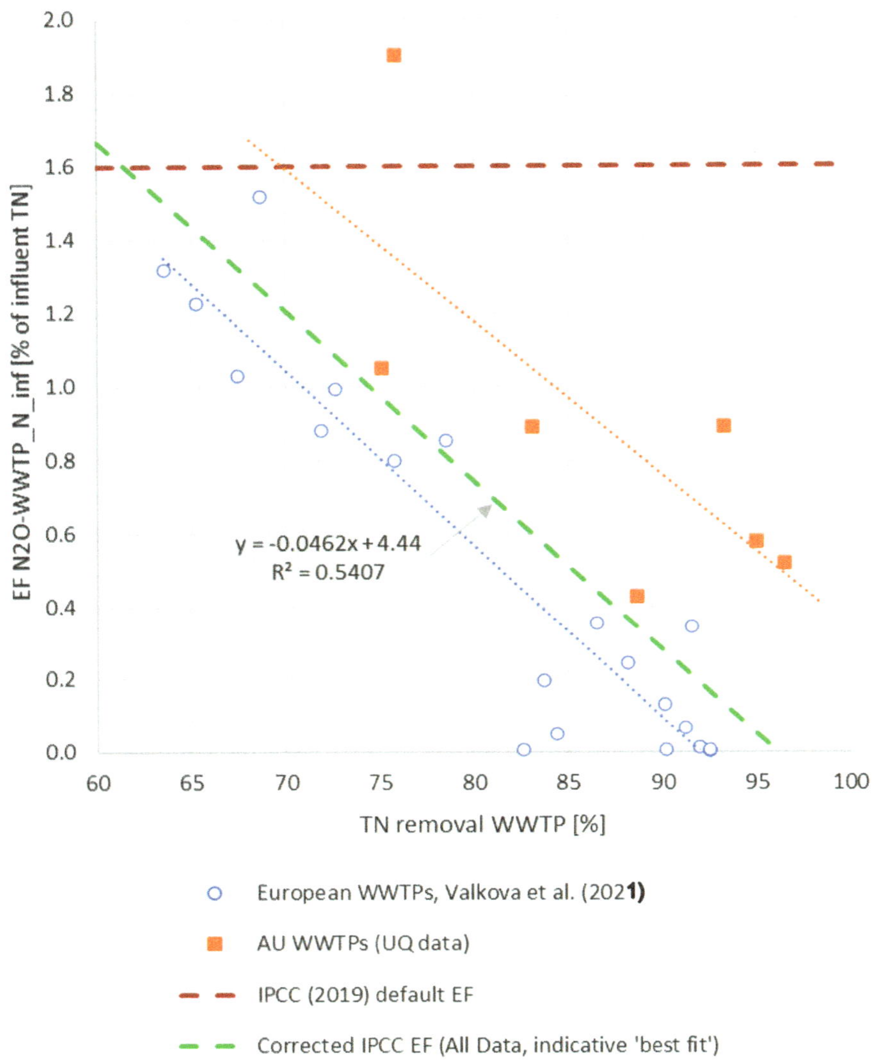

Figure 11.1 Comparison of average nitrous oxide emission factor for wastewater treatment ($EF_{N_2O\text{-}WWTP}$) from actual (measured) data showing suggested corrected EF trendline from 'best fit' of pooled actual datasets, related on an equivalent basis of influent TN load (de Haas and Ye 2021).

in terms of the UWS emissions. If not, they should be, as often these plants are even less monitored and controlled compared to their municipal counterparts. They can also be a significant source of emissions as their organic and nutrient loading can often be much higher than in municipal wastewater.

11.2.2 Reporting guidelines

One of the biggest issues of the current GHG accounting methodologies is that they provide a fixed emission factor (EF) for water utilities to report their N_2O and CH_4 emissions. In contrast, the actual

N_2O emission factor measured from different WWTPs can range from below one tenth of to five times greater than the IPCC fixed generic EF and vary between 0.001% and 12% of the incoming total nitrogen (TN) load (Valkova *et al.*, 2021). Selecting one EF within this extremely wide range for all WWTPs, as the existing protocols do, is just not scientifically sound. Eventually, the derivation of a process or treatment performance-based (de Haas & Ye, 2021) N_2O EF benchmarks would be an improvement over the existing GHG accounting guideline. However, challenges to be overcome reside in (i) differences in the N_2O generation triggered by the site-specific operational and environmental conditions; and (ii) the sensitivity of the quantified EF to differences in monitoring strategies and duration of monitoring campaigns. However, as more cases become available, more patterns will be found and these issues will become less of a challenge. In fact, a few studies have proposed that N_2O EF appears to be related to the plant's total nitrogen (TN) removal performance.

de Haas and Ye (2021) reported the measured average N_2O emissions and average TN removal data from 8 WWTPs in Australia collected by the University of Queensland (UQ) and compared results with the 10 WWTPs in Europe as reported by Valkova *et al.* (2021), and then compared the results of both studies with emission factors applied in the IPCC (2019) and the Australia National Greenhouse and Energy Reporting (NGER) (2020) protocols (shown in Figure 11.1). These datasets represent a total of 20 measurement campaigns in the period 2012–2018, including seasonal repetition. It can be clearly seen that the N_2O emission factor for wastewater treatment reduces on a percentage basis with improved total N removal based on the 'best fit' linear regression for the pooled EU and AU datasets. The linear correlation for the pooled datasets is relatively weak ($r^2 = 0.55$), as expected, given that the EU and AU datasets are undertaken with different equipment sets, durations and seasons. Further work is required to understand the underlying reasons for such dataset differences, be they process related (e.g., temperature, type and configuration of bioreactors) or methodological (e.g., around N_2O measurement campaigns, and/or TN removal calculation for the bioreactors or the WWTP as a whole). Furthermore, it is unclear whether it is the level of nitrogen removal itself that has more of an influence on N_2O emissions or how the process is being operated (at high risk or low risk of N_2O) to achieve that same removal.

The current reporting guidelines, therefore, urgently need further improvement. Data from further studies of actual emissions will help to confirm how correlation of emission factors to plant operation can be applied. It is recommended that GHG reporting protocols be updated to reflect such trends in the future. If any trends can be elucidated by future work, this may also provide useful guidance on mitigating N_2O and CH_4 emissions from WWTPs.

11.3 ISSUES, KNOWLEDGE GAPS AND PERSPECTIVES ON GHG MODELLING

Although important steps towards the development of new models which can predict N_2O emissions from WWTPs under different scenarios have been taken, there are still some challenges to overcome so they can be of more value for the sector in practice. Numerous full-scale datasets are starting to become available for suspended growth systems and although models have been successfully calibrated for lab-scale systems, they often require extensive calibration efforts to be able to predict full-scale emissions. This can be several reasons for this: kinetics are not completely understood, pathways are missing, certain environmental effects (T, pH, etc.) are not well understood or complex mixing phenomena are not captured in the simplified tanks-in-series models. All of these avenues deserve further investigation. With respect to the last one, Porro *et al.* (2019) have illustrated this by means of the coupling of an integrated CFD model and biokinetic activated sludge model extended with the knowledge-based AI to predict risk of N_2O production/emissions. This study showed that N_2O risk heavily depends on local conditions, which can be very different in large reactors as they are not completely mixed. If using a mechanistic model and looking at the N_2O related kinetic expressions, these are indeed full of half saturation indices, which are typically the ones that need calibration. Arnaldos *et al.* (2015) illustrated that this actually hints in the direction of poor description of advection. That would also explain why the models work at small volume scale and tend to fail for larger volume reactors. Next to using CFD, compartmental models should also be explored in conjunction with N_2O models. As mechanistic modelling can help to further unravel the actual mechanisms, there are other type of models that can be of use and that have not been explored extensively in this context: data-driven models. The reason for the lesser attention is likely the lack of data up to now. But as more data are becoming available, models in the AI realm can be further explored. We already touched on soft sensors which can be empirical models trained with full-scale data in search of patterns and correlations. But data-driven models can also be augmented to mechanistic models to yield so-called hybrid models. It is an interesting avenue to build models that can exploit the best of both worlds and lead to models with a better predictive power within the range of conditions they were trained for. This can be a fast lane to models with a high predictive power without having to wait until all mechanistic details have been completely revealed. Hence, it is a parallel road to the further development of mechanistic models. An already existing example in this hybrid AI realm is the N_2O Risk Model developed by Porro *et al.* (2014) which combines plant data, mechanistic model predictions and fuzzy logic to predict a risk value for N_2O emissions due to the various influencing factors and pathways, but this can also include prediction of actual N_2O emission with machine learning (ML) in combination with mechanistic model state variables as the ML model inputs.

Another avenue to investigate is to what extent data-driven models of a certain plant type, configuration or technology can be used to make predictions of N_2O for other plants. ML algorithms could be used for this and the feasibility would likely increase as more data are available to train these models. Porro *et al.* (2021) have shown this to be a promising approach as illustrated in Figure 11.2, which shows the comparison of N_2O emission estimates for a particular WWTP (Soerendonk, NL) based on the IPCC methodology and an ML model trained with data from a different WWTP, which yielded much more accurate results than using the IPCC EFs. This is mainly because the site-specific conditions were taken into account by the ML model, whereas the IPCC methodology is based on a generic emission factor.

A final remark is that most of the modelling efforts thus far have focused on suspended growth systems. The exploration of other systems (e.g., biofilms, granular sludge) has been initiated and might also be of use for describing the potential emissions coming from systems other than suspended activated sludge. It is likely that quite a bit of the gathered knowledge can be transferred, but similar problems are likely to arise for which similar avenues as described earlier can be explored.

And even further in the future are digital twins that are also capable of capturing GHG emissions. We are seeing the first examples of digital twins emerge, but focus is, for now, on system performance and cost. Once in place, the addition of GHG emission in order to make it a more complete decision support system is a logical and necessary next step.

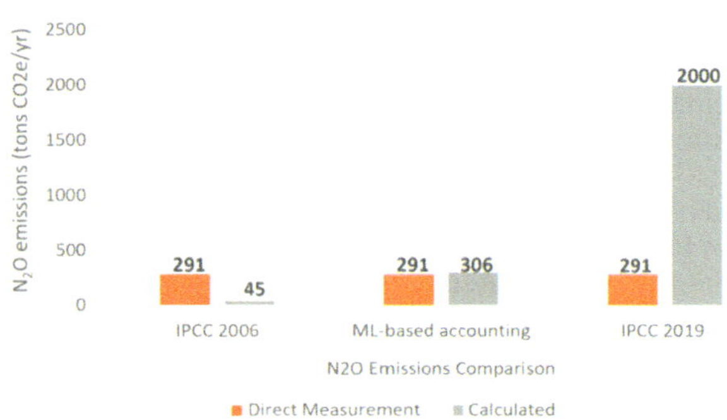

Figure 11.2 Comparison of N_2O accounting methods using IPCC EFs (2006, 2019) and a machine learning model from a different site versus direct measurements at Soerendonk WWTP (NL).

11.4 GHG MITIGATION STRATEGY AND PERSPECTIVES

11.4.1 N_2O mitigation

In many cases, N_2O emissions account for the majority of the carbon footprint of WWTPs. Research has focused more on full-scale quantification and lab-scale studies. Limited N_2O mitigation efforts have been carried out in full-scale WWTPs so far. There is an urgent need to demonstrate the feasibility of N_2O mitigation at full scale and investigate the long-term consequences of N_2O mitigation efforts.

While a large number of studies have shown the mitigation of N_2O can be feasibly achieved in WWTPs, the real implementation of N_2O mitigation in full-scale treatment plants is still rare. Recently, Duan *et al.* (2020) reported a full-scale N_2O mitigation work at an sequencing batch reactor (SBR) plant in Adelaide, Australia. To achieve this, the first monitoring campaign was carried out to monitor N_2O emission dynamics, nutrient performance and to identify the links between the plant operations and N_2O generation. N_2O mitigation strategies that were centred on the optimization of the aeration profiles were consequently proposed and evaluated using a multi-pathway N_2O production mathematical model before implementation. The second monitoring campaign was then carried out after implementing the mitigation strategy, the results showed that an approximate 35% reduction in N_2O emissions was achieved (shown in Figure 11.3). What is very important from this work is that it demonstrated that N_2O mitigation does NOT necessarily require additional operational cost, which was a misconception of many people. In contrast, the N_2O mitigation was achieved with reduced operational cost, due to savings in energy cost. In addition, the nutrient removal performance was not affected by N_2O mitigation. This was also the world's first N_2O mitigation work that has been permanently implemented at a full-scale plant.

Although the proposed mitigation strategy has been demonstrated to be effective in the studied plant, it is very difficult to generalize this to another plant. Applicability of one mitigation measure to another plant is always questionable. This is because different plant configurations, processes, and operational conditions (dissolved oxygen (DO), pH, loading rate, and so on.) would influence the N_2O production pathways and emission dynamics (Marques *et al.*, 2016), which all contribute to the high variability of N_2O generation from WWTPs. Consequently, a comprehensive guideline is urgently required to facilitate N_2O mitigation strategies to be developed on a case-by-case basis without a need for prior full-scale quantification.

By reviewing the existing N_2O mitigation studies in WWTPs, the essential knowledge to guide N_2O mitigations, and the logic behind N_2O mitigation strategies were presented by Duan *et al.* (2021).

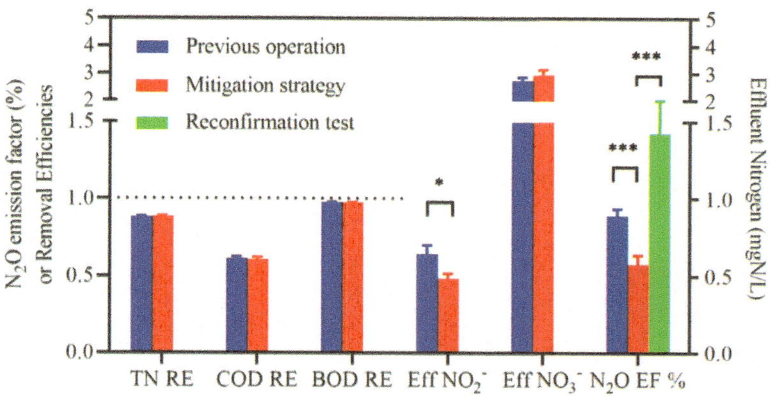

Figure 11.3 Comparison of nutrient removal performances, and N_2O emissions before and after implementing the mitigation. Standard errors are shown. RE: removal efficiency ($n = 61$); EF: emission factor ($n = 75$, 17 & 12); Eff: effluent ($n = 61$). *: $p < 0.05$; ***: $p < 0.001$ (Duan *et al.*, 2020.)

Table 11.1 N_2O mitigation strategies and associated typical studies summarized by Duan *et al.* (2021)

Strategies		Objectives and Mechanism	Environment	Outcomes	Status
Aeration control	Reduce dissolved oxygen setpoint	Reduce ammonium oxidation rate (AOR) and reduce nitrite accumulation. Reduced AOR will reduce N_2O emissions from the NH_2OH oxidation pathways. Low DO could also encourage simultaneous nitrification and denitrification and therefore minimize nitrite accumulation, leading to less N_2O emissions from the nitrifier denitrification pathway.	Mainstream full scale SBR reactors performing nitrification/ denitrification (N/D).	35% reductions in N_2O emissions.	Tested with mathematical model and permanently implemented at full-scale.
	Reduce aeration rate	Reduce the mass transfer of N_2O from liquid to gas phase (as long as oxygen limitation is prevented).	Full-scale WWTP receiving industrial wastewater and laboratory reactors in parallel.	75% reduction in aeration rate resulted in 53% reduction of N_2O emissions.	Tested at lab-scale for potential full-scale implementation.
	Change continuous aeration to intermittent aeration	Allowed heterotrophic denitrification between aeration intervals to consume N_2O and N_2O precursors (NO, NO_2^-).	A mainstream full scale SBR performing N/D.	About 90% reduction in N_2O emissions was observed during the short test run period.	Tested at full-scale.
	Change aeration scheme: reduce aeration period, avoid over-aeration	Minimize prolonged anoxic periods that were found to promote N_2O emissions.	Sidestream full-scale one-stage reactor performing partial nitritation-anammox (PN/A).	N/A	Proposed based on experimental observations.
Feed control	Supplying external chemical oxygen demand (COD) or improve COD utilization to prevent incomplete denitrification	Avoid insufficient denitrification, so that N_2O accumulation during denitrification can be prevented.	Lab-scale SBRs.	N_2O emissions reduced by 90% when COD/N ratio increased from 2.6 to 4.5	Tested at lab-scale.
	Flow equalization	Avoid sudden increase of COD or nitrogen loading. Increased ammonium could promote nitric oxide reduction (NOR) and thus N_2O emissions. Sudden increase of COD could reduce DO and lead to nitrite accumulation, which will stimulate N_2O emissions.	Mainstream full-scale WWTP occasionally receiving landfill leachate.	N/A	Proposed based on experimental observations.

	Step-feed/ intermittent feeding	Enhance the utilization of influent organic carbon to prevent N_2O and nitrite accumulation during denitrification. Maintain low levels of ammonium during the aerobic phase to limit the AOR. Maintain low level of nitrite during aerobic phase in order to reduce N_2O productions from the nitrifier denitrification pathway.	Laboratory reactors receiving real/ synthetic mainstream wastewater.	Up to 66% reductions reported.	Tested with lab reactors in a number of studies.
	Slow feeding and controlled supply of alkalinity	Control the pH within the window of 6.0–7.0. Under such non-optimal pH condition for ammonia oxidizing bacteria (AOB), the AOR can be controlled at a relatively low level, which would lead to lower N_2O emissions.	Lab scale SBR.	Reduced the N_2O production during aerobic phases by 75%.	Tested with lab reactors.
Process optimization	Applying long solids retention time	To ensure the sufficient growth of nitrite oxidizing bacteria (NOB), to minimize NO_2^- accumulation during nitrification.	Mainstream full-scale plug-flow WWTP.	N/A	Proposed based on experimental observation and literature.
	Operating one-stage PN/A or PN/ denitrification rather than two-stage processes	To reduce nitrite accumulation by anammox or denitrification.	Sidestream full-scale two-stage partial nitritation and anammox reactors.	N/A	Proposed based on comparison.
	Extending the anoxic phase	To allow complete denitrification for N_2O consumptions during anoxic phase.	Mainstream full-scale Carrousel reactors performing nitrification/ denitrification.	Extending anoxic phase by 33% and reducing DO set-point by 0.5 mg_{O2}/L led to 60% reduction of N_2O emissions.	Tested at full-scale for one week.
	Reduce or avoid anoxic period in nitritation reactor	Reduce N_2O formed during denitrification of anoxic period.	A sidestream full-scale SBR performing partial nitritation.	Resulted in 53% reduction of N_2O emissions.	Tested at full-scale.
	Rearrange the anoxic and aerobic phases in anammox SBR cycles or apply anammox without aeration	To allow for N_2O reduction during the anoxic phase and prevent strong stripping in the following aerobic phase.	Sidestream two-stage partial nitritation and anammox pilot plant (SBR).	N/A	Proposed based on experimental observation and literature.
	Change RAS return scheme	All RAS was returned to the first step of the reactor, so that the sludge concentration in the second step was lower, resulting in higher specific AOR in the second step. To reduce the high specific AOR of the second step of the reactor, a fraction of the RAS was proposed to be returned to the second step.	Mainstream full-scale step-feed plug-flow WWTP.	About 50% N_2O reduction, as simulated by a mathematical model.	Proposed and tested with mathematical model.

Table 11.1 summarizes the findings of Duan *et al.* (2021). In addition to the mechanistic fundamentals, from a practical perspective, it was found that current N_2O mitigation strategies can be categorized into three feasible types of implementation: aeration control, feed scheme optimization, and process optimization. In addition, three critical challenges, as well as opportunities, of N_2O mitigation were identified. It is proposed that (i) quantification methods for overall N_2O emissions and relative pathway contributions need improvement; (ii) a reliable while straightforward mathematical model is required to quantify benefits and compare mitigation strategies; and (iii) tailored risk assessment needs to be conducted for WWTPs, in which more long-term full-scale trials of N_2O mitigation are needed to enable robust assessments of the resulting operational costs and impact on nutrient removal performance.

When it comes to N_2O mitigation, it is imperative to understand the mechanisms that trigger its formation in order to counteract them. In that respect, it is noteworthy that these systems employ processes that can consume (part of) the produced N_2O and hence help in the mitigation. This means that mitigation is actually a matter of striking the right balance between allowing certain processes to proceed and making sure certain pathways that heavily contribute to emissions are not activated.

We already discussed the current shortcomings of mechanistic models in terms of predictive power at full-scale. But that does not mean that mitigation has to wait until all mechanisms have been completely understood. For sure these models are able to predict trends in the emissions resulting from changing operation. This means that they have the potential for pointing us in the direction of where mitigation can be achieved, but the exact amount will not be accurately predicted. However, since emissions will or should be physically measured anyway, this is not a road block by any means. In this respect, every successful mitigation and monitoring action is crucial in view of combating climate change and getting closer to net zero emissions. In terms of predicting more easily, and perhaps more accurately, the aforementioned data-driven and hybrid models can be incorporated into the practice of N_2O quantification and mitigation. Coupling the mechanistic insight provided by mechanistic models with knowledge-based AI insights into the contributing risk factors, and with data-driven predictions of N_2O, can provide the robustness needed to achieve the large-scale mitigation required on a global level.

Mitigation strategies will likely also not be fixed over seasons. Therefore, different mitigation measures might be required under different conditions. This too needs further exploration. Here the benchmark platform can prove to be very useful. It can test different control strategies over longer periods of time and also investigate supervisory control. Avenues to also explore are building different benchmarks using different models for the prediction of N_2O, including data-driven and hybrid models.

In the longer term, digital twins will become state of the art in UWSs and these will incorporate the gathered knowledge as well as being fed with real-time data and be able to continuously train themselves further and ensure that emissions are minimized at all times and at the lowest cost, while still ensuring system performance.

Increasingly stringent climate change policies may monetize the fugitive GHG emissions and drive more full-scale implementation of mitigation strategies. It should be noted that full-scale quantification of GHG emissions will continue to be essential in order to identify emission sources and to verify the effectiveness of mitigation strategies. However, these efforts are likely going to be at least partially replaced by digital twins, in conjunction with smarter sensor systems integrated with drone technology (e.g., remotely controlled mobile sensors).

11.4.2 CH_4 mitigation

Methane emissions also contribute a significant portion of the urban wastewater system carbon footprint. Fixing methane leaks from biogas piping systems at WWTPs are an obvious measure that will need to be carried out. Better managing the overall solids handling operations at WWTPs is another critical measure that needs to be implemented. However, the focus of this book was on modelling methane in the activated sludge units and in sewers.

In terms of modelling methane oxidation in the activated sludge process, this is helpful for understanding whether methane is oxidized or stripped to the atmosphere after being recycled back to the activated sludge process from sludge treatment, and whether changes can be made in the process to favour methane oxidation. In terms of modelling methane in sewers, this fills a large gap in the GHG accounting community as sewer emissions are completely overlooked in the accounting methodologies being followed by the water sector. Hydraulics can play a big role in sewer methane emissions, therefore, leveraging the hydraulic simulation outputs from the existing sewer network models that are widely available will be useful for running the sewer methane models and determining whether pumping can be optimized or whether real-time controls can be implemented in the networks to minimize methane emissions. And as digital twins are implemented these insights can be obtained more regularly. Furthermore, mitigating methane emissions via chemical dosing represents a big opportunity as it is already being done to minimize hydrogen sulfide emissions by changing the substrate composition that favours hydrogen sulfide and methane production in sewers. The modelling will again prove to be invaluable for assessing dosing schemes and their effectiveness in reducing methane emissions.

11.5 OVERALL CONCLUSION

In summary, a lot of progress has been made with regard to understanding, monitoring and modelling of GHG emissions from urban wastewater systems. This book provides the water community with the knowledge and tools needed to start monitoring and mitigating methane and nitrous oxide emissions and immediately bring the water industry closer to net zero. Now it is up to the water community to start applying the knowledge and tools the last decade of research has given us. Further work is still needed and will likely be identified as we learn from reducing GHG emissions in practice. However, it is of utmost importance to not wait. The urgency in taking climate action has never been greater. The time to quantify, model, and mitigate GHG emissions in the water sector is now.

ACKNOWLEDGEMENT

Liu Ye acknowledges the funding support through Australian Research Council Discovery Project (180103369), and the University of Queensland Foundation Research Excellence Award.

REFERENCES

American Society of Civil Engineers. (1991). Measurement of Oxygen Transfer in Clean Water, ASCE 2-91. American Society of Civil Engineers, Reston, Virginia.

Arnaldos M., Amerlinck Y., Rehman U., Van Hoey S., Naessens W. and Nopens I. (2015). From the affinity constant to the half-saturation index: Understanding conventional modeling concepts in novel wastewater treatment processes. *Water Research*, **70**, 458–470.

Australian Govt. (2020). National Greenhouse and Energy Reporting (Measurement) Determination 2008 made under subsection 10(3) of the National Greenhouse and Energy Reporting Act 2007- Compilation No. 12, 1 July 2020. Dept. of Industry, Science, Energy and Resources, Australian Government, Canberra.

Chandran K. (2011). Protocol for the Measurement of Nitrous Oxide Fluxes from Biological Wastewater Treatment Plants. *Methods in Enzymology*, ISSN: 0076-6879, **486**(C), 369–385. https://doi.org/10.1016/B978-0-12-381294-0.00016-X

Daelman M. R., De Baets B., van Loosdrecht M. C. and Volcke E. I. (2013). Influence of sampling strategies on the estimated nitrous oxide emission from wastewater treatment plants. *Water Research*, **47**(9), 3120–3130, https://doi.org/10.1016/j.watres.2013.03.016

Daelman M. R. J., van Voorthuizen E. M., van Dongen U. G. J. M., Volcke E. I. P. and van Loosdrecht M. C. M. (2015). Seasonal and diurnal variability of N_2O emissions from a full-scale municipal wastewater treatment plant. *Science of the Total Environment*, **536**, 1–11, https://doi.org/10.1016/j.scitotenv.2015.06.122

de Haas D. and Ye L. (2021). Nitrous oxide emissions from wastewater treatment: a case for variable emission factors. *Water E-Journal (Online Journal of the Australian Water Association)*, **6**(2), 2021, https://doi.org/10.21139/wej.2021.008

Duan H., van den Akker B., Thwaites B. J., Peng L., Herman C. E., Pan Y., Ni B., Watt S., Yuan Z. and Ye L. (2020). Mitigating nitrous oxide emissions at a full-scale wastewater treatment plant. *Water Research*, **185**, 116196, https://doi.org/10.1016/j.watres.2020.116196

Duan H., Zhao Y., Koch K., Wells G. F., Zheng M., Yuan Z. and Ye L. (2021). Insights into nitrous oxide mitigation strategies in wastewater treatment and challenges for wider implementation. *Environmental Science and Technology*, **55**(11), 7208–7224, https://doi.org/10.1021/acs.est.1c00840

Foley J., de Haas D., Yuan Z. and Lant P. (2010). Nitrous oxide generation in full-scale biological nutrient removal wastewater treatment plants. *Water Research*, **44**(3), 831–844, https://doi.org/10.1016/j.watres.2009.10.033

Gruber W., Villez K., Kipf M., Wunderlin P., Siegrist H., Vogt L. and Joss A. (2019). N_2O emission in full-scale wastewater treatment: proposing a refined monitoring strategy. *Science of the Total Environment*, **699**, 134157, https://doi.org/10.1016/j.scitotenv.2019.134157

IPCC. (2006). Guidelines for National Greenhouse Gas Inventories, (Chapter 5, Waste).

IPCC. (2019). Wastewater treatment and discharge. Bartram, D., Short, M.D., Ebie, Y., Farkăs, J., Gueguen, C., Peters, G.M., Zanzottera, N.M., Karthik, M., Masuda, S. In: 2019 Refinement to the 2006 IPCC Guidelines for National Greenhouse Gas Inventories, F. B. Demirok and A. Herold (eds.), Vol. **5** (Chapter 6), 6.39p, https://www.ipcc-nggip.iges.or.jp/public/2019rf/vol5.html.

Kosonen H., Heinonen M., Mikola A., Haimi H., Mulas M., Corona F. and Vahala R. (2016). Nitrous oxide production at a fully covered wastewater treatment plant: results of a long-term online monitoring campaign. *Environmental Science & Technology*, **50**(11), 5547–5554, https://doi.org/10.1021/acs.est.5b04466

Lim Y. and Kim D.-J. (2014). Quantification method of N_2O emission from full-scale biological nutrient removal wastewater treatment plant by laboratory batch reactor analysis. *Bioresource Technology*, **165**, 111–115, https://doi.org/10.1016/j.biortech.2014.03.021

Mannina G., Butler D., Benedetti L., Deletic A., Fowdar H., Fu G., Kleidorfer M., McCarthy D., Steen Mikkelsen P., Rauch W., Sweetapple C., Vezzaro L., Yuan Z. and Willems P. (2018). Greenhouse gas emissions from integrated urban drainage systems: where do we stand? *Journal of Hydrology*, **559**, 307–314, https://doi.org/10.1016/j.jhydrol.2018.02.058

Marques R., Rodríguez-Caballero A., Oehmen A. and Pijuan M. (2016). Assessment of online monitoring strategies for measuring N_2O emissions from full-scale wastewater treatment systems. *Water Research*, **99**, 171–179, https://doi.org/10.1016/j.watres.2016.04.052

Massara T. M., Malamis S., Guisasola A., Baeza J. A., Noutsopoulos C. and Katsou E. (2017). A review on nitrous oxide (N_2O) emissions during biological nutrient removal from municipal wastewater and sludge reject water. *Science of the Total Environment*, **596**, 106–123, https://doi.org/10.1016/j.scitotenv.2017.03.191

Porro J., Milleri C., Comas J., Rodriguez-Roda I., Pijuan M., Corominas L., Guo L. S., Daelman M. R. J., Volcke E. I. P., van Loosdrecht M. C. M., Vanrolleghem P. A. and Nopens I. (2014). Risk assessment modelling of nitrous oxide in activated sludge systems: quality not quantity. 4th IWA/WEF Wastewater Treatment Modelling Seminar, (WWTmod 2014), Spa, Belgium.

Porro J., Rehman U., Flameling T., Bellandi G., Deeke A., Audenaert W., Weijers S. and Nopens I. (2019). An integrated CFD/Biokinetic/N_2O risk modelling approach for mitigating N_2O emissions and optimal WWTP design. 10th IWA Symposium on Modelling and Integrated Assessment, (Watermatex 2019), Copenhagen, Denmark.

Porro J., Tessier M., Deeke A. and Weijers S. (2021). Using artificial intelligence for water utility-level nitrous oxide reduction and net zero emissions targets. 5th IWA Specialized International Conference of Ecotechnologies for Wastewater Treatment (EcoSTP 2021), Milan, Italy.

Rosso D., Iranpour R. and Stenstrom M. K. (2005). Fifteen years of off-gas transfer efficiency measurements on fine pore aerators: key role of sludge age and normalized air flux. *Water Environment Research*, **77**(3), 266–273, https://doi.org/10.2175/106143005X41843

Valkova T., Parravicini V., Saracevic E., Tauber J., Svardal K. and Krampe J. (2021). A method to estimate the direct nitrous oxide emissions of municipal wastewater treatment plants based on the degree of nitrogen removal. *Journal of Environmental Management*, **279**(111563), 1–10, https://doi.org/10.1016/j.jenvman.2020.111563

Vasilaki V., Danishvar S., Mousavi A. and Katsou E. (2020). Data-driven versus conventional N_2O EF quantification methods in wastewater; how can we quantify reliable annual EFs? *Computers & Chemical Engineering*, **141**, 106997, https://doi.org/10.1016/j.compchemeng.2020.106997

Ye L., Ni B.-J., Law Y., Byers C. and Yuan Z. (2014). A novel methodology to quantify nitrous oxide emissions from full-scale wastewater treatment systems with surface aerators. *Water Research*, **48**, 257–268, https://doi.org/10.1016/j.watres.2013.09.037

NOMENCLATURE

AI	Artificial intelligence
CH_4	Methane
DO	Dissolved oxygen
EF	Emission factor
ML	Machine learning
N_2O	Nitrous oxide

A note from the IWA Task Group GHG

By Jose Porro, Chair of IWA Task Group on the use of water quality and process models for minimizing wastewater utility greenhouse gas footprints.

I would like to thank each of the members of the *IWA Task Group on the use of water quality and process models for minimizing wastewater utility greenhouse gas footprints*, all of whom helped drive the knowledge on modelling GHG emissions from wastewater systems forward and make this book possible. As Chair of the Task Group, I was able to work closely with each of these all-star researchers and practitioners, and as individuals, I can say they are just world class people. It was truly a pleasure working with them. The urgency for climate action was not nearly what it is today in 2010, when we founded the Task Group, yet each member dedicated days and weeks of their personal time to the Task Group effort, not because it was urgently needed, but because they knew that one day it would be, exemplifying vision and passion for advancing knowledge for a better planet. If this coordinated research effort had not started when it did, we would not be in the position that we are in today, providing guidance on how we can use this knowledge to take climate action today. So a sincere thank you to each of the following Task Group members: Ingmar Nopens (Ghent University), Co-Chair, for his master organizational and leadership skills, and his exceptional modelling knowledge to help drive things forward, make connections, and make things happen around key IWA events; Kartik Chandran (Columbia University) and Marlies Kampschreur (Waterboard Aa en Maas) for their pioneering work in the modern N_2O era and for their early, critical input on pathways and processes that should be considered in model structures; Peter Vanrolleghem (Université Laval), a true environmental engineering modeller, for his unmatched passion for modelling and generous hosting of the Task Group at ModelEAU at Université Laval; Imre Takacs (Dynamita), Andy Shaw (Black & Veatch), and Bernhard Wett (Dynamita) for their invaluable modelling knowledge and practical insights on how we should be modelling greenhouse gas emissions; Mathieu Sperandio (INSA Toulouse) for helping us to not forget about NO and helping to push the limit of what we can include in the models; Eveline Volcke (Ghent University) for not only contributing to the N_2O models but also models for the other critical Scope 1 emission source from wastewater treatment, methane; Lisha Guo for her measurement experience and razor sharp modelling skills; Xavi Flores-Alsina (DTU) and Lluís Corominas (ICRA) for jump starting the benchmarking work; Maite Pijuan (ICRA) for contributing both her valuable N_2O field measurements and pathways knowledge; Vanessa Parravicini (TU Wien) for her N_2O survey work and insights; Oriol Gutierrez (ICRA) and Keshab Sharma (The University of Queensland) for the sewer methane expertise; Zhiguo Yuan (The University of Queensland), one of the hardest working people I know and who brought the same work ethic to the Task Group; and the last, but not least, member Sudhir Murthy (NEWHub) for championing our efforts to the industry, contributing interesting perspectives on how ammonia-oxidizing bacteria eat and produce N_2O, hosting a Task Group sleepover, and securing funding for us to keep meeting and pushing the envelope of what we could do with modelling GHG emissions from wastewater systems.

I would also like to thank the following affiliated members and Task Group supporters: Stefan Weijers (Waterboard De Dommel) and Mark van Loosdrecht (TU Delft) for continued support and collaboration; Barth Smets (DTU) for his collaboration; Youri Amerlinck (Aquafin), Lorenzo Benedetti (Waterways), and Stijn Van Hulle (Howest University) for their various inputs; Diego Rosso (UC Irvine) for his field measurements/gas transfer knowledge and collaboration; Bing Jie-Ni (UTS) for helping to take N_2O models to the next level; Peter Dold (Envirosim) for his collaboration; Ahlem Filali (INRAE) and Gaby Dotro (Cranfield University) for their collaboration on measurements; Matthijs Daelman (Trevi) for his N_2O and methane field measurements knowledge; Magnus Arnell (Lund University) for his benchmarking help; Dwight Houweling (Dynamita) for his collaboration; Yvonne Schneider (Ruhrverband) for her collaboration and Hannover Workshop; Lauren Fillmore (WERF, retired) for her excellent collaboration and support; Frank Rogalla (Aqualia), Eduardo Ayesa (CEIT), Alejandro Vargas (UNAM), Bruce Beck (FASresearch), Steven Kenway (The University of Queensland), and Ulf Jeppsson (Lund University) for their support through the IWA Specialist Groups; Sylvie Gillot (INRAE), Leiv Reiger (InCTRL Solutions), Lina Belia (Primodal), and Damien Batstone (The University of Queensland) for their support from other IWA Task Groups, Bill Hiatt (Clemson University) for pioneering denitrification N_2O modelling and sharing his insights; Juan Lema (University of Santiago de Compostela) for his generous collaboration with the NOVEDAR project; Albert Guisasola (Autonomous University of Barcelona) for his early sewer methane modelling feedback; Jeff Foley (GHD) for his early links of operational parameters to risk of N_2O emissions; Laura Snip (Hoogheemraadschap de Stichtse Rijnlanden), Ramesh Saagi (Lund University), and Celia Castro (Cetaqua) for their collaboration during the SANITAS Project; and Joaquim Comas (University of Girona) and Ignasi Rodriguez-Roda (University of Girona) for their knowledge of knowledge-based systems and helping to create one for N_2O.

I would also like to thank Gustaf Olsson (Lund University) and Glen Daigger (University of Michigan) for their thoughtful support, guidance, and encouragement.

We also want to thank the following organizations: IWA for their support and providing us with an excellent platform; WERF (now WRF) and Arcadis for their support.

And last but not least, I would like to thank Liu Ye (The University of Queensland) for bringing her leadership, knowledge and passion for N_2O research to help take this book effort across the finish line.

IWA Publishing's authorised EU representative for General Product Safety Regulations is Diane D'Arras, 15 rue Duret, 75116 Paris, France, e-mail: safety@iwap.co.uk.

Printed and bound by CPI Group (UK) Ltd, Croydon, CR0 4YY
16/06/2026
02140386-0001